T0185764

Universitext

Universitext

Universitext is a series of textbooks that presents material from a wide variety of mathematical disciplines at master's level and beyond. The books, often well class-tested by their author, may have an informal, personal, even experimental approach to their subject matter. Some of the most successful and established books in the series have evolved through several editions, always following the evolution of teaching curricula, to very polished texts.

Thus as research topics trickle down into graduate-level teaching, first textbooks written for new, cutting-edge courses may make their way into *Universitext*.

For further volumes:
www.springer.com/series/223

Anton Deitmar • Siegfried Echterhoff

Principles of Harmonic Analysis

Second Edition

Anton Deitmar
Universität Tübingen Institut für Mathematik
Tübingen
Baden-Württemberg
Germany

Siegfried Echterhoff
Universität Münster Mathematisches Institut
Münster
Germany

ISSN 0172-5939 ISSN 2191-6675 (electronic)
ISBN 978-3-319-37904-3 ISBN 978-3-319-05792-7 (eBook)
DOI 10.1007/978-3-319-05792-7
Springer Cham Heidelberg New York Dordrecht London

Printed on acid-free paper

Springer is part of Springer Science+Business Media (www.springer.com)

Preface

The thread of this book is formed by two fundamental principles of Harmonic Analysis: the Plancherel Formula and the Poisson Summation Formula. We first prove both for locally compact abelian groups. For non-abelian groups we discuss the Plancherel Theorem in the general situation for type-I groups. The generalization of the Poisson Summation Formula to non-abelian groups is the Selberg Trace Formula, which we prove for arbitrary groups admitting uniform lattices. As examples for the application of the Trace Formula we treat the Heisenberg group and the group $SL_2(\mathbb{R})$. In the former case the trace formula yields a decomposition of the L^2-space of the Heisenberg group modulo a lattice. In the case $SL_2(\mathbb{R})$, the trace formula is used to derive results like the Weil asymptotic law for hyperbolic surfaces and to provide the analytic continuation of the Selberg zeta function. We finally include a chapter on the applications of abstract Harmonic Analysis on the theory of wavelets, and we include a chapter on p-adic and adelic groups, which are important examples, as they are used in number theory.

The present book is a text book for a graduate course on abstract harmonic analysis and its applications. The book can be used as a follow up of the *First Course in Harmonic Analysis*, [Dei05], or independently, if the students have required a modest knowledge of Fourier Analysis already. In this book, among other things, proofs are given of Pontryagin Duality and the Plancherel Theorem for LCA groups, which were mentioned but not proved in [Dei05]. Using Pontryagin duality, we also obtain various structure theorems for locally compact abelian groups.

Knowledge of set theoretic topology, Lebesgue integration, and functional analysis on an introductory level will be required in the body of the book. For the convenience of the reader we have included all necessary ingredients from these areas in the appendices.

Differences to the first edition: Many details have been changed, new and better proofs have been found, some assertions have been sharpened and a few are even new to this book. Section 1.8 and Chap. 13 have not been part of the first edition. Whilst fitting in the changes, we tried to preserve the numbering of Theorems etc. We apologize for inconveniences that arise at those places where this was not possible.

Acknowledgments

The authors thank the following people for corrections and comments on the book: *Ralf Beckmann, Wolfgang Bertram, Robert Burckel, Cody Gunton, Linus Kramer, Yi Li, Jonas Morrissey, Michael Mueger, Kenneth Ross, Alexander Schmidt, Christian Schmidt, Vahid Shirbisheh, Frank Valckenborgh, Fabian Werner, Dana Williams.*

Chapters 3 and 4 are partly based on written notes of a course given by Prof. Eberhard Kaniuth on duality theory for abelian locally compact groups. The authors are grateful to Prof. Kaniuth for allowing us to use this material.

Chapter Dependency

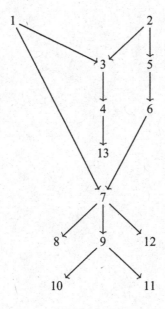

Notation

We write $\mathbb{N} = \{1,2,3,\ldots\}$ for the set of *natural numbers*. The sets of integer, real, and complex numbers are denoted as $\mathbb{Z}, \mathbb{R}, \mathbb{C}$. For a set A we write $\mathbf{1}_A$ for the *characteristic function* of A, i.e., $\mathbf{1}_A(x)$ is 1 if $x \in A$ and zero otherwise. The *Kronecker-delta* function is defined to be

$$\delta_{i,j} \stackrel{\text{def}}{=} \begin{cases} 1 & \text{if } i=j, \\ 0 & \text{otherwise.} \end{cases}$$

The word *positive* will always mean ≥ 0. For > 0, we use the words *strictly positive*.

Contents

Chapter 1
Haar Integration

In this chapter, topological groups and invariant integration are introduced. The existence of a translation invariant measure on a locally compact group, called Haar measure, is a basic fact that makes it possible to apply methods of analysis to study such groups. The Harmonic Analysis of a group is basically concerned with spaces of measurable functions on the group, in particular the spaces $L^1(G)$ and $L^2(G)$, both taken with respect to Haar measure. The invariance of this measure allows to analyze these function spaces by some generalized Fourier Analysis, and we shall see in further chapters of this book how powerful these techniques are.

In this book, we will freely use concepts of set-theoretic topology. For the convenience of the reader we have collected some of these in Appendix A.

1.1 Topological Groups

A *topological group* is a group G, together with a topology on the set G such that the group multiplication and inversion,

$$G \times G \to G \qquad G \to G$$
$$(x, y) \mapsto xy, \qquad x \mapsto x^{-1},$$

are both continuous maps.

Remark 1.1.1 It suffices to insist that the map $\alpha : (x, y) \to x^{-1}y$ is continuous. To see this, assume that α is continuous and recall that the map $G \to G \times G$, that maps x to (x, e) is continuous (Example A.5.3), where e is the unit element of the group G. We can thus write the inversion as a composition of continuous maps as follows $x \mapsto (x, e) \mapsto x^{-1}e = x^{-1}$. The multiplication can be written as the map $(x, y) \mapsto (x^{-1}, y)$ followed by the map α, so is continuous as well, if α is.

A. Deitmar, S. Echterhoff, *Principles of Harmonic Analysis*, Universitext,
DOI 10.1007/978-3-319-05792-7_1, © Springer International Publishing Switzerland 2014

Examples 1.1.2

- Any given group becomes a topological group when equipped with the *discrete topology*, i.e., the topology, in which every subset is open. In this case we speak of a *discrete group*

- The additive and multiplicative groups $(\mathbb{R}, +)$ and $(\mathbb{R}^{\times}, \times)$ of the field of real numbers are topological groups with their usual topologies. So is the group $GL_n(\mathbb{R})$ of all real invertible $n \times n$ matrices, which inherits the $\mathbb{R}^{n^2}$-topology from the inclusion $GL_n(\mathbb{R}) \subset M_n(\mathbb{R}) \cong \mathbb{R}^{n^2}$, where $M_n(\mathbb{R})$ denotes the space of all $n \times n$ matrices over the reals. As for the proofs of these statements, recall that in analysis one proves that if the sequences a_i and b_i converge to a and b, respectively, then their difference $a_i - b_i$ converges to $a - b$, and this implies that $(\mathbb{R}, +)$ is a topological group. The proof for the multiplicative group is similar. For the matrix groups recall that matrix multiplication is a polynomial map in the entries of the matrices, and hence continuous. The determinant map also is a polynomial and so the inversion of matrices is given by rational maps, as for an invertible matrix A one has $A^{-1} = \det(A)^{-1} A^{\#}$, where $A^{\#}$ is the adjugate matrix of A; entries of the latter are determinants of sub-matrices of A, therefore the map $A \mapsto A^{-1}$ is indeed continuous.

Let $A, B \subset G$ be subsets of the group G. We write

$$AB = \{ab : a \in A, b \in B\} \qquad \text{and} \qquad A^{-1} = \{a^{-1} : a \in A\},$$

as well as $A^2 = AA$, $A^3 = AAA$ and so on.

Lemma 1.1.3 *Let G be a topological group.*

(a) *For $a \in G$ the translation maps $x \mapsto ax$ and $x \mapsto xa$, as well as the inversion $x \mapsto x^{-1}$ are homeomorphisms of G. A set $U \subset G$ is a neighborhood of $a \in G$ if and only if $a^{-1}U$ is a neighborhood of the unit element $e \in G$. The same holds with Ua^{-1}.*

(b) *If U is a neighborhood of the unit, then $U^{-1} = \{u^{-1} : u \in U\}$ also is a neighborhood of the unit. We call U a symmetric unit-neighborhood if $U = U^{-1}$. Every unit-neighborhood U contains a symmetric one, namely $U \cap U^{-1}$.*

(c) *For a given unit-neighborhood U there exists a unit-neighborhood V with $V^2 \subset U$.*

(d) *If $A, B \subset G$ are compact subsets, then AB is compact.*

(e) *If A,B are subsets of G and A or B is open, then so is AB.*

(f) *For $A \subset G$ the topological closure $\overline{A}$ equals $\overline{A} = \bigcap_V AV$, where the intersection runs over all unit-neighborhoods V in G.*

Proof (a) follows from the continuity of the multiplication and the inversion. (b) follows from the continuity of the inversion. For (c), let U be an open unit-neighborhood, and let $A \subset G \times G$ be the inverse image of U under the continuous map $m : G \times G \to G$ given by the group multiplication. Then A is open in the product topology of $G \times G$. Any set, which is open in the product topology, is a union of sets of the form $W \times X$, where W, X are open in G. Therefore there are unit-neighborhoods W, X with $(e, e) \in W \times X \subset A$. Let $V = W \cap X$. Then V is a unit-neighborhood as well and $V \times V \subset A$, i.e., $V^2 \subset U$. For (d) recall that the set AB is the image of the compact set $A \times B$ under the multiplication map; therefore it is compact. (e) Assume A is open, then $AB = \bigcup_{b \in B} Ab$ is open since every set Ab is open. For (f) let $x \in \overline{A}$, and let V be a unit-neighborhood. Then xV^{-1} is a neighborhood of x, and so $xV^{-1} \cap A \neq \emptyset$. Let $a \in xV^{-1} \cap A$. Then $a = xv^{-1}$ for some $v \in V$, so $x = av \in AV$, which proves the first inclusion. For the other way round let x be in the intersection of all AV as above. Let W be a neighborhood of x. Then $V = x^{-1}W$ is a unit-neighborhood and so is V^{-1}. Hence $x \in AV^{-1}$, so there is $a \in A$, $v \in V$ with $x = av^{-1}$. It follows $a = xv \in xV = W$. This means $W \cap A \neq \emptyset$. As W was arbitrary, this implies $x \in \overline{A}$. $\square$

Lemma 1.1.4 *Let H be a subgroup of the topological group G. Then its closure $\overline{H}$ is also a subgroup of G. If H is normal, then so is $\overline{H}$.*

Proof Let $H \subset G$ be a subgroup. To show that the closure $\overline{H}$ is a subgroup, it suffices to show that $x, y \in \overline{H}$ implies $xy^{-1} \in \overline{H}$. Let m denote the continuous map $\overline{H} \times \overline{H} \to G$ given by $m(x, y) = xy^{-1}$. The pre-image $m^{-1}(\overline{H})$ must be closed and contains the dense set $H \times H$; therefore it contains the whole of $\overline{H} \times \overline{H}$, which proves the first claim. Next assume that H is normal, then for every $g \in G$ the set $g\overline{H}g^{-1}$ is closed and contains $gHg^{-1} = H$; therefore $\overline{H} \subset g\overline{H}g^{-1}$. Conjugating by g one gets $g^{-1}\overline{H}g \subset \overline{H}$. As g varies, the second claim of the lemma follows. $\square$

In functional analysis, people like to use *nets* in topological arguments. These have the advantage of providing very intuitive proofs. We refer the reader to Sect. A.6 for further details on nets and convergence in general. The next lemma is an example, how nets provide intuitive proofs.

Lemma 1.1.5 *Let G be a topological group. Let $A \subset G$ be closed and $K \subset G$ be compact. Then AK is closed.*

Proof Let $\left(x_j = a_j k_j\right)_{j \in J}$ be a net in AK, convergent in G. As K is compact, one can replace it with a subnet so that (k_j) converges in K. Since the composition in G and the inversion are continuous, the net $a_j = x_j k_j^{-1}$ converges too, with limit in $\overline{A} = A$. Therefore the limit of $x_j = a_j k_j$ lies in AK, which therefore is closed. $\square$

Lemma 1.1.6 *Let G be a topological group and $K \subset G$ a compact subset. Let U be an open set containing K. Then there exists a neighborhood V of the unit in G*

such that $KV \cup VK \subset U$. In particular, if U is open and compact, then there exist a neighborhood V of e such that $UV = VU = U$.

Proof For each $x \in K$ choose a unit-neighborhood V_x such that $x V_x^2 \subset U$. By compactness of K we may find $x_1, \ldots, x_l \in K$ such that $K \subset \bigcup_{i=1}^l x_i V_{x_i}$ and $K \subset \bigcup_{i=1}^l V_{x_i} x_i$. Set $V = \bigcap_{i=1}^l V_{x_i}$. Then $KV \subset \bigcup_{i=1}^l x_i V_{x_i} V \subset \bigcup_{i=1}^l x_i V_{x_i}^2 \subset U$ and similarly $VK \subset U$. $\qquad\square$

Recall (Appendix A) that a topological space X is a T_1-space if for $x \neq y$ in X there are neighborhoods U_x, U_y of x and y, respectively, such that y is not contained in U_x and x is not contained in U_y. So X is T_1 if and only if all singletons $\{x\}$ are closed. The space is called a T_2-space or Hausdorff space if the neighborhoods U_x and U_y can always be chosen disjoint.

Lemma 1.1.7 *Let G be a locally compact group.*

(a) *Let $H \subset G$ be a subgroup. Equip the left coset space $G/H = \{xH : x \in G\}$ with the quotient topology. Then the canonical projection $\pi : G \to G/H$, which sends $x \in G$ to the coset xH, is an open mapping. The space G/H is a T_1-space if and only if the group H is closed in G. If H is normal in G, then the quotient group G/H is a topological group.*

(b) *For any open symmetric unit-neighborhood V the set $H = \bigcup_{n=1}^\infty V^n$ is an open subgroup.*

(c) *Every open subgroup of G is closed as well.*

Proof (a) Let $U \subset G$ be open, then $\pi^{-1}(\pi(U)) = UH$ is open by Lemma 1.1.3 (e). As a subset of G/H is open in the quotient topology if and only if its inverse image under π is open in G, the map π is indeed open. So, for every $x \in G$ the set $G \smallsetminus xH$ is mapped to an open set if and only if H is closed. This proves that singletons are closed in G/H, if and only if H is closed.

Now suppose that H is normal in G. One has a canonical group isomorphism $(G \times G)/(H \times H) \to G/H \times G/H$ and one realizes that this map also is a homeomorphism, where the latter space is equipped with the product topology. Consider the map $\alpha : G \times G \to G$ and likewise for G/H. One gets a commutative diagram

$$
\begin{array}{ccc}
G \times G & \xrightarrow{\ \alpha\ } & G \\
\downarrow & & \downarrow \\
G/H \times G/H & \xrightarrow{\ \overline{\alpha}\ } & G/H.
\end{array}
$$

As $G/H \times G/H \cong (G \times G)/(H \times H)$, the map $\overline{\alpha}$ is continuous if and only if the map $G \times G \to G/H$ is continuous, which it is, as α and the projection are continuous.

(b) Let V be a symmetric unit-neighborhood. For $x \in V^n$ and $y \in V^m$ one has $xy \in V^{n+m}$ and as V is symmetric, one also has $x^{-1} \in V^n$, so H is an open subgroup.

(c) Let H be an open subgroup. Writing G as union of left cosets we get $G \smallsetminus H = \bigcup_{g \in G \smallsetminus H} gH$. As H is open, so is gH for every $g \in G$. Hence the complement $G \smallsetminus H$, being the union of open sets, is open, so H is closed. $\qquad\qquad\square$

Proposition 1.1.8 *Let G be a topological group. Let H be the closure of the set $\{1\}$.*

(a) *The set H is the smallest closed subgroup of G. The group H is a normal subgroup and the quotient G/H with the quotient topology is a T_1- space.*

(b) *Every continuous map of G to a T_1-space factors over the quotient G/H.*

(c) *Every topological group, which is T_1, is already T_2, i.e., a Hausdorff space.*

Proof We prove part (a). The set H is a normal subgroup by Lemma 1.1.4. The last assertion follows from Lemma 1.1.7 (a).

For part (b) let $x \in G$. As the translation by x is a homeomorphism, the closure of the set $\{x\}$ is the set $xH = Hx$. So, if $A \subset G$ is a closed set, then $A = AH = HA$. Let $f : G \to Y$ be a continuous map into a T_1-space Y. For $y \in Y$ the singleton $\{y\}$ is closed, so $f^{-1}(\{y\})$ is closed, hence of the form AH for some set $A \subset G$. This implies that $f(gh) = f(g)$ for every $g \in G$ and every $h \in H$.

To show part (c), let G be a topological group that is T_1. Let $x \neq y$ in G and set $U = G \smallsetminus \{xy^{-1}\}$. Then U is an open neighborhood of the unit. Let V be a symmetric unit-neighborhood with $V^2 \subset U$. Then $V \cap Vxy^{-1} = \emptyset$, for otherwise there would be $a, b \in V$ with $a = b^{-1}xy^{-1}$, so $xy^{-1} = ab \in V^2$, a contradiction. So it follows that $Vx \cap Vy = \emptyset$, i.e., Vx and Vy are disjoint neighborhoods of x and y, which means that G is a Hausdorff space. $\qquad\qquad\square$

The following observation is often useful.

Lemma 1.1.9 *Suppose that $\phi : G \to H$ is a homomorphism between topological groups G and H. Then ϕ is continuous if and only if it is continuous at the unit 1_G.*

Proof Assume that ϕ is continuous at 1_G. Let $x \in G$ be arbitrary and let (x_j) be a net with $x_j \to x$ in G. Then $x^{-1}x_j \to x^{-1}x = 1_G$ and we have $\phi(x)^{-1}\phi(x_j) = \phi(x^{-1}x_j) \to \phi(1_G) = 1_H$, which then implies $\phi(x_j) \to \phi(x)$. Thus ϕ is continuous. $\qquad\square$

Notation In the preceding proof we have used the notation $x_j \to x$ indicating that the net (x_j) converges to the point x.

1.2   Locally Compact Groups

A topological space is called *locally compact* if every point possesses a compact neighborhood. A topological group is called a *locally compact group* if it is Hausdorff and locally compact.

Note that by Proposition 1.1.8, every topological group has a biggest Hausdorff quotient group, and every continuous function to the complex numbers factors through that quotient. So, as far as continuous functions are concerned, a topological group is indistinguishable from its Hausdorff quotient. Thus it makes sense to restrict the attention to Hausdorff groups.

A subset $A \subset X$ of a topological space X is called *relatively compact* if its closure $\overline{A}$ is compact in X. Note that in a locally compact Hausdorff space X, every point has a neighborhood base consisting of compact sets. A subset S of G is called σ-*compact* if it can be written as a countable union of compact sets.

Proposition 1.2.1 *Let G be a locally compact group.*

(a) *For a closed subgroup H the quotient space G/H is a locally compact Hausdorff space.*

(b) *The group G possesses an open subgroup, which is σ-compact.*

(c) *The union of countably many open σ-compact subgroups generates an open σ-compact subgroup.*

Proof For (a) let $xH \neq yH$ in G/H. Choose an open, relatively compact neighborhood $U \subset G$ of x with $\overline{U} \cap yH = \emptyset$. The set $\overline{U}H$ is closed by Lemma 1.1.3, so there is an open, relatively compact neighborhood V of y such that $V \cap UH = \emptyset$. This implies $VH \cap UH = \emptyset$, and we have found disjoint open neighborhoods of xH and yH, which means that G/H is a Hausdorff space. It is locally compact, as for given $x \in H$, and a compact neighborhood U of x the set $UH \subset G/H$ is the image of the continuous map $G \to G/H$ of the compact set U; therefore it is a compact neighborhood of xH in G/H.

To show (b), let V be a symmetric, relatively compact open unit-neighborhood. For every $n \in \mathbb{N}$ one has $\overline{V}^n = \overline{V}^n \subset V \cdot V^n = V^{n+1}$. Therefore $H = \bigcup_n \overline{V}^n = \bigcup_n V^n$. An iterated application of Lemma 1.1.3 (d) shows that $\overline{V}^n$ is compact, so H is σ-compact. By Lemma 1.1.7 (b), H is an open subgroup.

Finally, for (c) let L_n be a sequence of σ-compact open subgroups. Then each L_n is the union of a sequence $(K_{n,j})_j$ of compact sets. The group L generated by all L_n is also generated by the family $(K_{n,j})_{n,j \in \mathbb{N}}$ and is therefore σ-compact. It is also open since it contains the open subgroup L_n for any n. $\square$

1.3   Haar Measure

For a topological space X, we naturally have a σ-algebra $\mathcal{B}$ on X, the smallest σ-algebra containing all open sets. This σ-algebra also contains all closed sets and is generated by either class. It is called the *Borel σ-algebra*. Any element of this σ-algebra is called a *Borel set*.

Fix a measure space $(X, \mathcal{A}, \mu)$, so $\mathcal{A} \subset \mathcal{P}(X)$ is a σ-algebra and $\mu : \mathcal{A} \to [0, \infty]$ is a measure. One calls μ a *complete measure* if every subset of a μ-null-set is an element of $\mathcal{A}$. If μ is not complete, one can extend μ in a unique way to the σ-algebra $\overline{\mathcal{A}}$ generated by $\mathcal{A}$ and all subsets of μ-null-sets; this is called the *completion* of $\mathcal{A}$ with respect to μ. A function $f : X \to \mathbb{C}$ will be called μ-*measurable* if $f^{-1}(S)$ lies in $\overline{\mathcal{A}}$ for every Borel set $S \subset \mathbb{C}$.

Any measure $\mu : \mathcal{A} \to [0, \infty]$ defined on a σ-algebra $\mathcal{A} \supset \mathcal{B}$ is called a *Borel measure*. Unless specified otherwise, we will always assume $\mathcal{A}$ to be the completion of $\mathcal{B}$ with respect to μ. A Borel measure μ is called *locally finite* if every point $x \in X$ possesses a neighborhood U with $\mu(U) < \infty$.

Example 1.3.1 The Lebesgue measure on $\mathbb{R}$ is a Borel measure. So is the counting measure #, which for any set A is defined by

$$\#(A) \overset{\text{def}}{=} \begin{cases} \text{cardinality of } A & \text{if } A \text{ is finite} \\ \infty & \text{otherwise.} \end{cases}$$

The Lebesgue measure is locally finite; the counting measure is not.

Definition A locally finite Borel measure μ on $\mathcal{B}$ is called an *outer Radon measure* if

- $\mu(A) = \inf_{U \supset A} \mu(U)$ holds for every $A \in \mathcal{B}$, where the infimum is taken over all open sets U containing A, and

- $\mu(U) = \sup_{K \subset U} \mu(K)$ holds for every open set U, where the supremum is extended over all compact sets K contained in U.

For the first property one says that an outer Radon measure is *outer regular*. The second says that an outer Radon measure is *weakly inner regular*. For simplicity, we will use the term *Radon measure* for an outer Radon measure. In the literature, one will sometimes find the notion of Radon measure used for what we call an inner Radon measure; see Appendix B.2 for a discussion.

Note that for an outer Radon measure μ one has $\mu(A) = \sup_{K \subset A} \mu(K)$ for every measurable A with $\mu(A) < \infty$, where the supremum is taken over all subsets of A which are compact in X. This is proved in Lemma B.2.1.

Example 1.3.2

- The Lebesgue measure on the Borel sets of $\mathbb{R}$ is a Radon measure.

- A locally finite measure, which is not outer regular, is given by the following example. Let X be an uncountable set equipped with the *cocountable topology*, i.e., a non-empty set A is open if and only if its complement $X \setminus A$ is countable. The Borel σ-algebra consists of all sets that are either countable or have a countable complement. On this σ-algebra define a measure μ by $\mu(A) = 0$ if A is countable and $\mu(A) = 1$ otherwise. Then μ is finite, but not outer regular, since every open subset U of X is either empty or satisfies $\mu(U) = 1$.

The following assertion is often used in the sequel.

Proposition 1.3.3 *Let μ be an outer Radon measure on a locally compact Hausdorff space X. Then the space $C_c(X)$ is dense in $L^p(\mu)$ for every $1 \leq p < \infty$.*

Proof Fix p as in the lemma and let $V \subset L^p(\mu)$ be the closure of $C_c(X)$ inside $L^p = L^p(\mu)$. We have to show $V = L^p$. By integration theory, the space of Lebesgue step functions is dense in L^p and any such is a linear combination of functions of the form 1_A, where $A \subset X$ is of finite measure. So we have to show $1_A \in V$. By outer regularity, there exists a sequence $U_n \supset A$ of open sets such that 1_{U_n} converges to 1_A in L^p. So it suffices to assume that A is open. By weak inner regularity we similarly reduce to the case when A is compact. For given $\varepsilon > 0$ there exists an open set $U \supset A$ with $\mu(U \smallsetminus A) < \varepsilon$. By Urysohn's Lemma (A.8.1) there is $g \in C_c(X)$ with $0 \leq g \leq 1$, the function vanishes outside U and is constantly equal to 1 on A. Then the estimate

$$\|1_A - g\|_p^p = \int_{U \smallsetminus A} |g(x)|^p \, dx \leq \mu(U \smallsetminus A) < \varepsilon$$

shows the claim. □

Let G be a locally compact group. A measure μ on the Borel σ-algebra of G is called a *left-invariant measure*, or simply *invariant* if $\mu(xA) = \mu(A)$ holds for every measurable set $A \subset G$ and every $x \in G$. Here xA stands for the set of all xa, where a ranges over A.

Examples 1.3.4

- The counting measure is invariant on any group.

- For the group $(\mathbb{R}, +)$ the Lebesgue measure dx is invariant under translations, so it is invariant in the sense above.

- For the multiplicative group $(\mathbb{R}^\times, \cdot)$ the measure $\frac{dx}{|x|}$ is invariant as follows from the change of variables rule.

Theorem 1.3.5 *Let G be a locally compact group. There exists a non-zero left-invariant outer Radon measure on G. It is uniquely determined up to positive multiples. Every such measure is called a* Haar *measure. The corresponding integral is called* Haar-integral.

The existence of an invariant measure can be made plausible as follows. Given an open set U in a topological group G one can measure the *relative size* of a set $A \subset G$

by the minimal number $(A : U)$ of translates xU needed to cover A. This relative measure is clearly invariant under left translation, and is finite, if A is compact. One can compare the sizes of sets and the quotient $\frac{(A:U)}{(K:U)}$, where K is a given fixed compact, should converge as U shrinks to a point. The limit is the measure in question. It is, however, hard to verify that the limit exists and defines a measure. We circumvent this problem by considering functionals on continuous functions of compact support instead of measures. Before giving the proof of the theorem, we will draw a few immediate conclusions.

For a function f on a topological space X the *support* is the closure of the set $\{x \in X : f(x) \neq 0\}$.

Corollary 1.3.6 *Let μ be a Haar measure on the locally compact group G.*

(a) *Every non-empty open set has strictly positive (> 0) measure.*

(b) *Every compact set has finite measure.*

(c) *Every continuous positive function $f \geq 0$ with $\int_G f(x)\,d\mu(x) = 0$ vanishes identically.*

(d) *Let f be a measurable function on G, which is integrable with respect to a Haar measure. Then the support of f is contained in a σ-compact open subgroup of G.*

Proof For (a) assume there is a non-empty open set U of measure zero. Then every translate xU of U has measure zero by invariance. As every compact set can be covered by finitely many translates of U, every compact set has measure zero. Being a Radon measure, μ is zero, a contradiction.

For (b) recall that the local-finiteness implies the existence of an open set U of finite measure. Then every translate of U has finite measure. A given compact set can be covered by finitely many translates, hence has finite measure.

For (c) let f be as above, then the measure of the open set $f^{-1}(0, \infty)$ must be zero, so it is empty by part (a).

To show (d), let f be an integrable function. It suffices to show that the set $A = \{x \in X : f(x) \neq 0\}$ is contained in an open σ-compact subgroup L, as the closure will then also be in L, which is closed by Lemma 1.1.7 (c). The set A is the union of the sets $A_n = \{x \in X : |f(x)| > 1/n\}$ for $n \in \mathbb{N}$, each of which is of finite measure. By Proposition 1.2.1 (c), it suffices to show that a set A of finite measure is contained in an open σ-compact subgroup L. By the outer regularity there exists an open set $U \supset A$ with $\mu(U) < \infty$. It suffices to show that U lies in a σ-compact open subgroup. Let $H \subset G$ be any open σ-compact subgroup of G, which exists by Proposition 1.2.1 (b). Then G is the disjoint union of the open cosets xH, where $x \in G$ ranges over a set of representatives of G/H. The set U can only meet countably many cosets xH, since for every coset one has either $xH \cap U = \emptyset$ or $\mu(xH \cap U) > 0$ by part (a) of this corollary. Let L be the group generated by H and the countably many cosets xH with $xH \cap U \neq \emptyset$. Then $L \supset U \supset A$ and L is σ-compact and open by Proposition 12.1 (c).   □

Proof of the theorem Let $C_c(G)$ denote the space of all continuous functions from G to $\mathbb{C}$ of compact support.

Definition We say that a map $f : G \to X$ to a metric space (X, d) is *uniformly continuous*, if for every $\varepsilon > 0$ there exists a unit-neighborhood U such that for $x^{-1}y \in U$ or $yx^{-1} \in U$ one has d$(f(x), f(y)) < \varepsilon$.

Lemma 1.3.7 *Any function $f \in C_c(G)$ is uniformly continuous.*

Proof We only show the part with $x^{-1}y \in U$ because the other part is proved similarly and to obtain both conditions, one simply intersects the two unit-neighborhoods. Let K be the support of f. Fix $\varepsilon > 0$ and a compact unit-neighborhood V. As f is continuous, for every $x \in G$ there exists an open unit-neighborhood $V_x \subset V$ such that $y \in xV_x \Rightarrow |f(x) - f(y)| < \varepsilon/2$. Let U_x be a symmetric open unit-neighborhood with $U_x^2 \subset V_x$. Then the sets xU_x, for $x \in KV$, form an open covering of the compact set KV, so there are $x_1, \ldots x_n \in KV$ such that $KV \subset x_1U_1 \cap \cdots \cap x_nU_n$, where we have written U_j for U_{x_j}. Let $U = U_1 \cap \cdots \cap U_n$. Then U is a symmetric open unit-neighborhood. Let now $x, y \in G$ with $x^{-1}y \in U$. If $x \notin KV$, then $y \notin K$ as $x \in yU^{-1} = yU \subset yV$. So in this case we conclude $f(x) = f(y) = 0$. It remains to consider the case when $x \in KV$. Then there exists j with $x \in x_jU_j$, and so $y \in x_jU_jU \subset x_jV_j$. It follows that

$$|f(x) - f(y)| \leq |f(x) - f(x_j)| + |f(x_j) - f(y)| < \frac{\varepsilon}{2} + \frac{\varepsilon}{2} = \varepsilon$$

as claimed.   □

In order to prove Theorem 1.3.5, we use Riesz's Representation Theorem B.2.2. It suffices to show that up to positive multiples there is exactly one positive linear map $I : C_c(G) \to \mathbb{C}$, $I \neq 0$, which is invariant in the sense that $I(L_xf) = I(f)$ holds for every $x \in G$ and every $f \in C_c(G)$, where the *left translation* is defined by $L_xf(y) \overset{\text{def}}{=} f(x^{-1}y)$. Likewise, the *right translation* is defined by $R_xf(y) \overset{\text{def}}{=} f(yx)$. Note that $L_{xy}f = L_xL_yf$ and likewise for R.

Definition We say that a function f on G is a *positive function* if $f(x) \geq 0$ for every $x \in G$. We then write $f \geq 0$. Write $C_c^+(G)$ for the set of all positive functions $f \in C_c(G)$. For any two functions $f, g \in C_c^+(G)$ with $g \neq 0$ there are finitely many elements $s_j \in G$ and positive numbers c_j such that for every $x \in G$ one has $f(x) \leq \sum_{j=1}^{n} c_j g(s_j^{-1}x)$. We can also write this inequality without arguments as $f \leq \sum_{j=1}^{n} c_j L_{s_j}g$. Put

$$(f : g) \overset{\text{def}}{=} \inf \left\{ \sum_{j=1}^{n} c_j : \begin{array}{c} \text{there are } s_j \in G \\ \text{such that } f \le \sum_{j=1}^{n} c_j L_{s_j} g \end{array} \right\}.$$

Lemma 1.3.8 *For* $f, f_1, f_2, g, h \in C_c^+(G)$ *with* $g, h \ne 0, c > 0$ *and* $y \in G$ *one has*

(a) $(L_y f : g) = (f : g)$, *so the index is translation-invariant,*

(b) $(f_1 + f_2 : g) \le (f_1 : g) + (f_2 : g)$, *sub-additive,*

(c) $(cf : g) = c(f, g)$, *homogeneous,*

(d) $f_1 \le f_2 \Rightarrow (f_1 : g) \le (f_2 : g)$, *monotonic,*

(e) $(f : h) \le (f : g)(g : h)$,

(f) $(f : g) \ge \frac{\max f}{\max g}$, *where* $\max f \overset{\text{def}}{=} \max\{f(x) : x \in G\}$.

Proof We only prove (e) and (f), as the other assertions are easy exercises. For (e) assume $f \le \sum_j c_j L_{s_j} g$ and $g \le \sum_l d_l L_{t_l} h$. Then $f \le \sum_j \sum_l c_j d_l L_{s_j t_l} h$, which implies the claim. For (f) choose $x \in G$ with $\max f = f(x)$. Then $\max f = f(x)$ is less than or equal to $\sum_j c_j g \left(s_j^{-1} x \right) \le \sum_j c_j \max g$. $\qquad\square$

Fix a non-zero $f_0 \in C_c^+(G)$. For $f, \phi \in C_c^+(G)$ with $\phi \ne 0$ let

$$J(f, \phi) = J_{f_0}(f, \phi) = \frac{(f : \phi)}{(f_0 : \phi)}.$$

Lemma 1.3.9 *For* $f, g, \phi \in C_c^+(G)$ *with* $f, \phi \ne 0, c > 0$ *and* $s \in G$ *one has*

(a) $\frac{1}{(f_0 : f)} \le J(f, \phi) \le (f : f_0)$,

(b) $J(L_s f, \phi) = J(f, \phi)$,

(c) $J(f + g, \phi) \le J(f, \phi) + J(g, \phi)$,

(d) $J(cf, \phi) = c J(f, \phi)$.

Proof This follows from Lemma 1.3.8. $\qquad\square$

The map $J(\cdot, \phi)$ will approximate the Haar-integral as the support of ϕ shrinks to $\{e\}$. Directly from Lemma 1.3.9 we only get sub-additivity, but in the limit this function will become additive as the following lemma shows. This is the central point of the proof of the existence of the Haar integral.

Lemma 1.3.10 *Let* $f_1, f_2 \in C_c^+(G)$ *and* $\varepsilon > 0$. *Then there is a unit-neighborhood* V *in* G *such that*

$$J(f_1, \phi) + J(f_2, \phi) \le J(f_1 + f_2, \phi)(1 + \varepsilon)$$

holds for every $\phi \in C_c^+(G) \smallsetminus \{0\}$ with support in V.

Proof Choose $f' \in C_c^+(G)$ such that $f' \equiv 1$ on the support of $f_1 + f_2$. Let $\varepsilon, \delta > 0$ be arbitrary. Set

$$f = f_1 + f_2 + \delta f', \quad h_1 = \frac{f_1}{f}, \quad h_2 = \frac{f_2}{f},$$

where we set $h_j(x) = 0$ if $f(x) = 0$. Then $h_j \in C_c^+(G)$ for $j = 1, 2$.

According to Lemma 1.3.7, every function in $C_c(G)$ is left uniformly continuous, so in particular, for h_j this means that there is a unit-neighborhood V such that for $x, y \in G$ with $x^{-1}y \in V$ and $j = 1, 2$ one has $|h_j(x) - h_j(y)| < \frac{\varepsilon}{2}$. Let $\phi \in C_c^+(G) \smallsetminus \{0\}$ with support in V. Choose finitely many $s_k \in G, c_k > 0$ with $f \le \sum_k c_k L_{s_k} \phi$. Then $\phi(s_k^{-1}x) \neq 0$ implies $|h_j(x) - h_j(s_k)| < \frac{\varepsilon}{2}$, and for all x one has

$$f_j(x) = f(x)h_j(x) \le \sum_k c_k \phi \left(s_k^{-1}x\right) h_j(x)$$

$$\le \sum_k c_k \phi \left(s_k^{-1}x\right) \left(h_j(s_k) + \frac{\varepsilon}{2}\right),$$

so that $(f_j : \phi) \le \sum_k c_k(h_j(s_k) + \frac{\varepsilon}{2})$, implying that $(f_1 : \phi) + (f_2 : \phi)$ is less than or equal to $\sum_k c_k(1 + \varepsilon)$, which yields

$$J(f_1, \phi) + J(f_2, \phi) \le J(f, \phi)(1 + \varepsilon)$$

$$\le (J(f_1 + f_2, \phi) + \delta J(f', \phi))(1 + \varepsilon).$$

For $\delta \to 0$ we get the claim.   □

Lemma 1.3.8(e) together with $(f : f) = 1$ implies $\frac{1}{(f_0:f)} \le (f : f_0)$. For $f \in C_c^+(G) \smallsetminus \{0\}$ let S_f be the compact interval $\left[\frac{1}{(f_0:f)}, (f : f_0)\right]$. The space $S \stackrel{\text{def}}{=} \prod_{f \neq 0} S_f$, where the product runs over all non-zero $f \in C_c^+(G)$, is compact by Tychonov's Theorem A.7.1. Recall from Lemma 1.3.9 (a) that for every $\phi \in C_c^+(G) \smallsetminus \{0\}$ we get an element $J(f, \phi) \in S_f$ and hence an element $(J(f, \phi))_f$ of the product space S. For a unit-neighborhood V let L_V be the closure in S of the set of all $(J(f, \phi))_f$ where ϕ ranges over all ϕ with support in V. As S is compact, the intersection $\bigcap_V L_V$ over all unit-neighborhoods V is non-empty. Choose an element $(I_{f_0}(f))_f$ in this intersection. By Lemma 1.3.9 and 1.3.10, it follows that $I = I_{f_0}$ is a positive invariant homogeneous and additive map on $C_c^+(G)$. Any real valued function $f \in C_c(G)$ can be written as the difference $f_+ - f_-$ of two positive functions. Setting $I(f) = I(f_+) - I(f_-)$, and for complex-valued functions $I(f) = I(\text{Re}(f)) + iI(\text{Im}(f))$, one gets a well-defined positive linear map that is invariant. This proves the existence of the Haar integral. For the proof of the uniqueness we need the following lemma.

Lemma 1.3.11 *Let v be a Haar measure on G. Then for every $f \in C_c(G)$ the function $s \mapsto \int_G f(xs)\,dv(x)$ is continuous on G.*

Proof We have to show that for a given $s_0 \in G$ and given $\varepsilon > 0$ there exists a neighborhood U of s_0 such that for every $s \in U$ one has $\left|\int_G f(xs) - f(xs_0)\,dv(x)\right| < \varepsilon$. Replacing f by $R_{s_0} f(x) = f(xs_0)$, we are reduced to the case $s_0 = e$. Let K be the support of f, and let V be a compact symmetric unit-neighborhood. For $s \in V$ one has $\mathrm{supp}(R_s f) \subset KV$. Let $\varepsilon > 0$. As f is uniformly continuous, there is a symmetric unit-neighborhood W such that for $s \in W$ one has $|f(xs) - f(x)| < \frac{\varepsilon}{v(KV)}$. For $s \in U = W \cap V$ one therefore gets

$$\left|\int_G f(xs) - f(x)\,dv(x)\right| \leq \int_{KV} |f(xs) - f(x)|\,dv(x)$$
$$< \frac{\varepsilon}{v(KV)} v(KV) = \varepsilon.$$

The lemma is proven. $\qquad\square$

Suppose now that μ, v are two non-zero invariant Radon measures. We have to show that there is $c > 0$ with $v = c\mu$. For $f \in C_c(G)$ with $\int_G f(t)\,d\mu(t) = I_\mu(f) \neq 0$ set $D_f(s) \overset{\text{def}}{=} \int_G f(ts)\,dv(t)\frac{1}{I_\mu(f)}$. Then the function D_f is continuous by the lemma.

Let $g \in C_c(G)$. Using Fubini's Theorem (B.3.3) and the invariance of the measures μ, v we get

$$I_\mu(f)I_v(g) = \int_G \int_G f(s)g(t)\,dv(t)\,d\mu(s)$$
$$= \int_G \int_G f(s)g(s^{-1}t)\,dv(t)\,d\mu(s) = \int_G \int_G f(ts)g(s^{-1})\,d\mu(s)\,dv(t)$$
$$= \int_G \int_G f(ts)g(s^{-1})\,dv(t)\,d\mu(s) = \int_G \int_G f(ts)\,dv(t)\,g(s^{-1})\,d\mu(s)$$
$$= I_\mu(f) \int_G D_f(s)g(s^{-1})\,d\mu(s).$$

Since $I_\mu(f) \neq 0$ one concludes $I_v(g) = \int_G D_f(s)g(s^{-1})\,d\mu(s)$. Let f' be another function in $C_c(G)$ with $I_\mu(f') \neq 0$, so it follows $\int_G (D_f(s) - D_{f'}(s))g(s^{-1})\,d\mu(s) = 0$ for every $g \in C_c(G)$. Replacing g with the function $\tilde{g}$ given by $\tilde{g}(s) = |g(s)|^2 (\overline{D_f(s^{-1}) - D_{f'}(s^{-1})})$ one gets $\int_G |(D_f(s) - D_{f'}(s))g(s^{-1})|^2\,d\mu(s) = 0$. Corollary 1.3.6 (c) implies that $(D_f(s) - D_{f'}(s))g(s^{-1}) = 0$ holds for every $s \in G$. As g is arbitrary, one gets $D_f = D_{f'}$. Call this function D. For every f with $I_\mu(f) \neq 0$ one has $\int_G f(t)\,d\mu(t)D(e) = \int_G f(t)\,dv(t)$. By linearity, it follows that this equality holds everywhere. This finishes the proof of the theorem. $\qquad\square$

Example 1.3.12 Let B be the subgroup of $GL_2(\mathbb{R})$ defined by

$$B = \left\{ \begin{pmatrix} 1 & x \\ & y \end{pmatrix} : x, y \in \mathbb{R}, \ y \neq 0 \right\}.$$

Then $I(f) = \int_{\mathbb{R}^\times} \int_{\mathbb{R}} f\left(\begin{pmatrix} 1 & x \\ & y \end{pmatrix} \right) dx \frac{dy}{y}$ is a Haar-integral on B. (See Exercise 1.8.)

1.4   The Modular Function

From now on, if not mentioned otherwise, for a given locally compact group G, we will always assume a fixed choice of Haar measure. For the integral we will then write $\int_G f(x) \, dx$, and for the measure of a set $A \subset G$ we write $\mathrm{vol}(A)$. If the group G is compact, any Haar measure is finite, so, if not mentioned otherwise, we will then assume the measure to be the *normalized Haar measure*, i.e., we assume $\mathrm{vol}(G) = 1$ in that case. Also, for $p \geq 1$ we write $L^p(G)$ for the L^p-space of G with respect to a Haar measure, see Appendix B.4. Note that this space does not depend on the choice of a Haar measure.

Definition Let G be a locally-compact group, and let μ be a Haar measure on G. For $x \in G$ the measure μ_x, defined by $\mu_x(A) = \mu(Ax)$, is a Haar measure again, as for $y \in G$ one has $\mu_x(yA) = \mu(yAx) = \mu(Ax) = \mu_x(A)$. Therefore, by the uniqueness of the Haar measure, there exists a number $\Delta(x) > 0$ with $\mu_x = \Delta(x)\mu$. In this way one gets a map $\Delta : G \to \mathbb{R}_{>0}$, which is called the *modular function* of the group G. If $\Delta \equiv 1$, then G is called a *unimodular group*. In this case every left Haar measure is right invariant as well.

Let X be any set, and let $f : X \to \mathbb{C}$ be a function. The *sup-norm* or *supremum-norm* of f is defined by

$$\|f\|_X \overset{\text{def}}{=} \sup_{x \in X} |f(x)|.$$

Note that some authors use $\|\cdot\|_\infty$ to denote the sup-norm. This, however, is in conflict with the equally usual and better justified notation for the norm on the space L^∞ (See Appendix B.4).

Theorem 1.4.1

(a) *The modular function $\Delta : G \to \mathbb{R}_{>0}^\times$ is a continuous group homomorphism.*

(b) *One has $\Delta \equiv 1$ if G is abelian or compact.*

(c) *For $y \in G$ and $f \in L^1(G)$ one has $R_y f \in L^1(G)$ and*

$$\int_G R_y f(x) \, dx = \int_G f(xy) \, dx = \Delta(y^{-1}) \int_G f(x) \, dx.$$

(d) *The equality* $\int_G f(x^{-1}) \Delta(x^{-1}) \, dx = \int_G f(x) \, dx$ *holds for every integrable function* f.

Proof   Part (c) is clear if f is the characteristic function $\mathbf{1}_A$ of a measurable set A. It follows generally by the usual approximation argument.

We now prove part (a) of the theorem. For $x, y \in G$ and a measurable set $A \subset G$, one computes

$$\Delta(xy)\mu(A) = \mu_{xy}(A) = \mu(Axy) = \mu_y(Ax)$$
$$= \Delta(y)\mu(Ax) = \Delta(y)\Delta(x)\mu(A).$$

Choose A with $0 < \mu(A) < \infty$ to get $\Delta(xy) = \Delta(x)\Delta(y)$. Hence Δ is a group homomorphism.

Continuity: Let $f \in C_c(G)$ with $c = \int_G f(x) \, dx \neq 0$. By part (c) we have

$$\Delta(y) = \frac{1}{c} \int_G f\left(xy^{-1}\right) dx = \frac{1}{c} \int_G R_{y^{-1}} f(x) \, dx.$$

So the function Δ is continuous in y by Lemma 1.3.11.

For part (b), if G is abelian, then every right translation is a left translation, and so every left Haar measure is right-invariant.

If G is compact, then so is the image of the continuous map Δ. As Δ is a group homomorphism, the image is also a subgroup of $\mathbb{R}_{>0}$. But the only compact subgroup of the latter is the trivial group $\{1\}$, which means that $\Delta \equiv 1$.

Finally, for part (d) of the theorem let $f \in C_c(G)$ and set $I(f) = \int_G f\left(x^{-1}\right) \Delta\left(x^{-1}\right) dx$. Then, by part (c),

$$I(L_z f) = \int_G f(z^{-1}x^{-1}) \Delta(x^{-1}) \, dx = \int_G f((xz)^{-1}) \Delta(x^{-1}) \, dx$$
$$= \Delta(z^{-1}) \int_G f(x^{-1}) \Delta((xz^{-1})^{-1}) \, dx = \int_G f(x^{-1}) \Delta(x^{-1}) \, dx = I(f).$$

It follows that I is an invariant integral; hence there is $c > 0$ with $I(f) = c \int_G f(x) \, dx$. To show that $c = 1$ let $\varepsilon > 0$ and choose a symmetric unit-neighborhood V with $|1 - \Delta(s)| < \varepsilon$ for every $s \in V$. Choose a nonzero symmetric function $f \in C_c^+(V)$. Then

$$|1 - c| \int_G f(x) \, dx = \left| \int_G f(x) \, dx - I(f) \right| \leq \int_G |f(x) - f(x^{-1})\Delta(x^{-1})| \, dx$$

$$= \int_V f(x) |1 - \Delta(x^{-1})| \, dx < \varepsilon \int_G f(x) \, dx.$$

So one gets $|1 - c| < \varepsilon$, and as ε was arbitrary, it follows $c = 1$ as claimed. The proof of the theorem is finished. $\qquad\square$

Lemma 1.4.2 *For given $1 \leq p < \infty$, and $g \in L^p(G)$ the maps $y \mapsto L_y g$ and $y \mapsto R_y g$ are continuous maps from G to $L^p(G)$. In particular, for every $\varepsilon > 0$ there exists a neighborhood U of the unit such that*

$$y \in U \quad \Rightarrow \quad \begin{array}{l} \|L_y g - g\|_p < \varepsilon, \\ \|R_y g - g\|_p < \varepsilon. \end{array}$$

The proof will show that L is even uniformly continuous and in case that G is unimodular, so is R.

Proof   Note that by invariance of the Haar integral we have

$$\|L_y g - L_x g\|_p = \|L_{x^{-1}y} g - g\|_p,$$

so uniform continuity as claimed follows from continuity at 1, which is the displayed formula in the lemma. Likewise, for the right translation we have $\|R_y g - R_x g\|_p = \Delta(x^{-1})^{1/p} \|R_{x^{-1}y} g - g\|_p$ as follows from part (c) of the theorem. It remains to show continuity at the unit element. We first assume that $g \in C_c(G)$. Choose $\varepsilon > 0$. Let K be the support of g. Then the support of $L_y g$ is yK. Let U_0 be a compact symmetric unit-neighborhood. Then for $y \in U_0$ one has $\mathrm{supp} L_y g \subset U_0 K$.

Let $\delta > 0$. By Lemma 1.3.7 there exists a unit-neighborhood $U \subset U_0$ such that for $y \in U$, the sup-norm $\|L_y g - g\|_G$ is less than δ.

In particular, for every $y \in U$ one has

$$\|L_y g - g\|_p = \left(\int_G |g(y^{-1}x) - g(x)|^p \, dx \right)^{\frac{1}{p}} < \delta \, \mathrm{vol}(U_0 K)^{\frac{1}{p}}.$$

By setting δ equal to $\varepsilon/\mathrm{vol}(U_0 K)^{1/p}$, one gets the claim for $g \in C_c(G)$.

For general g, choose $f \in C_c(G)$ such that $\|f - g\|_p < \varepsilon/3$. Choose a unit-neighborhood U with $\|f - L_y f\|_p < \varepsilon/3$ for every $y \in U$. Then for $y \in U$ one has

$$\|g - L_y g\|_p \leq \|g - f\|_p + \|f - L_y f\|_p + \|L_y f - L_y g\|_p < \frac{\varepsilon}{3} + \frac{\varepsilon}{3} + \frac{\varepsilon}{3} = \varepsilon.$$

In the last step we have used $\|L_y f - L_y g\|_p = \|f - g\|_p$, i.e., the left-invariance of the p-norm. This implies the claim for the left translation. The proof for the right-translation R_y is similar, except for the very last step, where instead of the invariance we use the continuity of the modular function and the equality $\|R_y f - R_y g\|_p = \Delta(y^{-1})^{1/p} \|f - g\|_p$, which follows from part (c) of the theorem. $\qquad\square$

Example 1.4.3 Let B be the group of real matrices of the form $\left(\begin{smallmatrix} 1 & x \\ 0 & y \end{smallmatrix}\right)$ with $y \neq 0$. Then the modular function Δ is given by $\Delta\left(\begin{smallmatrix} 1 & x \\ 0 & y \end{smallmatrix}\right) = |y|$ (See Exercise 1.8).

Proposition 1.4.4 *Let G be a locally compact group. The following assertions are equivalent.*

(a) *There exists $x \in G$ such that the singleton $\{x\}$ has non-zero measure.*

(b) *The set $\{1\}$ has non-zero measure.*

(c) *The Haar measure is a multiple of the counting-measure.*

(d) *G is a discrete group.*

Proof The equivalence of (a) and (b) is clear by the invariance of the measure. Assume (b) holds. Let $c > 0$ be the measure of $\{1\}$. Then for every finite set $E \subset G$ one has $\mathrm{vol}(E) = \sum_{e \in E} \mathrm{vol}(\{e\}) = c\#E$. Since the measure is monotonic, every infinite set gets measure ∞, and so the Haar measure equals c times the counting measure.

To see that (c) implies (d) recall that every compact set has finite measure, and by locally compactness, there exists an open set of finite measure, i.e., a finite set U that is open. By the Hausdorff axiom one can separate the elements of U by open sets, so the singletons in U are open; hence every singleton, and so every set, is open, i.e., G is discrete. Finally, if G is discrete, then each singleton is open, hence has strictly positive measure by Corollary 1.3.6. $\square$

Proposition 1.4.5 *Let G be a locally compact group. Then G has finite volume under a Haar measure if and only if G is compact.*

Proof If G is compact, it has finite volume by Corollary 1.3.2. For the other direction suppose G has finite Haar measure. Let U be a compact unit-neighborhood. As the Haar measure of G is finite, there exists a maximal number $n \in \mathbb{N}$ of pairwise disjoint translates xU of U. Let $z_1 U, \ldots, z_n U$ be such pairwise disjoint translates, and set K equal to the union of these finitely many translates. Then K is compact, and for every $x \in G$ one has $K \cap xK \neq \emptyset$. This means that $G = KK^{-1}$, which is a compact set. $\square$

1.5 The Quotient Integral Formula

Let G be a locally compact group and let H be a closed subgroup. Then G/H is a locally compact Hausdorff space by Proposition 1.2.1. For $f \in C_c(G)$ let $f^H(x) \stackrel{\text{def}}{=} \int_H f(xh)\, dh$. For any x the function mapping h to $f(xh)$ is continuous of compact support, so the integral exists.

Lemma 1.5.1 *The function f^H lies in $C_c(G/H)$, and the support of f^H is contained in* $(\mathrm{supp}(f)H)/H$. *The map $f \mapsto f^H$ from $C_c(G)$ to $C_c(G/H)$ is surjective.*

Proof Let K be the support of f. Then KH/H is compact in G/H and contains the support of f^H, which therefore is compact. To prove continuity, let $x_0 \in G$ and U a compact neighborhood of x_0. For every $x \in U$, the function $h \mapsto f(xh)$ is supported in the compact set $U^{-1}K \cap H$. Put $d = \mu_H(U^{-1}K \cap H)$, where μ_H denotes the given Haar measure on H. Given $\varepsilon > 0$ it follows from uniform continuity of f (Lemma 1.3.7) that there exists a neighborhood $V \subseteq U$ of x_0 such that $|f(xh) - f(x_0h)| < \frac{\varepsilon}{d}$ for every $x \in V$, from which it follows that

$$|f^H(x) - f^H(x_0)| \leq \int_{U^{-1}K \cap H} |f(xh) - f(x_0h)| \, dh \; < \; \varepsilon$$

for every $x \in V$, which proves continuity of f^H.

Write π for the natural projection $G \to G/H$. To show surjectivity of the map $f \mapsto f^H$, we first show that for a given compact subset C of the quotient G/H there exists a compact subset K of G such that $C \subset \pi(K)$. To this end choose a pre-image $y_c \in G$ to every $c \in C$ and an open, relatively compact neighborhood $U_c \subset G$ of y_c. As π is open, the images $\pi(U_c)$ form an open covering of C, so there are $c_1, \ldots c_n \in C$ such that $C \subset \pi(K)$ with K being the compact set $\overline{U}_{c_1} \cup \cdots \cup \overline{U}_{c_n}$.

Apply this construction to the set C being the support of a given $g \in C_c(G/H)$. Let $\phi \in C_c(G)$ be such that $\phi \geq 0$ and $\phi \equiv 1$ on K, which exists by Urysohn's Lemma (Lemma A.8.1). Then set $f = g\phi/\phi^H$ where g is non-zero and $f = 0$ otherwise. This definition makes sense as $\phi^H > 0$ on the support of g. One gets $f \in C_c(G)$ and $f^H = g\phi^H/\phi^H = g$. □

Remark 1.5.2 For later use we note that in the proof of the above lemma we also showed that for any compact set $C \subset G/H$ there exists a compact set $K \subset G$ such that $C \subseteq \pi(K)$. By passing to $\pi^{-1}(C) \cap K$ if necessary, we can even choose K such that $\pi(K) = C$.

A measure ν on the Borel σ-algebra of G/H is called an *invariant measure* if $\nu(xA) = \nu(A)$ holds for every $x \in G$ and every measurable $A \subset G/H$. Let Δ_G be the modular function of G and Δ_H the modular function of H.

Theorem 1.5.3 (Quotient Integral Formula) *Let G be a locally compact group, and let H be a closed subgroup. There exists an invariant Radon measure $\nu \neq 0$ on the quotient G/H if and only if the modular functions Δ_G and Δ_H agree on H. In this case, the measure ν is unique up to a positive scalar factor. Given Haar measures on G and H, there is a unique choice for ν, such that for every $f \in C_c(G)$ one has the quotient integral formula*

$$\int_G f(x) \, dx = \int_{G/H} \int_H f(xh) \, dh \, dx.$$

We will always assume this normalization and call the ensuing measure on G/H the quotient measure.

The quotient integral formula is valid for every $f \in L^1(G)$.

The last assertion says that if f is an integrable function on G, then the integral $f^H(x) = \int_H f(xh) \, dx$ exists almost everywhere in x and defines a ν-measurable, indeed integrable function on G/H, such that the integral formula holds.

See Exercise 1.10 for a generalization of the quotient integral formula.

Proof Assume first that there exists an invariant Radon measure $\nu \neq 0$ on the quotient space G/H. Define a linear functional I on $C_c(G)$ by $I(f) = \int_{G/H} f^H(x) \, d\nu(x)$. Then $I(f)$ is a non-zero invariant integral on G, so it is given by a Haar measure. We write $I(f) = \int_G f(x) \, dx$. For $h_0 \in H$ one gets

$$\Delta_G(h_0) \int_G f(x) \, dx = \int_G f\left(xh_0^{-1}\right) dx = \int_G R_{h_0^{-1}} f(x) \, dx$$

$$= \int_{G/H} \int_H f\left(xhh_0^{-1}\right) dh \, d\nu(x)$$

$$= \Delta_H(h_0) \int_{G/H} \int_H f(xh) \, dh \, d\nu(x)$$

$$= \Delta_H(h_0) \int_G f(x) \, dx.$$

As f can be chosen with $\int_G f(x) \, dx \neq 0$, it follows that $\Delta_G|_H = \Delta_H$.

For the converse direction assume $\Delta_G|_H = \Delta_H$, and let $f \in C_c(G)$ with $f^H = 0$. We want to show that $\int_G f(x) \, dx = 0$. For let ϕ be another function in $C_c(G)$. We use Fubini's Theorem to get

$$0 = \int_G \int_H f(xh)\phi(x) \, dh \, dx = \int_H \int_G \phi(x) f(xh) \, dx \, dh$$

$$= \int_H \Delta_G(h^{-1}) \int_G \phi\left(xh^{-1}\right) f(x) \, dx \, dh$$

$$= \int_G \int_H \Delta_H\left(h^{-1}\right) \phi\left(xh^{-1}\right) dh \, f(x) \, dx$$

$$= \int_G \int_H \phi(xh) \, dh \, f(x) \, dx = \int_G \phi^H(x) f(x) \, dx.$$

As we can find ϕ with $\phi^H \equiv 1$ on the support of f, it follows that $\int_G f(x) \, dx = 0$. This means that we can unambiguously define a non-zero invariant integral on G/H by $I(g) = \int_G f(x) \, dx$, whenever $g \in C_c(G/H)$ and $f \in C_c(G)$ with $f^H = g$. By Riesz's Theorem, this integral comes from an invariant Radon measure. In particular, it follows that the quotient integral formula is valid for every $f \in C_c(G)$. All but the last assertion of the theorem is proven.

We want to prove the quotient integral formula for an integrable function f on G. It suffices to assume $f \geq 0$. Then f is a monotone limit of step-functions, so by monotone convergence one may assume f is a step-function itself and by linearity one reduces to the case of f being the characteristic function of a measurable set A with finite Haar measure. We have to show that $\mathbf{1}_A^H$ is measurable on G/H and that its integral equals $\int_G \mathbf{1}_A(x)\,dx$. We start with the case of $A = U$ being open. Note that by Lemma B.3.2 the function $g(xH) = \sup_{\substack{\phi \in C_c(G) \\ 0 \leq \phi \leq \mathbf{1}_U}} \int_H \phi(xh)\,dh$ is measurable on G/H and coincides with $\mathbf{1}_U^H$. A repeated use of the Lemma of Urysohn and Lemma B.3.2 shows the claim for $A = U$,

$$\int_{G/H} \int_H \mathbf{1}_U(xh)\,dh\,dx = \int_{G/H} \int_H \sup_{0 \leq \phi \leq \mathbf{1}_U} \phi(xh)\,dh\,dx$$

$$= \sup_{0 \leq \phi \leq \mathbf{1}_U} \int_{G/H} \int_H \phi(xh)\,dh\,dx$$

$$= \sup_{0 \leq \phi \leq \mathbf{1}_U} \int_G \phi(x)\,dx = \int_G \sup_{0 \leq \phi \leq \mathbf{1}_U} \phi(x)\,dx$$

$$= \int_G \mathbf{1}_U(x)\,dx.$$

If $A = K$ is a compact set, then let V be a relatively compact open neighborhood of K. Then $\mathbf{1}_K = \mathbf{1}_V - \mathbf{1}_{V \smallsetminus K}$. The claim follows for $A = K$. For general A of finite measure and given $n \in \mathbb{N}$, by regularity and Lemma B.2.1, there are a compact set K_n and an open set U_n such that $K_n \subset A \subset U_n$ and $\mu(U_n \smallsetminus K_n) < 1/n$. We can further assume that the sequence U_n is decreasing and K_n is increasing. Let g be the pointwise limit of the increasing sequence $\mathbf{1}_{K_n}^H$ and let h be the limit of $\mathbf{1}_{U_n}$. Then g and h are integrable on G/H, one has $0 \leq g \leq \mathbf{1}_A^H \leq h$ and $h - g$ is a positive function of integral zero, hence a nullfunction. This means that $\mathbf{1}_A^H$ coincides with g up to a nullfunction and thus is measurable. One has

$$\int_{G/H} \mathbf{1}_A^H(x)\,dx = \int_{G/H} g(x)\,dx = \lim_n \int_{G/H} \mathbf{1}_{K_n}^H(x)\,dx$$

$$= \lim_n \int_G \mathbf{1}_{K_n}(x)\,dx = \int_G \mathbf{1}_A(x)\,dx. \qquad \square$$

The quotient integral formula should be understood as a one-sided version of Fubini's Theorem for product spaces. As for Fubini, it has a partial converse, which we give now. Let μ be a measure on a set X. Recall that a measurable subset $A \subset X$ is called σ-finite if A can be written as a countable union $A = \bigcup_{j=1}^{\infty} A_j$ of sets with $\mu(A_j) < \infty$ for every j. If X itself is σ-finite, one also says that the measure μ is σ-finite.

Corollary 1.5.4 *Suppose that H is a closed subgroup of G such that there exists an invariant Radon measure $\neq 0$ on G/H. Let $f : G \to \mathbb{C}$ be a measurable function such that the set $A = \{x \in G : f(x) \neq 0\}$ is σ-finite. If the iterated integral $\int_{G/H} \int_H |f(xh)|\,dh\,dx$ exists, then f is integrable.*

Proof It suffices to show that $|f|$ is integrable. So choose a sequence $(A_n)_{n\in\mathbb{N}}$ of measurable sets in G with finite Haar measure such that $A = \bigcup_{n=1}^{\infty} A_n$, and define $f_n : G \to \mathbb{C}$ by $f_n = \min(|f| \cdot \mathbf{1}_{A_n}, n)$. Then $(f_n)_{n\in\mathbb{N}}$ is an increasing sequence of integrable functions that converges point-wise to $|f|$. It follows from Theorem 1.5.3 that $\int_G f_n(x)\,dx = \int_{G/H} \int_H f_n(xh)\,dh\,dx \leq \int_{G/H} \int_H |f(xh)|\,dh\,dx$ for every $n \in \mathbb{N}$. The result follows then from the Monotone Convergence Theorem. □

Corollary 1.5.5

(a) *If H is a normal closed subgroup of G, then the modular functions Δ_G and Δ_H agree on H.*

(b) *Let H be the kernel of Δ_G. Then H is unimodular.*

Proof (a) The Haar measure of the group G/H is an invariant Radon measure, so (a) follows from the theorem. Part (b) follows from part (a). □

Proposition 1.5.6 *Let G be a locally compact group, $K \subset G$ a compact subgroup and $H \subset G$ a closed subgroup such that $G = HK$. Then one can arrange the Haar measures on G, H, K in a way that for every $f \in L^1(G)$ one has*

$$\int_G f(x)\,dx = \int_H \int_K f(hk)\,dk\,dh.$$

Proof The group $H \times K$ acts on G by $(h, k).g = hgk^{-1}$. As this operation is transitive, G can be identified with $H \times K / H \cap K$, where we embed $H \cap K$ diagonally into $H \times K$. The group $H \cap K$ is compact; therefore it has trivial modular function and the modular function of $H \times K$ is trivial on this subgroup. By Theorem 1.5.3 there is a unique $H \times K$-invariant Radon measure on G up to scaling. We show that the Haar measure on G also is $H \times K$-invariant, so the uniqueness implies our claim. Obviously, the Haar measure is invariant under the action of H as the latter is the left multiplication. As K is compact, we have $\Delta_G(k) = 1$ for every $k \in K$ and so $\int_G f(xk)\,dx = \int_G f(x)\,dx$ for every $f \in C_c(G)$ by Theorem 1.4.1 (c). □

Lemma 1.5.7 *Let H be a closed subgroup of the locally compact group G such that there exists a G-invariant Radon measure on G/H. Fix such a measure. For given $1 \leq p < \infty$, and $g \in L^p(G/H)$ the map $y \mapsto L_y g$ is a uniformly continuous map from G to $L^p(G/H)$. In particular, for every $\varepsilon > 0$ there exists a neighborhood U of the unit such that*

$$y \in U \quad \Rightarrow \quad \|L_y g - g\|_p < \varepsilon.$$

Proof The lemma is a generalization of Lemma 1.4.2 and the proof of the latter extends to give a proof of the current lemma. □

1.6   Convolution

An *algebra* over $\mathbb{C}$ is a complex vector space $\mathcal{A}$ together with a map $\mathcal{A} \times \mathcal{A} \to \mathcal{A}$, called product or multiplication and written $(a, b) \mapsto ab$, which is bilinear, i.e., it satisfies

$$a(b + c) = ab + ac, \quad (a + b)c = ab + ac, \quad \lambda(ab) = (\lambda a)b = a(\lambda b)$$

for $a, b, c \in \mathcal{A}$ and $\lambda \in \mathbb{C}$, and it is *associative*, i.e., one has

$$a(bc) = (ab)c$$

for all $a, b, c \in \mathcal{A}$. The algebra $\mathcal{A}$ is called a *commutative algebra* if in addition for all $a, b \in \mathcal{A}$ one has $ab = ba$.

Example 1.6.1

- The vector space $\mathcal{A} = M_n(\mathbb{C})$ of complex $n \times n$ matrices forms an algebra with matrix multiplication as product. This algebra is not commutative unless $n = 1$.
- For a set S the vector space $\mathrm{Map}(S, \mathbb{C})$ of all maps from S to $\mathbb{C}$ forms a commutative algebra with the point-wise product, i.e., for $f, g \in \mathrm{Map}(S, \mathbb{C})$ the product fg is the function given by $(fg)(s) = f(s)g(s)$ for $s \in S$.

Definition Let G be a locally-compact group. For two measurable functions $f, g : G \to \mathbb{C}$ define the *convolution product* as

$$f * g(x) = \int_G f(y)g(y^{-1}x)\,dy,$$

whenever the integral exists.

Theorem 1.6.2 *If $f, g \in L^1(G)$, then the integral $f * g$ exists almost everywhere in x and defines a function in $L^1(G)$. The L^1-norm satisfies $\|f * g\|_1 \leq \|f\|_1 \|g\|_1$. The convolution product endows $L^1(G)$ with the structure of an algebra.*

Proof Let f, g be integrable functions on G. Then f and g are measurable in the sense that pre-images of Borel-sets are in the completed Borel-σ-algebra. Let the function ψ be defined by $\psi(y, x) = f(y)g(y^{-1}x)$. We write ψ as a composition of the map $\alpha : G \times G \to G \times G; (y, x) \mapsto (y, y^{-1}x)$ followed by $f \times g$ and multiplication, which are measurable. We show that α is measurable. Recall that we need measurability here with respect to the completion of the Borel σ-algebra. Since α is continuous, it is measurable with respect to the Borel σ-algebra, so we need to know that the pre-image of a null-set is a null-set. This however is clear, as α preserves the Haar measure on $G \times G$, as follows from the formula

$$\int_G \int_G \phi(y, x)\, dx\, dy = \int_G \int_G \phi\left(y, y^{-1}x\right) dx\, dy, \quad \phi \in C_c(G \times G),$$

and Fubini's Theorem. Being a composition of measurable maps, ψ is measurable. Let $S(f)$ and $S(g)$ be the supports of f and g, respectively. Then the sets $S(f)$ and $S(g)$ are σ-compact by Corollary 1.3.6 (d). The support of ψ is contained in the σ-compact set $S(f) \times S(f)S(g)$, and therefore is σ-compact itself. We can apply the Theorem of Fubini to calculate

$$\|f * g\|_1 \le \int_G \int_G |f(y)g\left(y^{-1}x\right)|\, dy\, dx = \int_G \int_G |f(y)g\left(y^{-1}x\right)|\, dx\, dy$$

$$= \int_G \int_G |f(y)g(x)|\, dx\, dy = \|f\|_1 \|g\|_1 < \infty.$$

The function $\psi(x, \cdot)$ is therefore integrable almost everywhere in x, and the function $f * g$ exists and is measurable. Further, the norm $\|f * g\|_1$ is less than or equal to $\int_{G \times G} |\psi(x, y)|\, dx\, dy = \|f\|_1 \|g\|_1$. Associativity and distributivity are proven by straightforward calculations. $\square$

Recall that for a function $f : G \to \mathbb{C}$ and $y \in G$ we have defined

$$R_y(f)(x) = f(xy) \quad \text{and} \quad L_y(f)(x) = f(y^{-1}x).$$

Lemma 1.6.3 *For $f, g \in L^1(G)$ and $y \in G$ one has $R_y(f * g) = f * (R_y g)$ and $L_y(f * g) = (L_y f) * g$.*

Proof We compute

$$R_y(f * g)(x) = \int_G f(z)g\left(z^{-1}xy\right) dz = \int_G f(z)R_y g\left(z^{-1}x\right) dz = f * (R_y g)(x),$$

and likewise for L. $\square$

Theorem 1.6.4 *The algebra $L^1(G)$ is commutative if and only if G is abelian.*

Proof Assume $L^1(G)$ is commutative. Let $f, g \in L^1(G)$. For $x \in G$ we have

$$0 = f * g(x) - g * f(x) = \int_G f(xy)g\left(y^{-1}\right) - g(y)f\left(y^{-1}x\right) dy$$

$$= \int_G g(y)\left(\Delta(y^{-1})f(xy^{-1}) - f(y^{-1}x)\right) dy.$$

Since this is valid for every g, one concludes that $\Delta\left(y^{-1}\right) f\left(xy^{-1}\right) = f\left(y^{-1}x\right)$ holds for every $f \in C_c(G)$. For $x = 1$ one gets $\Delta \equiv 1$, so G is unimodular and $f\left(xy^{-1}\right) = f\left(y^{-1}x\right)$ for every $f \in C_c(G)$ and all $x, y \in g$. This implies that G is abelian. The converse direction is trivial. $\square$

Definition By a *Dirac function* we mean a function $\phi \in C_c(G)$, which

- is positive, i.e., $\phi \geq 0$,
- has integral equal to one, $\int_G \phi(x)\,dx = 1$, and
- is symmetric, $\phi(x^{-1}) = \phi(x)$.

A *Dirac family* is a family $(\phi_U)_U$ of Dirac functions indexed by the set $\mathcal{U}$ of all unit-neighborhoods U such that ϕ_U has support inside U. Note that the set $\mathcal{U}$ can be partially ordered by reversed inclusion which makes it a directed set. So a Dirac-family is a net, which we also refer to as a *Dirac net*

Lemma 1.6.5 *If ϕ and ψ are Dirac functions, then so is their convolution product $\phi * \psi$. To every unit neighborhood U their exists a Dirac function ϕ_U such that ϕ_U as well $\phi_U * \phi_U$ have support inside U.*

Proof If ϕ and ψ are positive, then so is their convolution product. For the integral we have $\int_G \phi * \psi(x)\,dx = \int_G \phi(x)\,dx \int_G \psi(x)\,dx = 1$ and symmetry is preserved by convolution. For the second assertion, let U be a given unit neighborhood. Then their exists a symmetric unit neighborhood $W \subset U$ such that $W^2 \subset U$ as well. The Lemma of Urysohn (A.8.1) yields a function $h \in C_c(G)$ with $0 \neq h \geq 0$ and supp$(h) \subset W$. Set $\phi_U(x) = h(x) + h(x^{-1})$ and scale this function so it has integral one. Then supp$(\phi_U * \phi_U) \subset$ supp(ϕ_U)supp$(\phi_U) \subset W^2 \subset U$, so ϕ_U satisfies the claim. □

Lemma 1.6.6 *Let $\varepsilon > 0$. For every $f \in L^1(G)$ there exists a unit-neighborhood U such that for every Dirac function ϕ_U with support in U one has*

$$\|f * \phi_U - f\|_1 < \varepsilon, \quad \|\phi_U * f - f\|_1 < \varepsilon.$$

For every continuous function f on G and every compact set $K \subset G$ there exists a unit-neighborhood U such that for every Dirac function ϕ_U with support in U one has

$$\|f * \phi_U - f\|_K < \varepsilon, \quad \|\phi_U * f - f\|_K < \varepsilon,$$

where $\|g\|_K = \sup_{x \in K} |g(x)|$.

In other words this means that the net $(\phi_U * f)_U$ indexed by the set of all unit-neighborhoods, converges to f in the L^1 norm if $f \in L^1(G)$ and compactly uniformly, if f is continuous.

Proof It suffices to consider $\phi_U * f$, as the other side is treated similarly. We compute

$$
\|\phi_U * f - f\|_1 = \int_G \left| \int_G \phi_U(y)(f\left(y^{-1}x\right) - f(x))\,dy \right| dx
$$

$$
\leq \int_G \int_G \phi_U(y)|f\left(y^{-1}x\right) - f(x)|\,dy\,dx
$$

$$
= \int_G \phi_U(y)\|L_y f - f\|_1\,dy.
$$

The claim nor follows from Lemma 1.4.2.

For the last statement let f be continuous and let $K \subset G$ be compact. Since a continuous function on a compact set is uniformly continuous, for every $\varepsilon > 0$ there exists a unit-neighborhood U, such that for $x \in K$, $y^{-1}x \in U$ one has $|f(y) - f(x)| < \varepsilon$. Let now ϕ_U be a Dirac function with support in U. Then $|\phi_U * f(x) - f(x)| \leq \int_G \phi_U(y^{-1}x)|f(y) - f(x)|\,dy < \varepsilon$. □

1.7 The Fourier Transform

A locally compact abelian group will be called an *LCA-group* for short. A *character* of an LCA-group A is a continuous group homomorphism

$$
\chi : A \to \mathbb{T},
$$

where $\mathbb{T}$ is the *circle group*, i.e., the multiplicative group of all complex numbers of absolute value one.

Example 1.7.1

- The characters of the group $\mathbb{Z}$ are the maps $k \mapsto e^{2\pi i kx}$, where x varies in $\mathbb{R}/\mathbb{Z}$ (See [Dei05] Sect. 7.1).

- The characters of $\mathbb{R}/\mathbb{Z}$ are the maps $x \mapsto e^{2\pi i kx}$, where k varies in $\mathbb{Z}$ (See [Dei05] Sect. 7.1).

Definition The set of characters forms a group under point-wise multiplication, called the *dual group* and denoted $\widehat{A}$. Later, we will equip the group $\widehat{A}$ with a topology that makes it an LCA-group.

Let $f \in L^1(A)$ and define its *Fourier transform* to be the map $\hat{f} : \widehat{A} \to \mathbb{C}$ given by

$$
\hat{f}(\chi) \overset{\text{def}}{=} \int_A f(x)\overline{\chi(x)}\,dx.
$$

This integral exists as χ is bounded.

Lemma 1.7.2 *For $f, g \in L^1(A)$ and $\chi \in \widehat{A}$ one has $|\hat{f}(\chi)| \leq \|f\|_1$ and $\widehat{f * g} = \hat{f}\hat{g}$.*

Proof The first assertion is clear. Using Fubini's Theorem, one computes,

$$\widehat{f * g}(\chi) = \int_A f * g(x)\overline{\chi(x)}\, dx = \int_A \int_A f(y)g\left(y^{-1}x\right)\, dy\, \overline{\chi(x)}\, dx$$

$$= \int_A f(y) \int_A g\left(y^{-1}x\right)\, \overline{\chi(x)}\, dx\, dy$$

$$= \int_A f(y)\overline{\chi(y)}\, dy \int_A g(x)\,\overline{\chi(x)}\, dx = \hat{f}(\chi)\hat{g}(\chi). \qquad \square$$

1.8   Limits

In this section we shall give a construction principle for locally compact groups as *limits*, i.e., direct and projective limits. The reader mostly interested in developing the theory may proceed to the next chapter.

We first recall the notion of a *partial order* on a set I. This is a relation $\leq$ such that for all $a, b, c \in I$ one has $a \leq a$, (reflexivity); $a \leq b$ and $b \leq a$ implies $a = b$ (antisymmetry); $a \leq b$ and $b \leq c$ implies $a \leq c$ (transitivity).

Definition A *directed set* is a tuple $(I, \leq)$ consisting of a non-empty set I and a partial order $\leq$ on I, such that any two elements of I have a common upper bound, which means that for any two $a, b \in I$ there exists an element $c \in I$ with $a \leq c$ and $b \leq c$.

Examples 1.8.1

- The set $\mathbb{N}$ of natural numbers is an example with the natural order $\leq$. In this case the order is even *linear*, which means that any two elements on $\mathbb{N}$ can be compared. Every linear order is directed.

- Let Ω be an infinite set and let I be the set of all finite subsets of Ω, ordered by inclusion, so $A \leq B \Leftrightarrow A \subset B$. Then I is directed, as for $A, B \in I$ the union $C = A \cup B$ is an upper bound.

- Directed sets are precisely the index sets of nets $(x_i)_{i \in I}$.

Direct Limits

A *direct system* of groups consists of the following data

- a directed set $(I, \leq)$,
- a family $(G_i)_{i \in I}$ of groups and
- a family of group homomorphisms $\phi_i^j : G_i \to G_j$, if $i \leq j$,

such that the following axioms are satisfied:

$$\phi_i^i = \mathrm{Id}_{G_i} \quad \text{and} \quad \phi_j^k \circ \phi_i^j = \phi_i^k, \text{ if } i \leq j \leq k.$$

Examples 1.8.2

- Let G be a group and let $(G_i)_{i \in I}$ be a family of subgroups, such that for any two indices $i, j \in I$ there exists an index $k \in I$, such that $G_i, G_j \subset G_k$. Then the G_i form a direct system, if on I one installs the partial order

$$i \leq j \quad \Leftrightarrow \quad G_i \subset G_j,$$

and if for group homomorphisms ϕ_i^j one takes the inclusions.

- Let X be a topological space, fix $x_0 \in X$ and let I be the set of all neighborhoods of x_0 in X. For $U \in I$ let G_U be the group $C(U)$ of all continuous functions from U to $\mathbb{C}$. We order I by the *inverse inclusion*, i.e., $U \leq V \Leftrightarrow U \supset V$. The *restriction homomorphisms*

$$\phi_U^V : C(U) \to C(V), \quad \phi_U^V(f) = f|_V$$

form a direct system.

Definition Let $\left((G_i)_{i \in I}, (\phi_i^j)_{i \leq j} \right)$ be a direct system of groups. The *direct limit* of the system is the set

$$\varinjlim_{i \in I} G_i \stackrel{\text{def}}{=} \coprod_{i \in I} G_i \; / \sim,$$

where $\coprod$ denotes the disjoint union and $\sim$ the following equivalence relation: For $a \in G_i$ and $b \in G_j$ we say $a \sim b$, if there is $k \in I$ with $k \geq i, j$ and $\phi_i^k(a) = \phi_j^k(b)$.

On the set $G = \varinjlim G_i$ we define a group multiplication as follows. Let $a \in G_i$ and $b \in G_j$ and let $[a]$ and $[b]$ denote their equivalence classes in G. Then there is $k \in I$ with $k \geq i$ and $k \geq j$. We define $[a][b]$ to be the equivalence class of the element $\phi_i^k(a)\phi_j^k(b)$ in G_k, so $[a][b] = [\phi_i^k(a)\phi_j^k(b)]$. Some authors also use the notion *inductive limit* instead of direct limit.

Proposition 1.8.3 *The multiplication is well-defined and defines a group structure on the set G. This group is called the* direct limit *of the system* (G_i, ϕ_j^i). *For every $i \in I$ the map*

$$\psi_i : G_i \hookrightarrow \coprod_{j \in I} G_j \to G$$

is a group homomorphism.

The direct limit has the following universal property: Let Z be a group and for every
$i \in I$ *let a group homomorphism* $\alpha_i : G_i \to Z$ *be given, such that* $\alpha_i = \alpha_j \circ \phi_i^j$
holds if $i \leq j$. *Then there exists exactly one group homomorphism* $\alpha : G \to Z$
making all diagrams

commutative.

Note that in this construction the word "group" can be replaced with other algebraic
structures, like rings. Then one assumes that the structure homomorphisms ϕ_i^j are
ring homomorphisms and gets a ring as direct limit.

Proof   To show well-definedness, we need to show that the product is independent
of the choice of k. If k' is another element of I with $k' \geq i, j$, there exists a common
upper bound l for k and k', so $l \geq k, k'$. We show that the construction gives the same
element with l as with k. Then we apply the same argument to k' and l. Note that by
definition for every $c \in G_k$ one has $[c] = [\phi_k^l(c)]$. Gs ϕ_k^l is a group homomorphism,
it follows that

$$\left[\phi_i^k(a)\phi_j^k(b)\right] = \left[\phi_k^l\left(\phi_i^k(a)\phi_j^k(b)\right)\right] = \left[\phi_k^l(\phi_i^k(a))\phi_k^l\left(\phi_j^k(b)\right)\right] = \left[\phi_i^l(a)\phi_j^l(b)\right].$$

This proves well-definedness. The rest is left as an exercise to the reader.   □

Examples 1.8.4

- In the case of the direct system $(C(U))_U$, where U runs through all neighborhoods
  of a point in a topological space, one calls the elements of $\varinjlim C(U)$ *germs of*
  continuous functions.

- A special example of a direct limit is the *direct sum* of groups. So let $S \neq \emptyset$ be
  an index set and for each $s \in S$ let G_s be a group. Let I be the directed set of all
  finite subsets of S. For each $E \in I$ we let G_E be the finite product of groups,

$$G_E = \prod_{s \in E} G_s.$$

For $E \subset F$ in I we have the natural group homomorphism $\phi_E^F : G_E \to G_F$
sending x to $(x, 1, \ldots, 1)$. The direct limit constructed in this way is called the
direct sum of the groups G_s and is denoted as

$$\bigoplus_{s \in S} G_s.$$

Since all groups G_E can also be embedded into the product $\prod_{s\in S} G_s$ we find that the direct sum is isomorphic to the subgroup of $\prod_{s\in S} G_s$ consisting of those elements x with $x_s = 1$ for almost all $s \in S$.

Definition We say that a direct system $((G_i)_{i\in I}, (\phi_i^j)_{i\le j})$ is a *Mittag-Leffler direct system*, if the kernel of the homomorphism ϕ_i^k stabilizes as k grows. More precisely, if for every $i \in I$ there is a $k_0 \ge i$ such that for every $k \ge k_0$ one has

$$\ker(\phi_i^k) = \ker(\phi_i^{k_0}).$$

In particular, it follows that $\ker(\psi_i) = \ker(\phi_i^{k_0})$.

Examples 1.8.5

- The case of a family of subgroups provides an example of a Mittag-Leffler system, as here the structure homomorphisms are indeed injective.

- The system of germs of continuous functions at a point is in general not a Mittag-Leffler system.

Definition Suppose that (G_i, ϕ_i^j) is a direct system of topological groups, i.e., each G_i is a topological group and each ϕ_i^j is continuous. Then one defines the *direct product topology* on the limit $G = \varinjlim G_i$ to be the topology generated by the maps $\psi_i : G_i \to G$, i.e., it is the finest topology that makes all maps $G_i \to G$ continuous. Recall that a map $f : G \to X$ into some topological space is continuous if and only if all compositions $f \circ \psi_i$ are continuous (see Appendix A.5).

Proposition 1.8.6 *Let (G_i, ϕ_i^j) be a direct system of topological groups with limit G. Assume that all structure homomorphisms ϕ_i^j are open maps.*

(a) *The limit G is a topological group, when equipped with the inductive limit topology. The natural homomorphisms $\psi_i : G_i \to G$ are open maps.*

(b) *Suppose that all the groups G_i are Hausdorff, then the limit G is Hausdorff if and only if each of the kernels of the maps $\psi_i : G_i \to G$ is closed.*

(c) *If the system is Mittag-Leffler and all G_i are Hausdorff, then G is Hausdorff.*

(d) *If all G_i are locally compact groups and $\ker(\psi_i)$ is closed for each $i \in I$, then G is a locally compact group.*

Proof (a) A subset U of G is open if and only if the pre-image $\psi_i^{-1}(U) \subset G_i$ is open in G_i for every $i \in I$. Since the structure homomorphisms are open, the maps $\psi_i : G_i \to G$ are open as well and a set $U \subset G$ is open if and only if it can be written as $U = \bigcup_{i\in I} \psi_i(U_i)$ for some open sets $U_i \subset G_i$. We use this to show that the natural continuous bijection

$$\lim_{\substack{\rightarrow \\ i,j}} G_i \times G_j \to G \times G$$

is also open, hence a homeomorphism. As any open subset of the left hand side is a union of images of open subsets of $G_i \times G_j$, it suffices to show that the image of an open subset of $G_i \times G_j$ in $G \times G$ is open. For this it suffices to assume that the open set be a rectangle, i.e., of the form $U_i \times U_j$ for open sets $U_i \subset G_i$ and $U_j \subset G_j$. But then the images of U_i and U_j in G are open, hence the image of $U_i \times U_j$ in $G \times G$ is open.

We need to show continuity of the multiplication map $G \times G \to G$. As $G \times G$ is homeomorphic with the direct limit of the $G_i \times G_j$, it suffices to show that the composite map $\alpha : G_i \times G_j \to G \times G \to G$ is continuous, where the second map is multiplication. Choose some $k \in I$ with $k \geq i, j$. Then α also equals the map $G_i \times G_j \to G_k \times G_k \to G_k \to G$. In the second description the continuity follows from the continuity of the multiplication map of G_k. The inversion is dealt with in a similar way. This shows that G is a topological group if all G_i are.

(b) Suppose that G is Hausdorff. Then for given $i \in I$ the map $\psi_i : G_i \to G$ is continuous, hence its kernel is closed, as it is the pre-image of the closed set $\{1\}$. Conversely, assume all kernels $\ker(\psi_i)$ are closed and let $y \neq 1$ in G. Then there exists $i \in I$ and $y_i \in G_i$ such that $y = \psi_i(y_i)$. Now ψ_i is open, and so $U = \psi_i(G_i \smallsetminus \ker(\psi_i))$ is an open neighborhood of y which does not contain 1. Therefore, G is Hausdorff.

(c) Now suppose the system is a Mittag-Leffler direct system and that all G_i are Hausdorff. Let $i \in I$ and fix $k_0 \in I$ such that $H_i = \ker(\phi_i^k) = \ker\left(\phi_i^{k_0}\right)$ holds for every $k \geq k_0$. Then the closed subgroup H_i is also the kernel of ψ_i, so G is Hausdorff by part (b).

(d) Finally, suppose that all G_i are locally compact groups and the kernels $\ker(\psi_i)$ are closed. Then G is Hausdorff by (b), further, as each $\psi_i : G_i \to G$ is open as well, a compact unit neighborhood U inside G_i maps to a compact unit neighborhood in G, which therefore is locally compact. $\square$

Examples 1.8.7

- If all G_i are open subgroups of a given topological group H with their subspace topology, then the limit is their union and the limit topology is the subspace topology as well.

- If $\left(G_i, \phi_i^j\right)$ is a direct system of discrete groups, then the limit is a discrete, hence a locally compact group.

- This example shows that the Hausdorff property in the direct limit can fail if the system does not satisfy the Mittag-Leffler condition. Let V be an infinite-dimensional Hilbert space and let $D \neq V$ be a dense subspace of V. Let I be the set of all finite subsets of D, for each $\alpha \in I$ let V_α denote the (finite-dimensional) linear span of α and set

$$G_\alpha = V/V_\alpha.$$

We order I by set inclusion, then if $\alpha \leq \beta$ there is a natural projection $\phi_\alpha^\beta : G_\alpha \to G_\beta$. This family of maps forms a direct system. Each V_α is a Hilbert space, hence a Hausdorff topological group and the structure maps are open, but the direct limit, which can be identified with V/D is no longer Hausdorff, indeed, it carries the trivial topology.

Projective Limits

There is a dual construction to the direct limit, called the projective limit.

Definition A *projective system* of groups consists of the following data

- a directed set $(I, \leq)$,
- a family $(G_i)_{i \in I}$ of groups and
- a family of group homomorphisms

$$\pi_i^j : G_j \to G_i, \qquad \text{if } i \leq j,$$

such that the following axioms are met:

$$\pi_i^i = \mathrm{Id}_{G_i} \quad \text{and} \quad \pi_i^j \circ \pi_j^k = \pi_i^k, \text{ if } i \leq j \leq k.$$

Note that, in comparison to a direct system, the homomorphisms now run in the opposite direction.

Example 1.8.8 Let p be a prime number. Let $I = \mathbb{N}$ with the usual order. For $n \in \mathbb{N}$ let $G_n = \mathbb{Z}/p^n\mathbb{Z}$ and for $m \geq n$ let $\pi_n^m : \mathbb{Z}/p^m\mathbb{Z} \to \mathbb{Z}/p^n\mathbb{Z}$ be the canonical projection. Then (G_n, π_n^m) form a projective system of groups.

Definition Let (G_i, π_i^j) be a projective system of groups. The *projective limit* of the system is the set

$$G = \varprojlim G_i$$

of all $a \in \prod_{i \in I} G_i$ such that $a_i = \pi_i^j(a_j)$ holds for every pair $i \leq j$ in I.

Proposition 1.8.9 *The projective limit G of the system (G_i) is a subgroup of the product $\prod_{i \in I} G_i$. Let $\pi_i : G \to G_i$ be the map given by the projection to the i-th coordinate. Then π_i is a group homomorphism. The projective limit has the following* *universal property: If Z is a group with group homomorphisms* $\alpha_i : Z \to G_i$, *such that* $\alpha_i = \pi_i^j \circ \alpha_j$ *holds for all* $i \leq j$ *in* I, *then there exists exactly one group homomorphism* $\alpha : Z \to G$, *such that all diagrams*

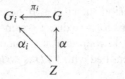

commute.

As in the case of direct limits, one can replace the word "group" with, say, the word "ring". Then one assumes that the structure homomorphisms π_i^j are ring homomorphisms and gets a ring as projective limit.

Proof The proof is left to the reader.   □

Definition Again assume that the groups G_i in a given projective system are topological groups and that all structure homomorphisms π_i^j are continuous. Then one equips $G = \lim\limits_{\leftarrow} G_i$ with the topology induced by the projections $p_i : G \to G_i$ and calls this the *projective limit topology*.

Since the topology of the product $\prod_i G_i$ is induced by the projections as well, the projective limit G carries the subspace topology of the product. Hence the question of locally compactness is connected to the same question for products.

Lemma 1.8.10 *Let I be an index set and for every* $i \in I$ *let there be given a nonempty locally compact space* X_i. *Then the product space* $X = \prod_{i \in I} X_i$ *is locally compact if and only if almost all the spaces* X_i *are compact.*

Proof Let $E \subset I$ be a finite subset and for each $i \in E$ let $U_i \subset X_i$ be a subset. These data define a *rectangle*

$$R = R((U_i)_{i \in E}) = \prod_{i \in E} U_i \times \prod_{i \in I \setminus E} X_i.$$

A rectangle is open if and only if every U_i is open.

By definition of the product topology, every open set is a union of open rectangles. The intersection of two open rectangles is again an open rectangle. If $X \neq \emptyset$ is locally compact, there therefore exists a non-empty open rectangle with compact closure. The closure of a rectangle $R((U_i)_i)$ is the rectangle $R((\overline{U_i})_i)$, and for this to be compact, almost all X_i must be compact.

The converse direction follows from Tychonov's Theorem and the simple observation that a finite product of topological spaces is locally compact if and only if all factors are locally compact.   □

Proposition 1.8.11 *Let $\left(G_i, \pi_i^j\right)$ be a projective system of topological groups with limit G. Then G is a closed subgroup of the product $\prod_i G_i$ and carries the subspace topology, hence it is a topological group. If all G_i are Hausdorff, then G is Hausdorff. If all G_i are locally compact and all but finitely many are compact, then G is locally compact.*

Proof The assertions of this propositions are clear by what has been said above. □

Definition A *profinite group* is a locally compact group isomorphic to a projective limit of finite groups.

Example 1.8.12 Let p be a prime number. The profinite group

$$\mathbb{Z}_p = \varprojlim_n \mathbb{Z}/p^n\mathbb{Z}$$

is called the group of *p-adic integers*, see Sect. 14.1.

1.9 Exercises

Exercise 1.1 Determine the Haar measures of the groups $\mathbb{Z}, \mathbb{R}, (\mathbb{R}^\times, \cdot), \mathbb{T}$.

Exercise 1.2 Give an example of a locally compact group G and two closed subsets A, B of G such that AB is not closed.

(Hint: There is an example with $G = \mathbb{R}$.)

Exercise 1.3 Let G be a topological group and suppose there exists a compact subset K of G such that $xK \cap K \neq \emptyset$ for every $x \in G$. Show that G is compact.

Exercise 1.4 Let G be a locally compact group with Haar measure μ, and let $S \subset G$ be a measurable subset with $0 < \mu(S) < \infty$. Show that the map $x \mapsto \mu(S \cap xS)$ from G to $\mathbb{R}$ is continuous.

(Hint: Note that $\mathbf{1}_S \in L^2(G)$. Write the map as $\langle \mathbf{1}_S, L_{x^{-1}}\mathbf{1}_S \rangle$ and use the Cauchy-Schwarz inequality.)

Exercise 1.5 Let G be a locally compact group with Haar measure μ, and let S be a measurable subset with $0 < \mu(S) < \infty$. Show that the set K of all $k \in G$ with $\mu(S \cap kS) = \mu(S)$ is a closed subgroup of G.

Exercise 1.6 Let G be a locally compact group, H a dense subgroup, and μ a Radon measure on G such that $\mu(hA) = \mu(A)$ holds for every measurable set $A \subset G$ and every $h \in H$. Show that μ is a Haar measure.

Exercise 1.7 Let G be a locally compact group, H a dense subgroup, and μ a Haar measure. Let $S \subset G$ be a measurable subset such that for each $h \in H$ the sets

$$hS \cap (G \smallsetminus S) \quad \text{and} \quad S \cap (G \smallsetminus S)$$

are both null-sets. Show that either S or its complement $G \smallsetminus S$ is a null-set.

(Hint: Show that the measure $v(A) = \mu(A \cap S)$ is invariant.)

Exercise 1.8 Let B be the subgroup of $GL_2(\mathbb{R})$ defined by

$$B = \left\{ \begin{pmatrix} 1 & x \\ & y \end{pmatrix} : x, y \in \mathbb{R}, \ y \neq 0 \right\}.$$

Show that $I(f) = \int_{\mathbb{R}^\times} \int_{\mathbb{R}} f\begin{pmatrix} 1 & x \\ & y \end{pmatrix} dx \frac{dy}{y}$ is a Haar-integral on B. Show that the modular function Δ of B satisfies: $\Delta \begin{pmatrix} 1 & x \\ & y \end{pmatrix} = |y|$.

Exercise 1.9 Let G be a locally-compact group. Show that the convolution satisfies $f * (g * h) = (f * g) * h$, $f * (g + h) = f * g + f * h$.

Exercise 1.10 Let G be a locally compact group, and let $\chi : G \to \mathbb{R}_{>0}^\times$ be a continuous group homomorphism.

(a) Show that there exists a unique Radon measure μ on G, which is χ-quasi-invariant in the sense that $\mu(xA) = \chi(x)\mu(A)$ holds for every $x \in G$ and every measurable subset $A \subset G$.

(b) Let $H \subset G$ be a closed subgroup. Show that there exists a Radon measure v on G/H with $v(xA) = \chi(x)v(A)$ for every $x \in G$ and every measurable $A \subset G/H$, if and only if for every $h \in H$ one has $\chi(h)\Delta_G(h) = \Delta_H(h)$.

(Hint: Verify that the measure μ is χ-quasi-invariant if and only if the corresponding integral J satisfies $J(L_x f) = \chi(x)J(f)$. If I is a Haar-integral, consider $J(f) = I(\chi f)$.)

Exercise 1.11 Let G, H be locally compact groups and assume that G acts on H by group homomorphisms $h \mapsto {}^g h$, such that the ensuing map $G \times H \to H$ is continuous.

(a) Show that the product $(h, g)(h', g') = (h \, {}^g h', gg')$ gives $H \times G$ (with the product topology) the structure of a locally compact group, called the *semi-direct product* $H \rtimes G$.

(b) Show that there is a unique group homomorphism $\delta : G \to (0, \infty)$ such that $\mu_H({}^g A) = \delta(g)\mu_H(A)$, where μ_H is a Haar measure on H and A is a measurable subset of H.

(c) Show that $\int_H f(^g x) d\mu_H(x) = \delta(g) \int_H f(x) d\mu_H(x)$ for $f \in C_c(H)$ and deduce that δ is continuous.

(d) Show that a Haar integral on $H \rtimes G$ is given by

$$\int_H \int_G f(h, g)\delta(g) d\mu_H(h) d\mu_G(g).$$

Exercise 1.12 For a finite group G define the *group algebra* $\mathbb{C}[G]$ to be a vector space of dimension equal to the group order $|G|$, with a special basis $(v_g)_{g \in G}$, and equipped with a multiplication $v_g v_{g'} \stackrel{\text{def}}{=} v_{gg'}$. Show that $\mathbb{C}[G]$ indeed is an algebra over $\mathbb{C}$. Show that the linear map $v_g \mapsto \mathbf{1}_{\{g\}}$ is an isomorphism of $\mathbb{C}[G]$ to the convolution algebra $L^1(G)$.

Chapter 2
Banach Algebras

In this chapter we present the basic concepts on Banach algebras and C^*-algebras, which are needed to understand many of the further topics in this book. In particular, we shall treat the basics of the Gelfand theory for commutative Banach algebras, and we shall give a proof of the Gelfand-Naimark theorem, which asserts that a commutative C^*-algebra is naturally isomorphic to the algebra of continuous functions vanishing at infinity on a locally compact Hausdorff space.

2.1 Banach Algebras

Recall the notion of an *algebra* from Sect. 1.6. A *Banach algebra* is an algebra $\mathcal{A}$ over the complex numbers together with a norm $\|\cdot\|$, in which $\mathcal{A}$ is complete, i.e., $\mathcal{A}$ is a Banach space, such that the norm is submultiplicative, i.e., the inequality

$$\|a \cdot b\| \leq \|a\| \|b\|$$

holds for all $a, b \in \mathcal{A}$. Note that this inequality in particular implies that the multiplication on $\mathcal{A}$ is a continuous map from $\mathcal{A} \times \mathcal{A} \to \mathcal{A}$, which means that if $(a_n)_{n \in \mathbb{N}}$ and $(b_n)_{n \in \mathbb{N}}$ are sequences in $\mathcal{A}$ converging to a and b, respectively, then the product sequence $a_n b_n$ converges to ab. This follows from the estimate

$$\|a_n b_n - ab\| = \|a_n b_n - a_n b + a_n b - ab\|$$
$$\leq \|a_n\| \|b_n - b\| + \|b\| \|a_n - a\|,$$

as the latter term tends to zero as $n \to \infty$.

Examples 2.1.1

- The algebra $M_n(\mathbb{C})$ equipped with the norm

$$\|a\| = \sum_{i,j=1}^{n} |a_{i,j}|$$

  is a Banach algebra.

A. Deitmar, S. Echterhoff, *Principles of Harmonic Analysis,* Universitext,
DOI 10.1007/978-3-319-05792-7_2, © Springer International Publishing Switzerland 2014

- For a topological space X let $C(X)$ denote the vector space of continuous functions $f : X \to \mathbb{C}$. If X is compact, the space $C(X)$ becomes a commutative Banach algebra if it is equipped with the sup-norm $\|f\|_X = \sup_{x \in X} |f(x)|$.

- If G is a locally compact group, then $L^1(G)$ equipped with $\|\cdot\|_1$ and the convolution product $(f, g) \mapsto f * g$ is a Banach algebra by Theorem 1.6.2, which is commutative if and only if G is abelian by Theorem 1.6.4.

- Let V be a Banach space. For a linear operator $T : V \to V$ define the *operator norm* by

$$\|T\|_{op} \overset{\text{def}}{=} \sup_{v \neq 0} \frac{\|Tv\|}{\|v\|}.$$

The operator T is called a *bounded operator* if $\|T\|_{\text{op}} < \infty$. By Lemma C.1.2 an operator is bounded if and only if it is continuous. The set $\mathcal{B}(V)$ of all bounded linear operators on V is a Banach algebra with the operator norm (see Exercise 2.1 below).

Definition An algebra $\mathcal{A}$ is *unital* if there exists an element $1_{\mathcal{A}} \in \mathcal{A}$ such that

$$1_{\mathcal{A}} a = a 1_{\mathcal{A}} = a \quad \text{for every } a \in \mathcal{A}.$$

The element $1_{\mathcal{A}}$ is then called the *unit* of $\mathcal{A}$. It is uniquely determined, for if $1'_{\mathcal{A}}$ is a second unit, one has $1_{\mathcal{A}} = 1_{\mathcal{A}} 1'_{\mathcal{A}} = 1'_{\mathcal{A}}$. We shall often write 1 for $1_{\mathcal{A}}$ if no confusion can arise.

Recall that two norms $\|\cdot\|$ and $\|\cdot\|'$ on a complex vector space V are called *equivalent norms* if there is $C > 0$ with

$$\frac{1}{C} \|\cdot\| \leq \|\cdot\|' \leq C \|\cdot\|.$$

In that case, V is complete in the norm $\|\cdot\|$ if and only if it is complete in the norm $\|\cdot\|'$ and both norms define the same topology on V.

Lemma 2.1.2 *Let $\mathcal{A}$ be a unital Banach algebra with unit 1. Then $\|1\| \geq 1$ and there is an equivalent norm $\|\cdot\|'$ such that $(\mathcal{A}, \|\cdot\|')$ is again a Banach algebra with $\|1\|' = 1$.*

With this lemma in mind, we will, when talking about a unital Banach algebra, always assume that the unit element is of norm one.

Proof In the situation of the lemma one has $\|1\|^2 \geq \|1^2\| = \|1\|$, so $\|1\| \geq 1$. For $a \in \mathcal{A}$ let $\|a\|'$ be the *operator norm* of the multiplication operator M_a, which sends x to ax, so $\|a\|' = \sup_{x \neq 0} \frac{\|ax\|}{\|x\|}$. Then $\|\cdot\|'$ is a norm with $\|1\|' = 1$. Since $\|ax\| \leq \|a\| \|x\|$, it follows that $\|a\|' \leq \|a\|$. On the other hand one has $\|a\|' = \sup_{x \neq 0} \frac{\|ax\|}{\|x\|} \geq \frac{\|a \cdot 1\|}{\|1\|} = \frac{\|a\|}{\|1\|}$. This shows that $\|\cdot\|$ and $\|\cdot\|'$ are equivalent. The inequality $\|ab\|' \leq \|a\|' \|b\|'$ is easy to show (See Exercise 2.1). $\square$

Proposition 2.1.3 *Let G be a locally compact group. The algebra $\mathcal{A} = L^1(G)$ is unital if and only if G is discrete.*

Proof If G is discrete, then the function $\mathbf{1}_{\{1\}}$ is easily seen to be a unit of $\mathcal{A}$. Conversely, assume that $\mathcal{A} = L^1(G)$ possesses a unit ϕ and G is non-discrete. The latter fact implies that any unit neighborhood U has at least two points. This implies by Urysohn's Lemma (A.8.1) that for every unit-neighborhood U there are two Dirac functions ϕ_U and ψ_U, both with support in U, such that the supports of ϕ_U and ψ_U are disjoint, hence in particular, $\|\phi_U - \psi_U\|_1 = 2$ for every $n \in \mathbb{N}$. The function ϕ being a unit means that we have $\phi * f = f * \phi = f$ for every $f \in L^1(G)$. There exists a unit-neighborhood U, such that one has $\|\phi_U * \phi - \phi\|_1 < 1$ and $\|\psi_U * \phi - \phi\|_1 < 1$. Hence $2 = \|\phi_U - \psi_U\|_1 \leq \|\phi_U - \phi\|_1 + \|\phi - \psi_U\|_1 < 2$, a contradiction! Hence the assumption is false and G must be discrete. $\qquad\square$

Definition Let $\mathcal{A}, \mathcal{B}$ be Banach algebras. A *homomorphism of Banach algebras* is by definition a continuous algebra homomorphism $\phi : \mathcal{A} \to \mathcal{B}$. This means that ϕ is continuous, $\mathbb{C}$-linear and multiplicative, i.e., satisfies $\phi(ab) = \phi(a)\phi(b)$. A *topological isomorphism of Banach algebras* is a homomorphism with continuous inverse, and an *isomorphism of Banach algebras* is an isomorphism ϕ, which is an isometry, i.e., which satisfies $\|\phi(a)\| = \|a\|$ for every $a \in \mathcal{A}$. For better distinction we will call an isomorphism of Banach algebras henceforth an *isometric isomorphism of Banach algebras*.

Example 2.1.4 Let $Y \subset X$ be a compact subspace of the compact topological space X. Then the restriction of functions is a homomorphism of Banach algebras from $C(X)$ to $C(Y)$. Note that this includes the special case when $Y = \{x\}$ consists of a single element. In this case $C(Y) \cong \mathbb{C}$, and the restriction is the *evaluation homomorphism* $\delta_x : C(X) \to \mathbb{C}$ mapping f to $f(x)$.

If $\mathcal{A}$ is a unital Banach algebra, we denote by $\mathcal{A}^\times$ the group of invertible elements of $\mathcal{A}$, i.e., the multiplicative group of all a in $\mathcal{A}$, for which there exists some $b \in \mathcal{A}$ with $ab = ba = 1$. This b then is uniquely determined, as for a second such b' one has $b' = b'ab = b$. Therefore it is denoted a^{-1} and called the *inverse* of a.

Recall that for $a \in \mathcal{A}$ we denote by $B_r(a)$ the open ball of radius $r > 0$ around $a \in \mathcal{A}$, in other words, $B_r(a)$ is the set of all $z \in \mathcal{A}$ with $\|a - z\| < r$.

Lemma 2.1.5 (Neumann series). *Let $\mathcal{A}$ be a unital Banach algebra, and let $a \in \mathcal{A}$ with $\|a\| < 1$. Then $1 - a$ is invertible with inverse*

$$(1 - a)^{-1} = \sum_{n=0}^{\infty} a^n.$$

The unit group $\mathcal{A}^\times$ is an open subset of $\mathcal{A}$. With the subspace topology, $\mathcal{A}^\times$ is a topological group.

Proof Since $\|a\| < 1$ one has $\sum_{n=0}^{\infty} \|a^n\| \leq \sum_{n=0}^{\infty} \|a\|^n < \infty$, so the series $b = \sum_{n=0}^{\infty} a^n$ converges absolutely in $\mathcal{A}$, and we get the first assertion by computing $(1-a)b = (1-a)\sum_{n=0}^{\infty} a^n = \sum_{n=0}^{\infty} a^n - \sum_{n=0}^{\infty} a^{n+1} = 1$, and likewise $b(1-a) = 1$.

For the second assertion let $y \in \mathcal{A}^\times$. As the multiplication on $\mathcal{A}$ is continuous, the map $x \mapsto yx$ is a homeomorphism. This implies that $yB_1(1) \subset \mathcal{A}^\times$ is an open neighborhood of y, so $\mathcal{A}^\times$ is indeed open.

To show that $\mathcal{A}^\times$ is a topological group, it remains to show that the inversion is continuous on $\mathcal{A}^\times$. Note that the map $a \mapsto \sum_{n=0}^{\infty} a^n = (1-a)^{-1}$ is continuous on $B_1(0)$, which implies that inversion is continuous on $B_1(1)$. But then it is continuous on the open neighborhood $yB_1(1) \subset \mathcal{A}^\times$ of any $y \in \mathcal{A}^\times$.   □

Examples 2.1.6

- Let $\mathcal{A} = M_n(\mathbb{C})$. Then the unit group $\mathcal{A}^\times$ is the group of invertible matrices, i.e., of those matrices $a \in \mathcal{A}$ with $\det(a) \neq 0$. The continuity of the determinant function in this case gives another proof that $\mathcal{A}^\times$ is open.

- Let $\mathcal{A} = C(X)$ for a compact Hausdorff space X. Then the unit group $\mathcal{A}^\times$ consists of all $f \in C(X)$ with $f(x) \neq 0$ for every $x \in X$.

2.2 The Spectrum $\sigma_{\mathcal{A}}(a)$

Let $\mathcal{A}$ be a unital Banach algebra. For $a \in \mathcal{A}$ we denote by

$$\mathrm{Res}\,(a) \stackrel{\mathrm{def}}{=} \{\lambda \in \mathbb{C} : \lambda 1 - a \text{ is invertible}\}$$

the *resolvent set* of $a \in \mathcal{A}$. Its complement,

$$\sigma_{\mathcal{A}}(a) \stackrel{\mathrm{def}}{=} \mathbb{C} \backslash \mathrm{Res}(a)$$

is called the *spectrum* of a. Since $\mathcal{A}^\times$ is open in $\mathcal{A}$ by Lemma 2.1.5 and since $\lambda \mapsto (\lambda 1 - a)$ is continuous, we see that $\mathrm{Res}(a)$ is open, and $\sigma_{\mathcal{A}}(a)$ is closed in $\mathbb{C}$.

Examples 2.2.1

- Let $\mathcal{A} = M_n(\mathbb{C})$. Then for $a \in \mathcal{A}$ the spectrum $\sigma(a)$ equals the set of eigenvalues of a.

- Let X be a compact topological space, and let $\mathcal{A} = C(X)$. For $f \in \mathcal{A}$ the spectrum $\sigma(f)$ equals the image of the map $f : X \to \mathbb{C}$.

Lemma 2.2.2 *Let $\mathcal{A}$ be a unital Banach algebra. Then for every $a \in \mathcal{A}$ the spectrum $\sigma(a)$ is a closed subset of the closed ball $\bar{B}_{\|a\|}(0)$ around zero of radius $\|a\|$, so in particular, $\sigma(a)$ is compact.*

Proof As the resolvent set is open, the spectrum is closed. Let $a \in \mathcal{A}$, and let $\lambda \in \mathbb{C}$ with $|\lambda| > \|a\|$. We have to show that $\lambda 1 - a$ is invertible. As $\|\lambda^{-1}a\| < 1$, by Lemma 2.1.5 one has $1 - \lambda^{-1}a \in \mathcal{A}^{\times}$; it follows that $\lambda \cdot 1 - a = \lambda(1 - \lambda^{-1}a) \in \mathcal{A}^{\times}$, so $\lambda \in \text{Res}(a)$. $\qquad\square$

Definition Let $D \subset \mathbb{C}$ be an open set, and let $f : D \to V$ be a map, where V is a Banach space. We say that f is *holomorphic* if for every $z \in D$ the limit

$$f'(z) = \lim_{h \to 0} \frac{1}{h}(f(z+h) - f(z))$$

exists in V. Note that if f is holomorphic and $\alpha : V \to \mathbb{C}$ is a continuous linear functional, then the function $z \mapsto \alpha(f(z))$ is a holomorphic function from D to $\mathbb{C}$. A holomorphic map is continuous (see Exercise 2.2).

Lemma 2.2.3 *Let $a \in \mathcal{A}$, then the map $f : \lambda \mapsto (\lambda - a)^{-1}$ is holomorphic on the resolvent set* $\text{Res}(a)$. *Here we have written λ for $\lambda 1 \in \mathcal{A}$.*

Proof Let $\lambda \in \text{Res}(a)$, and let h be a small complex number. Then $\frac{1}{h}(f(\lambda+h) - f(\lambda))$ equals

$$\frac{1}{h}\left((\lambda + h - a)^{-1} - (\lambda - a)^{-1}\right)$$
$$= \frac{1}{h}\left((\lambda - a) - (\lambda + h - a)\right)(\lambda + h - a)^{-1}(\lambda - a)^{-1}$$
$$= -(\lambda + h - a)^{-1}(\lambda - a)^{-1}.$$

This map is continuous at $h = 0$, since the inversion is a continuous map on the resolvent set by Lemma 2.1.5. $\qquad\square$

Theorem 2.2.4 *Let $\mathcal{A}$ be a unital Banach algebra, and let $a \in \mathcal{A}$. Then $\sigma_{\mathcal{A}}(a) \neq \emptyset$.*

Proof Assume there exists $a \in \mathcal{A}$ with empty spectrum. Let α be a continuous linear functional on $\mathcal{A}$, then the function $f_{\alpha} : \lambda \mapsto \alpha\left((a - \lambda)^{-1}\right)$ is entire. As α is continuous, hence bounded, there exists $C > 0$ such that $|\alpha(b)| \leq C\|b\|$ holds for every $b \in \mathcal{A}$. For $|\lambda| > 2\|a\|$ we get

$$|f_{\alpha}(\lambda)| = |\alpha\left((a - \lambda)^{-1}\right)| = \frac{1}{|\lambda|}|\alpha\left((1 - \lambda^{-1}a)^{-1}\right)|$$
$$= \frac{1}{|\lambda|}\left|\alpha\left(\sum_{n=0}^{\infty}(\lambda^{-1}a)^n\right)\right| \leq \frac{1}{|\lambda|}\sum_{n=0}^{\infty}|\alpha\left((\lambda^{-1}a)^n\right)|$$
$$\leq \frac{C}{|\lambda|}\sum_{n=0}^{\infty}\|\lambda^{-1}a\|^n < \frac{C}{|\lambda|}\sum_{n=0}^{\infty}\left(\frac{1}{2}\right)^n = \frac{2C}{|\lambda|}.$$

It follows that the entire function f_α is constantly zero by the Theorem of Liouville [Rud87]. This holds for every α, so the Hahn-Banach Theorem C.1.3 implies that $f : \lambda \mapsto (\lambda - a)^{-1}$ is the zero map as well, which is a contradiction. □

Corollary 2.2.5 (Gelfand-Mazur). *Let $\mathcal{A}$ be a unital Banach algebra such that all non-zero elements $a \in \mathcal{A}$ are invertible. Then $\mathcal{A} = \mathbb{C}1$.*

Proof If $a \in \mathcal{A} \setminus \mathbb{C}1$ we have $\lambda 1 - a$ invertible for every $\lambda \in \mathbb{C}$. But this means that $\sigma_\mathcal{A}(a) = \emptyset$, which contradicts Theorem 2.2.4. □

Definition For an element a of a unital Banach algebra $\mathcal{A}$ we define the *spectral radius $r(a)$* of a by

$$r(a) \overset{\text{def}}{=} sup\{|\lambda| : \lambda \in \sigma_\mathcal{A}(a)\}.$$

In what follows next we want to prove an important formula for the spectral radius $r(a)$.

Theorem 2.2.6 (Spectral radius formula). *Let $\mathcal{A}$ be a unital Banach algebra. Then $r(a) \leq \|a\|$ and*

$$r(a) = \lim_{n \to \infty} \|a^n\|^{\frac{1}{n}}.$$

Proof As $\|a^n\| \leq \|a\|^n$ the first assertion follows from the second. We shall show the inequalities

$$r(a) \leq \liminf \|a^n\|^{\frac{1}{n}} \leq \limsup \|a^n\|^{\frac{1}{n}} \leq r(a),$$

which clearly implies the theorem.

For $\lambda \in \sigma_\mathcal{A}(a)$, the equation $\lambda^n 1 - a^n = (\lambda 1 - a) \sum_{j=0}^{n-1} \lambda^j a^{n-1-j}$ implies that $\lambda^n \in \sigma_\mathcal{A}(a^n)$ and hence that $|\lambda|^n \leq \|a^n\|$ for every $n \in \mathbb{N}$. Thus $r(a) \leq \|a^n\|^{\frac{1}{n}}$ for every $n \in \mathbb{N}$, which gives the first inequality.

To see that $\limsup \|a^n\|^{\frac{1}{n}} \leq r(a)$ recall $(\lambda 1 - a)^{-1} = \lambda^{-1}(1 - \frac{a}{\lambda})^{-1} = \sum_{n=0}^{\infty} a^n \frac{1}{\lambda^{n+1}}$ for $|\lambda| > \|a\|$, and hence, as the function is holomorphic there, the series converges in the norm-topology for every $|\lambda| > r(a)$, as we derive from Corollary B.6.7 applied to $z = \frac{1}{\lambda}$.

For a fixed $|\lambda| > r(a)$, it follows that the sequence $a^n \frac{1}{\lambda^{n+1}}$ is bounded in $\mathcal{A}$, so that there exists a constant $C \geq 0$ such that $\|a^n\| \leq C|\lambda|^{n+1}$ for every $n \in \mathbb{N}$. Taking n-th roots on both sides and then applying $\limsup$ shows that $\limsup \|a^n\|^{\frac{1}{n}} \leq |\lambda|$. Since this holds for every $|\lambda| > r(a)$ we get $\limsup \|a^n\|^{\frac{1}{n}} \leq r(a)$. □

Lemma 2.2.7 *Suppose that $\mathcal{A}$ is a closed subalgebra of the unital Banach algebra $\mathcal{B}$ such that $1 \in \mathcal{A}$. Then*

$$\partial\sigma_{\mathcal{A}}(a) \subset \sigma_{\mathcal{B}}(a) \subset \sigma_{\mathcal{A}}(a)$$

for every $a \in \mathcal{A}$, where $\partial\sigma_{\mathcal{A}}(a)$ denotes the boundary of $\sigma_{\mathcal{A}}(a) \subset \mathbb{C}$.

Proof If $a \in \mathcal{A}$ is invertible in $\mathcal{A}$ it is invertible in $\mathcal{B} \supset \mathcal{A}$. So $\mathrm{Res}_{\mathcal{A}}(a) \subset \mathrm{Res}_{\mathcal{B}}(a)$, which is equivalent to the second inclusion.

To see the first inclusion let $\lambda \in \partial\sigma_{\mathcal{A}}(a) \subset \sigma_{\mathcal{A}}(a)$, and let $(\lambda_n)_{n \in \mathbb{N}}$ be a sequence in $\mathrm{Res}_{\mathcal{A}}(a)$ with $\lambda_n \to \lambda$. If $\lambda \in \mathrm{Res}_{\mathcal{B}}(a)$, then $\mathcal{A} \ni (\lambda_n 1 - a)^{-1} \to (\lambda 1 - a)^{-1} \in \mathcal{B}$. Since $\mathcal{A}$ is closed in $\mathcal{B}$ we get $(\lambda 1 - a)^{-1} \in \mathcal{A}$, which implies that $\lambda \in \mathrm{Res}_{\mathcal{A}}(a)$. This contradicts $\lambda \in \sigma_{\mathcal{A}}(a)$. $\square$

Example 2.2.8 Let $\mathbb{D} \subset \mathbb{C}$ be the closed disk of radius 1 around zero, and let $\mathring{\mathbb{D}}$ be its interior. The *disk-algebra* $\mathcal{A}$ is by definition the subalgebra of $C(\mathbb{D})$ consisting of all functions that are holomorphic inside $\mathring{\mathbb{D}}$. Since uniform limits of holomorphic functions are again holomorphic, the disk-algebra is a closed subalgebra of $C(\mathbb{D})$ and hence a Banach algebra. Let $\mathbb{T} = \partial\mathbb{D}$ be the circle group. By the maximum principle for holomorphic functions, every $f \in \mathcal{A}$ takes its maximum on $\mathbb{T}$. Therefore the restriction homomorphism $\mathcal{A} \to C(\mathbb{T})$ mapping $f \in \mathcal{A}$ to its restriction $f|_{\mathbb{T}}$ is an isometry. So $\mathcal{A}$ can be viewed as a sub Banach algebra of $\mathcal{B} = C(\mathbb{T})$. For $f \in \mathcal{A}$ and $\lambda \in \mathbb{C}$ the function $(\lambda - f)^{-1}$ is defined in $\mathcal{A}$ if and only if λ is not in the image of f. Therefore, the spectrum $\sigma_{\mathcal{A}}(f)$ equals the image $f(\mathbb{D})$. Likewise, considered as an element of $\mathcal{B}$, the spectrum of f equals $\sigma_{\mathcal{B}}(f) = f(\mathbb{T})$.

2.3   Adjoining a Unit

The results of the previous section always depended on the existence of a unit in the Banach algebra $\mathcal{A}$. But many important Banach algebras, like $L^1(G)$ for a non-discrete locally compact group G, do not have a unit. We solve this problem by adjoining a unit if needed. Indeed, if $\mathcal{A}$ is any Banach algebra (with or without unit), then the cartesian product,

$$\mathcal{A}^e \overset{\mathrm{def}}{=} \mathcal{A} \times \mathbb{C}$$

equipped with the obvious vector space structure and the multiplication

$$(a, \lambda)(b, \mu) = (ab + \lambda b + \mu a, \lambda\mu)$$

becomes an algebra with unit $(0, 1)$. If we define $\|(a, \lambda)\| = \|a\| + |\lambda|$, one easily checks that $\mathcal{A}^e$ becomes a Banach algebra containing $\mathcal{A} \cong \mathcal{A} \times \{0\}$ as a closed subalgebra of codimension 1. We call $\mathcal{A}^e$ the *unitization* of $\mathcal{A}$.

If $\mathcal{A}$ already has a unit $1_\mathcal{A}$, then the algebra $\mathcal{A}^e$ is isomorphic to the direct sum $\mathcal{A} \oplus \mathbb{C}$ of the algebras $\mathcal{A}$ and $\mathbb{C}$, where we define multiplication component-wise. The isomorphism from $\mathcal{A}^e$ to $\mathcal{A} \oplus \mathbb{C}$ is given by $(a, \lambda) \mapsto (a + \lambda 1_\mathcal{A}) \oplus \lambda$.

If $\mathcal{A}$ is a Banach algebra without unit then we define the spectrum of $a \in \mathcal{A}$ as

$$\sigma_\mathcal{A}(a) \overset{\text{def}}{=} \sigma_{\mathcal{A}^e}(a),$$

where we identify $\mathcal{A}$ with a subset of $\mathcal{A}^e$ via $a \mapsto (a, 0)$. With this convention, all results from the previous section, in particular the spectral radius formula, have natural analogues in the non-unital case.

A very important class of commutative Banach-Algebras without unit is given as follows:

Definition Suppose that X is a locally compact Hausdorff space. A function $f : X \to \mathbb{C}$ is said to *vanish at infinity* if for every $\varepsilon > 0$ there exists a compact set $K = K_\varepsilon \subset X$ such that $|f(x)| < \varepsilon$ holds for every $x \in X \setminus K$. Let $C_0(X)$ denote the vector space of all continuous functions on X that vanish at infinity. Then $C_0(X)$ is a Banach algebra with point-wise multiplication and the sup-norm $\|f\|_X = \sup_{x \in X} |f(x)|$. Note that $C_0(X)$ is unital if and only if X is compact in which case $C_0(X)$ equals $C(X)$.

Example 2.3.1 As a crucial example we want to compute the unitization $C_0(X)^e$. We recall the construction of the *one point compactification*, also called *Alexandrov compactification* X^∞, of the space X. Let ∞ denote a new point and $X^\infty = X \cup \{\infty\}$, so X^∞ is just a set that contains X as a subset plus one more element. On X^∞ one introduces the following topology. A set $U \subset X^\infty$ is open if it is either already contained in X and open in the topology of X, or if $\infty \in U$ and the set $X \setminus U$ is a compact subset of X. Note that X being a Hausdorff space implies that compact sets in X are closed in X. Every continuous function in $C(X^\infty)$ defines, by restriction, a continuous function on X. In this way one can identify $C_0(X)$ with the subspace of all continuous functions f on X^∞ with $f(\infty) = 0$. This ultimately justifies the notion "vanishing at infinity".

Lemma 2.3.2 *There is a canonical topological isomorphism of Banach algebras* $C(X^\infty) \cong C_0(X)^e$.

Proof Extending every $f \in C_0(X)$ by zero to X^∞, we consider $C_0(X)$ as a subspace of $C(X^\infty)$. Define $\psi : C_0(X)^e \to C(X^\infty)$ by $\psi(f, \lambda) = f + \lambda e$, where $e(x) = 1$ for every $x \in X^\infty$. Then ψ is an isomorphism of algebras. For the norms one has

$$\|\psi(f, \lambda)\|_{X^\infty} = \sup_{x \in X^\infty} |f(x) + \lambda| \leq \sup_{x \in X} |f(x)| + |\lambda| = \|(f, \lambda)\|.$$

This implies that ψ is continuous. On the other hand, one has

$$\sup_{x \in X^\infty} |f(x) + \lambda| \geq |f(\infty) + \lambda| = |\lambda|.$$

Since $|f(x)| \leq |\lambda| + |f(x) + \lambda|$ for every x, we then get

$$\|(f, \lambda)\| = \sup_{x \in X^\infty} |f(x)| + |\lambda| \leq 3\|\psi(f, \lambda)\|_{X^\infty}.$$

The lemma is proven. Note that ψ is not an isometry, but restricted to $C_0(X)$, it becomes one. □

2.4 The Gelfand Map

In this section we shall always assume that $\mathcal{A}$ is a *commutative* Banach algebra. In this case we define the *structure space* $\Delta_{\mathcal{A}}$ to be the set of all non-zero continuous algebra homomorphisms $m : \mathcal{A} \to \mathbb{C}$. This space is often called the *maximal ideal space*, which is justified by Theorem 2.5.2 below, that sets up a bijection between $\Delta_{\mathcal{A}}$ and the set of maximal ideals of $\mathcal{A}$. The elements of $\Delta_{\mathcal{A}}$ are also called *multiplicative linear functionals*, which is why we use the letter m to denote them. If $\mathcal{A}$ is unital, it follows automatically that $m(1) = 1$ for every $m \in \Delta_{\mathcal{A}}$, since $m(1) = m(1^2) = m(1)^2$ implies $m(1) = 0$ or $m(1) = 1$. Now $m(1) = 0$ implies $m = 0$, a case which is excluded.

Examples 2.4.1

- Let $\mathcal{A} = C_0(X)$ for a locally compact Hausdorff space X. For $x \in X$ one gets an element m_x of $\Delta_{\mathcal{A}}$ defined by $m_x(f) = f(x)$.

- Let $\mathcal{A} = L^1(A)$, where A is an LCA-group. Let $\chi \in \widehat{A}$, then the map $m_\chi : \mathcal{A} \to \mathbb{C}$ defined by

$$m_\chi(f) = \hat{f}(\chi) = \int_A f(x)\overline{\chi(x)}\, dx$$

is an element of $\Delta_{\mathcal{A}}$ as follows from Lemma 1.7.2.

For a given multiplicative functional $m \in \Delta_{\mathcal{A}}$, there exists precisely one extension of m to a multiplicative functional on $\mathcal{A}^e$ given by

$$m^e(a, \lambda) = m(a) + \lambda.$$

A multiplicative functional of $\mathcal{A}^e$ that is not extended from $\mathcal{A}$ must vanish on $\mathcal{A}$, and hence it must be equal to the *augmentation functional* m_∞ of $\mathcal{A}^e$ given by $m_\infty(a, \lambda) = \lambda$. Thus we get

$$\Delta_{\mathcal{A}^e} = \{m^e : m \in \Delta_{\mathcal{A}}\} \cup \{m_\infty\}.$$

We now want to equip $\Delta_{\mathcal{A}}$ with a natural topology that makes $\Delta_{\mathcal{A}}$ into a locally compact space (compact if $\mathcal{A}$ is unital). For a Banach space V let V' be the *dual space* consisting of all continuous linear maps $\alpha : V \to \mathbb{C}$. This is a Banach space with the norm $\|\alpha\| = \sup_{v \in V \setminus \{0\}} \frac{|\alpha(v)|}{\|v\|}$.

Lemma 2.4.2 *Suppose that $\mathcal{A}$ is a commutative Banach algebra. Let $m \in \Delta_{\mathcal{A}}$. Then m is continuous with $\|m\| \leq 1$. If $\mathcal{A}$ is unital, then $\|m\| = 1$.*

Proof We first consider the case when $\mathcal{A}$ is unital. If $a \in \mathcal{A}$, then $m(a - m(a)1) = 0$, which implies that $a - m(a)1$ is not invertible in $\mathcal{A}$, so that $m(a) \in \sigma(a)$. Since $\sigma(a) \subset B_{\|a\|}(0)$, this implies that $|m(a)| \leq \|a\|$ for every $a \in \mathcal{A}$. By $m(1) = 1$ we see that m is continuous with $\|m\| = 1$.

If $\mathcal{A}$ is not unital, the extension $m^e : \mathcal{A}^e \to \mathbb{C}$ is continuous with $\|m^e\| = 1$. But then the restriction $m = m^e|_{\mathcal{A}}$ is also continuous with $\|m\| \leq 1$. $\square$

It follows from the lemma that $\Delta_{\mathcal{A}} \subset \bar{B}' \subset \mathcal{A}'$, where $\bar{B}' = \{f \in \mathcal{A}' : \|f\| \leq 1\}$ is the closed ball of radius one. Recall that for any normed space V, the *weak-*-topology* on V' is defined as the initial topology on V' defined by the maps $\{\delta_v : v \in V\}$ with $\delta_v : V' \to \mathbb{C}; \alpha \mapsto \alpha(v)$. It is the topology of point-wise convergence, i.e., a net $(\alpha_j)_j$ in V' converges to $\alpha \in V'$ in the weak-* topology if and only if $\alpha_j(v) \to \alpha(v)$ for all $v \in V$.

We need the following important fact, which, as we shall see, is a consequence of Tychonov's Theorem.

Theorem 2.4.3 (Banach-Alaoglu). *Let V be a normed (complex) vector space. Then the closed unit ball*

$$\bar{B}' \overset{\text{def}}{=} \{f \in V' : \|f\| \leq 1\} \subset V'$$

equipped with the weak-*-topology is a compact Hausdorff space.

Proof Recall $\mathbb{D} = \{z \in \mathbb{C} : |z| \leq 1\}$. For $\alpha \in \bar{B}'$ and $v \in V$ one has $|\alpha(v)| \leq \|v\|$, so $\alpha(v)$ is an element of the compact set $\mathbb{D}\|v\|$. So one gets an injective map

$$\bar{B}' \to \prod_{v \in V} \mathbb{D}\|v\|$$

$$\alpha \to (\alpha(v))_v.$$

Note that the product space on the right is Hausdorff and compact by Tychonov's Theorem A.7.1. Since a net in the product space converges if and only if it converges in each component, the weak-*-topology on $\bar{B}'$ coincides with the subspace topology if one views $\bar{B}'$ as a subspace of the product space. Thus, all we need to show is that $\bar{B}'$ is closed in the product space. An element x of the product space lies in $\bar{B}'$ if and only if its coordinates satisfy $x_{v+w} = x_v + x_w$ and $x_{\lambda v} = \lambda x_v$ for all $v, w \in V$ and every $\lambda \in \mathbb{C}$. These conditions define a closed subset of the product. $\square$

If $\mathcal{A}$ is any commutative Banach algebra, we equip the structure space $\Delta_{\mathcal{A}} \subset \mathcal{A}'$ of the algebra $\mathcal{A}$ with the topology induced from the weak-*-topology on $\mathcal{A}'$.

Lemma 2.4.4 *Suppose that $\mathcal{A}$ is a commutative Banach algebra. Then the inclusion map $\Phi : \Delta_{\mathcal{A}} \to \Delta_{\mathcal{A}^e}$ that maps m to m^e is a homeomorphism onto its image.*

Proof Φ is clearly injective. Since the weak-*-topology is the topology of point-wise convergence, we only have to check that a net $(m_\nu)_\nu$ converges point-wise to $m \in \Delta_A$ if and only if $(m_\nu^e)_\nu$ converges point-wise to m^e. But it is clear that $m_\nu^e(a, \lambda) = m_\nu(a) + \lambda \to m(a) + \lambda = m^e(a, \lambda)$ if and only if $m_\nu(a) \to m(a)$. $\qquad \square$

Every $a \in A$ defines a function $\hat{a}$ on the structure space,

$$\hat{a} : \Delta_A \to \mathbb{C}, \qquad m \mapsto \hat{a}(m) = m(a).$$

The map $a \mapsto \hat{a}$ is an algebra homomorphism from A to the algebra of continuous functions on Δ_A. It is known as the *Gelfand transform*.

Theorem 2.4.5 (Gelfand Transform). *Let A be a commutative Banach algebra. Then the following hold.*

(a) Δ_A *is a locally compact Hausdorff space.*

(b) *If A is unital, then Δ_A is compact.*

(c) *For every $a \in A$ the function $\hat{a}$ on Δ_A is continuous and vanishes at ∞. The Gelfand transform*

$$A \to C_0(\Delta_A); \quad a \mapsto \hat{a}.$$

is an algebra homomorphism.

(d) *For every $a \in A$ one has $\|\hat{a}\|_{\Delta_A} \le \|a\|$, so the Gelfand transform is continuous.*

Proof We will show (a) and (b) in one go. The closure $\overline{\Delta_A}$ of Δ_A in A' is compact by the Banach-Alaoglu Theorem. So we show that it equals Δ_A if A is unital, and Δ_A or $\Delta_A \cup \{0\}$ if A is not unital. For this let $(m_\nu)_\nu$ be a net in Δ_A which converges point-wise to $f \in A'$, it follows that

$$f(ab) = \lim_\nu m_\nu(ab) = \lim_\nu m_\nu(a) \lim_\nu m_\nu(b) = f(a)f(b),$$

which shows that $f : A \to \mathbb{C}$ is an algebra homomorphism. If A is unital, then $f(1) = \lim_\nu m_\nu(1) = 1$, we have $f \ne 0$, and hence $f \in \Delta_A$. If A is not unital, it can happen that $f = 0$. This implies (a) and (b).

For (c) note that, as the topology of Δ_A is induced by the weak-*-topology, the point-evaluation $\hat{a}$ is continuous by definition. If Δ_A is compact, the claim is clear. Otherwise, the closure of Δ_A is $\Delta_A \cup \{0\}$ and coincides with the one-point-compactification of Δ_A. Clearly, every $\hat{a}$ vanishes at the point $0 \in \overline{\Delta_A}$.

Finally, for (d) note that for $a \in A$ one has $|\hat{a}(m)| = |m(a)| \le \|m\| \|a\| = \|a\|$ by Lemma 2.4.2. $\qquad \square$

Example 2.4.6 Consider the case $A = C_0(X)$ for a locally compact Hausdorff space X. For $x \in X$ let $m_x : A \to \mathbb{C}$ be defined by $m_x(f) = f(x)$. Then the map $x \mapsto m_x$ is a homeomorphism from X to the structure space Δ_A (See Exercise 2.14).

Lemma 2.4.7 *Suppose that* $\phi : \mathcal{A} \to \mathcal{B}$ *is an algebra homomorphism between commutative Banach algebras* $\mathcal{A}$ *and* $\mathcal{B}$ *such that* $m \circ \phi \neq 0$ *for every* $m \in \Delta_\mathcal{B}$. *Then*

$$\phi^* : \Delta_\mathcal{B} \to \Delta_\mathcal{A}$$

$$m \mapsto m \circ \phi$$

is continuous and is a homeomorphism if it is bijective.

Proof Since point-wise convergence of a net (m_i) in $\Delta_\mathcal{B}$ implies point-wise convergence of $(m_i \circ \phi)$ in $\Delta_\mathcal{A}$, the map ϕ^* is continuous. Assume now that it is bijective. Let $\phi^e : \mathcal{A}^e \to \mathcal{B}^e$ be the canonical extension of ϕ to the unitizations defined by $\phi^e(a, \lambda) = (\phi(a), \lambda)$. Then the map $(\phi^e)^* : \Delta_{\mathcal{B}^e} \to \Delta_{\mathcal{A}^e}$, defined by $m \mapsto m \circ \phi^e$, is a continuous bijection between the one-point compactifications $\Delta_{\mathcal{B}^e}$ and $\Delta_{\mathcal{A}^e}$ of $\Delta_\mathcal{A}$ and $\Delta_\mathcal{B}$, respectively, which restricts to the map $\phi^* : \Delta_\mathcal{B} \to \Delta_\mathcal{A}$. The result then follows from the fact that bijective continuous maps between compact Hausdorff spaces are homeomorphisms. $\qquad\square$

2.5 Maximal Ideals

If $\mathcal{A}$ is an algebra, a linear subspace $I \subset \mathcal{A}$ is called an *ideal* of $\mathcal{A}$ if for every $a \in \mathcal{A}$ one has $aI \subset I$ and $Ia \subset I$. Obviously, $\mathcal{A}$ itself is an ideal. Any ideal different from $\mathcal{A}$ will be called a *proper ideal*. A *maximal ideal* is a proper ideal M such that for any other proper ideal I with $I \supset M$, one has $I = M$.

Notice that if $\mathcal{A}$ is unital, then $\mathcal{A}^\times \cap I = \emptyset$ for any proper ideal $I \subset \mathcal{A}$, since if $a \in \mathcal{A}^\times \cap I$, then $b = ba^{-1}a \in I$ for every $b \in \mathcal{A}$, and hence $I = \mathcal{A}$. Since $\mathcal{A}^\times$ is open in $\mathcal{A}$, it follows that the closure $\overline{I}$ of any proper ideal in a unital Banach algebra $\mathcal{A}$ is again a proper ideal in the algebra $\mathcal{A}$.

Lemma 2.5.1 *If* $\mathcal{A}$ *is a unital Banach algebra, then every proper ideal of* $\mathcal{A}$ *is contained in a maximal ideal of* $\mathcal{A}$. *Every maximal ideal is closed. Finally, if* $\mathcal{A}$ *is commutative, then every non-invertible element of* $\mathcal{A}$ *lies in some maximal ideal.*

Proof The first assertion is an easy application of Zorn's Lemma. The second is clear by the above remarks. For the third take an element a of $\mathcal{A} \setminus \mathcal{A}^\times$. By the first assertion it suffices to show that a lies in some proper ideal. Since we assume $\mathcal{A}$ commutative, the set $I = a\mathcal{A}$ consisting of all elements of the form aa' for $a' \in \mathcal{A}$, is an ideal that contains a. It does not contain 1, as $\mathcal{A}$ is not invertible; therefore it is a proper ideal. $\qquad\square$

Definition If $\mathcal{A}$ is a Banach algebra and $I \subset \mathcal{A}$ is a closed ideal in $\mathcal{A}$, then the quotient vector space $\mathcal{A}/I \overset{\text{def}}{=} \{a + I : a \in \mathcal{A}\}$ is again a Banach algebra if we equip

$\mathcal{A}/I$ with the *quotient norm*

$$\|a + I\| = \inf\{\|a + d\| : d \in I\}$$

and with multiplication given by $(a + I)(b + I) = ab + I$ (See Proposition C.1.7).

Theorem 2.5.2 *Let $\mathcal{A}$ be a commutative unital Banach algebra.*

(a) *The map $m \mapsto \ker(m) = m^{-1}(0)$ is a bijection between $\Delta_{\mathcal{A}}$ and the set of all maximal ideals of $\mathcal{A}$.*

(b) *An element $a \in \mathcal{A}$ is invertible if and only if $m(a) \neq 0$ for every $m \in \Delta_{\mathcal{A}}$.*

(c) *For $a \in \mathcal{A}$ one has $\sigma(a) = \text{Im}(\hat{a})$.*

Proof (a) For the injectivity let m, n be two elements of $\Delta_{\mathcal{A}}$ with $\ker(m) = \ker(n) = I$. As I is a subspace of codimension one, every $a \in \mathcal{A}$ can be written as $a_0 + \lambda 1$ for some $a_0 \in I$. Then

$$m(a) = m(a_0 + \lambda 1) = \lambda = n(a_0 + \lambda 1) = n(a).$$

For the surjectivity let I be a maximal ideal. Then $\mathcal{B} = \mathcal{A}/I$ is a unital Banach algebra without any proper ideal. Let $b \in \mathcal{B}$ be non-zero. As $\mathcal{B}$ does not have any proper ideal, Lemma 2.5.1 implies that b is invertible. Since this holds for any b, Corollary 2.2.5 implies that $\mathcal{B} \cong \mathbb{C}$. So I is the kernel of the map $\mathcal{A} \to \mathcal{B} \cong \mathbb{C}$, which lies in $\Delta_{\mathcal{A}}$.

For (b) let $a \in \mathcal{A}$ be invertible. For $m \in \Delta_{\mathcal{A}}$ it holds $m(a)m(a^{-1}) = m(1) = 1$; therefore $m(a) \neq 0$. For the converse suppose a is not invertible. Then it is contained in a maximal ideal, which is the kernel of some $m \in \Delta_{\mathcal{A}}$ by part (a). Finally, (c) follows by putting things together as follows: A given $\lambda \in \mathbb{C}$ lies in $\sigma(a) \Leftrightarrow a - \lambda 1$ is not invertible $\Leftrightarrow$ there is $m \in \Delta_{\mathcal{A}}$ with $m(a - \lambda 1) = 0$, i.e., $\hat{a}(m) = m(a) = \lambda$, which is equivalent to $\lambda \in \text{Im}(\hat{a})$. Hence $\text{Im}(\hat{a}) \subseteq \sigma(a)$. The converse follows from $m(a - m(a)1) = 0$ for every $m \in \Delta_{\mathcal{A}}$, from which it follows that $a - m(a)1$ is not invertible, hence $\hat{a}(m) = m(a) \in \sigma(a)$. $\square$

2.6 The Gelfand-Naimark Theorem

Let $\mathcal{A}$ be an algebra over $\mathbb{C}$. An *involution* is a map $\mathcal{A} \to \mathcal{A}$, denoted $a \mapsto a^*$ such that for $a, b \in \mathcal{A}$ and $\lambda \in \mathbb{C}$ one has

$$(a + b)^* = a^* + b^*, \quad (\lambda a)^* = \bar{\lambda} a^*, \quad (ab)^* = b^* a^*,$$

as well as $(a^*)^* = a$. A *Banach-*-algebra* is a Banach algebra together with an involution * such that for every $a \in \mathcal{A}$ one has

$$\|a^*\| = \|a\|.$$

A Banach-*-algebra is called C^*-*algebra* if

$$\|a^*a\| = \|a\|^2$$

holds for every $a \in \mathcal{A}$. Equivalently, one can say $\|aa^*\| = \|a\|^2$.

An element a of a Banach-*-algebra $\mathcal{A}$ is called *self-adjoint* if $a^* = a$.

Examples 2.6.1

- Let H be a Hilbert space and $\mathcal{A} = \mathcal{B}(H)$ the Banach algebra of all bounded linear operators on H. Then the map $T \mapsto T^*$ is an involution, where T^* is the *adjoint* of T, i.e., the unique operator with $\langle Tv, w \rangle = \langle v, T^*w \rangle$ for all $v, w \in H$. Then $\mathcal{A}$ is a C^*-algebra, as for $v \in H$ the Cauchy-Schwarz inequality says that

$$\|Tv\|^2 = \langle Tv, Tv \rangle = \langle v, T^*Tv \rangle \le \|v\| \|T^*Tv\|.$$

So, for $v \ne 0$, one gets $\frac{\|Tv\|^2}{\|v\|^2} \le \frac{\|T^*Tv\|}{\|v\|}$ and so, for the operator norm it follows $\|T\|^2 \le \|T^*T\| \le \|T^*\| \|T\|$. This inequality implies $\|T\| \le \|T^*\|$, and by replacing T by T^* one gets equality, and it also follows that $\|T\|^2 = \|T^*T\|$.

  An operator $T \in \mathcal{B}(H)$ with $T^* = T$ is called a *self adjoint operator*.

- Let $\mathcal{A} = C_0(X)$ for a locally compact Hausdorff space X. The involution

$$f^*(x) \stackrel{\text{def}}{=} \overline{f(x)}$$

  makes this commutative Banach algebra into a C^*-algebra as $\|f^*f\| = \sup_{x \in X} |f^*(x)f(x)| = \sup_{x \in X} |f(x)|^2 = \|f\|^2$.

- A simple example of a Banach-*-algebra, which is not a C^*-algebra is given by the disk-algebra (Example 2.2.8) with the involution $f^*(z) = \overline{f(\bar{z})}$ (See Exercise 2.13).

Proposition 2.6.2 *Let G be a locally compact group, and let $\mathcal{A}$ be the Banach algebra $L^1(G)$. With the involution*

$$f^*(x) = \Delta_G\left(x^{-1}\right) \overline{f\left(x^{-1}\right)}$$

the algebra $\mathcal{A}$ is a Banach--algebra but not a C^*-algebra unless G is trivial, in which case $\mathcal{A} = \mathbb{C}$.*

Proof The axioms of an involution are easily verified, and so is the fact that $\|f^*\| = \|f\|$ for $f \in \mathcal{A}$. For the last assertion, we need a lemma.

Lemma 2.6.3 (a) *Let X be a locally compact Hausdorff space and let $x_1, \ldots, x_n \in X$ pairwise different points. Let $\lambda_1, \ldots, \lambda_n \in \mathbb{C}$ any given numbers. Then there exists $f \in C_c(G)$ with $f(x_j) = \lambda_j$ for each $j = 1, \ldots, n$.*

(b) *Let G be a locally compact group and $g \in C_c(G)$ with the property that $\left| \int_G g(y) \, dy \right| = \int_G |g(y)| \, dy$. Then there is $\theta \in \mathbb{T}$ such that $g(x) \in \theta[0, \infty)$ for every $x \in G$.*

Proof (a) By the Hausdorff property there are open sets $U_{i,j}$ and $V_{i,j}$ for $i \neq j$ with $x_i \in U_{i,j}$, $x_j \in V_{i,j}$ and $U_{i,j} \cap V_{i,j} = \emptyset$. Set $W_j = \bigcap_{j \neq i} U_{j,i} \cap V_{i,j}$. Then W_j is an open neighborhood of x_j and the sets $W_1, \ldots, W_n$ are pairwise disjoint. By the Lemma of Urysohn (A.8.1) there is $f_j \in C_c(X)$ with support in W_j and $f(x_j) = 1$. Set $f = \lambda_1 f_1 + \cdots + \lambda_n f_n$. Then f satisfies the assertion of the lemma.

(b) Let g be given as in the assumption. If $\int_G g(x) \, dx = 0$, then $\int_G |g(x)| \, dx = 0$, so $g = 0$. So we assume $\int_G g(x) \, dx \neq 0$. Replacing g by λg for some $\lambda \in \mathbb{T}$, we can assume $\int_G g(x) \, dx > 0$. Then

$$\int_G |g(x)| \, dx = \int_G g(x) \, dx = \operatorname{Re}\left(\int_G g(x) \, dx \right) = \int_G \operatorname{Re}(g(x)) \, dx.$$

So $\int_G (|g| - \operatorname{Re}(g))(x) \, dx = 0$. Since the continuous function $|g| - \operatorname{Re}(G)$ is positive, it vanishes, hence $|g| = \operatorname{Re}(g)$, which means $g \geq 0$. $\qquad\square$

To prove the proposition assume that $L^1(G)$ is a C^*-algebra. We show that $G = \{1\}$. By the C^*-property one has for every $f \in C_c(G)$ that

$$\int_G \left| \int_G \Delta(x^{-1}y) f(y) \overline{f(x^{-1}y)} \, dy \right| dx$$

$$= \| f * f^* \| = \| f \|^2$$

$$= \int_G \int_G \Delta(x^{-1}y) |f(y) \overline{f(x^{-1}y)}| \, dy \, dx.$$

The outer integrals on both sides have continuous integrands ≥ 0. The integrands satisfy the inequality $\leq$. As the integrals are equal, the integrands are equal, too. So for every $x \in G$ we have

$$\left| \int_G \Delta(x^{-1}y) f(y) \overline{f(x^{-1}y)} \, dy \right| = \int_G \Delta(x^{-1}y) |f(y) \overline{f(x^{-1}y)}| \, dy.$$

By the lemma there is, for given $x \in G$, a $\theta \in \mathbb{T}$ such that $f(y) \overline{f(x^{-1}y)} \in \theta[0, \infty)$ for every $y \in G$.

Assume now that G is non-trivial. Then there is $x_0 \neq 1$ in G. By the lemma there is a function $f \in C_c(G)$ with $f(x_0) = f(x_0^{-1}) = i$ and $f(1) = 1$. For $x = x_0$ we deduce

$$y = 1 \quad \Rightarrow \quad f(y)\overline{f(x^{-1}y)} = f(1)\overline{f(x_0^{-1})} = -i,$$
$$y = x_0 \quad \Rightarrow \quad f(y)\overline{f(x^{-1}y)} = f(x_0)\overline{f(1)} = i.$$

This is a contradiction! Therefore G is the trivial group. $\square$

Recall that for an element a in a Banach algebra $\mathcal{A}$, we denote by $r(a) = \max\{|\lambda| : \lambda \in \sigma(a)\}$ the spectral radius of a. By adjoining a unit if necessary (See Sect. 2.3) this makes sense also for non-unital algebras and the spectral radius formula $r(a) = \lim_{n\to\infty} \|a^n\|^{1/n}$ holds in general.

Lemma 2.6.4 *Suppose that $\mathcal{A}$ is a C^*-algebra and that $a \in \mathcal{A}$ is self adjoint, i.e., $a = a^*$. Then the spectral radius of a equals its norm, i.e., $r(a) = \|a\|$.*

Proof If $a = a^*$ the C^*-condition implies that $\|a^2\| = \|a^*a\| = \|a\|^2$. By induction we get $\|a^{2^n}\| = \|a\|^{2^n}$ for every $n \in \mathbb{N}$. This implies $r(a) = \lim_{n\to\infty} \|a^{2^n}\|^{\frac{1}{2^n}} = \|a\|$. $\square$

We should point out that the above result is remarkable, since it implies that the norm in any C^*-algebra $\mathcal{A}$ only depends on the purely algebraic properties of $\mathcal{A}$, since the spectral radius $r(a)$ only depends on the $*$-algebraic operations in $\mathcal{A}$. This fact is clear for a self-adjoint by the above lemma, and it follows for arbitrary $a \in \mathcal{A}$ by $\|a\|^2 = \|a^*a\| = r(a^*a)$. In particular, it follows that every C^*-algebra has a unique C^*-norm.

For later use we need to know that if $\mathcal{A}$ is a C^*-algebra, then $\mathcal{A}^e$ is also a C^*-algebra, meaning that there exists a complete C^*-norm on $\mathcal{A}^e$. It is easy to extend the involution to $\mathcal{A}^e$: Put $(a, \lambda)^* := (a^*, \overline{\lambda})$. If $\mathcal{A}$ is unital, then we saw above that $\mathcal{A}^e$ is isomorphic to the direct sum $\mathcal{A} \oplus \mathbb{C}$, and it is easily checked that the point-wise operations and the norm $\|(a, \lambda)\| := \max\{\|a\|, |\lambda|\}$ give $\mathcal{A} \oplus \mathbb{C}$ the structure of a C^*-algebra. When $\mathcal{A}$ has no unit, the proof becomes more complicated:

Lemma 2.6.5 *Suppose that $\mathcal{A}$ is a C^*-algebra. Then there exists a norm on $\mathcal{A}^e$ that makes $\mathcal{A}^e$ a C^*-algebra. Moreover, the embedding of $\mathcal{A}$ into $\mathcal{A}^e$ is isometric.*

Proof By the above discussion we may assume that $\mathcal{A}$ has no unit. Consider the homomorphism $L : \mathcal{A}^e \to \mathcal{B}(\mathcal{A})$ given by $L_{(a,\lambda)}b := ab + \lambda b$, and define $\|(a, \lambda)\| := \|L_{(a,\lambda)}\|_{\mathrm{op}}$. To see that this is a Banach algebra norm on $\mathcal{A}^e$ we only have to check that L is injective. So let $(a, \lambda) \in \mathcal{A}^e$ such that $ab + \lambda b = 0$ for every $b \in \mathcal{A}$. If $\lambda \neq 0$, it follows that $\left(-\frac{a}{\lambda}\right)b = b$ for every $b \in \mathcal{A}$, and hence that $-\frac{a}{\lambda}$ is a unit for $\mathcal{A}$ (See Exercise 2.4). This contradicts the assumption that $\mathcal{A}$ has no unit. If $\lambda = 0$, then $ab = 0$ for every $b \in \mathcal{A}$ implies that $aa^* = 0$. But then $\|a\|^2 = \|aa^*\| = 0$.

The inclusion $a \mapsto (a,0), \mathcal{A} \to \mathcal{A}^e$ is isometric: since $\|ab\| \le \|a\|\|b\|$ for all $a, b \in \mathcal{A}$ it follows that $\|L_{(a,0)}\|_{\mathrm{op}} \le \|a\|$, and the equation $\|aa^*\| = \|a\|^2$ implies that $\|a\| \le \|L_{(a,0)}\|_{\mathrm{op}}$. Since then $\mathcal{A}$ is a complete subspace of finite codimension in $\mathcal{A}^e$, this also implies that $\mathcal{A}^e$ is complete (See Lemma C.1.9).

It only remains to check that the norm on $\mathcal{A}^e$ satisfies the C^*- condition $\|(a,\lambda)^*(a,\lambda)\| = \|(a,\lambda)\|^2$. For this let $\varepsilon > 0$. By the definition of the operator norm there exists $b \in \mathcal{A}$ with $\|b\| = \|(b,0)\| = 1$ such that $\|ab + \lambda b\| \ge \|(a,\lambda)\|(1-\varepsilon)$. This implies

$$(1-\varepsilon)^2\|(a,\lambda)\|^2 \le \|ab + \lambda b\|^2 = \|(ab + \lambda b)^*(ab + \lambda b)\|$$

$$= \|(b^*,0)(a^*,\bar{\lambda})(a,\lambda)(b,0)\|,$$

and the latter is less than or equal to

$$\|(b^*,0)\|\|(a,\lambda)^*(a,\lambda)\|\|(b,0)\| = \|(a,\lambda)^*(a,\lambda)\|.$$

Since $\varepsilon > 0$ is arbitrary, we get the inequalities

$$\|(a,\lambda)\|^2 \le \|(a,\lambda)^*(a,\lambda)\| \le \|(a,\lambda)^*\|\|(a,\lambda)\|.$$

Replacing (a,λ) by $(a,\lambda)^*$, this implies equality everywhere. $\qquad\square$

Definition If $\mathcal{A}$ is a C*-algebra and $a \in \mathcal{A}$, we define the *real* and *imaginary part* of a as

$$\mathrm{Re}(a) = \frac{1}{2}(a + a^*) \quad \text{an} \quad \mathrm{Im}(a) = \frac{1}{2i}(a - a^*).$$

Then $\mathrm{Re}(a)$ and $\mathrm{Im}(a)$ are self-adjoint with $a = \mathrm{Re}(a) + i\,\mathrm{Im}(a)$.

Lemma 2.6.6 *If $\mathcal{A}$ is a commutative C*-algebra, then $m(a^*) = \overline{m(a)}$ for every $a \in \mathcal{A}$ and every $m \in \Delta_{\mathcal{A}}$.*

Proof Let $\mathcal{A}$ be a commutative C^*-algebra. By passing to $\mathcal{A}^e$ if necessary, we may assume that $\mathcal{A}$ is unital. Let $m \in \Delta_{\mathcal{A}}$ and let $a \in \mathcal{A}$. Decomposing $a \in \mathcal{A}$ in its real and imaginary parts, we may assume that $a = a^*$. We want to show $m(a) \in \mathbb{R}$. For this write $m(a) = x + iy$ with $x, y \in \mathbb{R}$. We have to show that $y = 0$. Put $a_t \overset{\mathrm{def}}{=} a + it$. Then we get $a_t^* a_t = (a - it)(a + it) = a^2 + t^2$ and it follows $m(a_t) = x + i(y+t)$. So for every $t \in \mathbb{R}$ we have $x^2 + (y+t)^2 = |m(a_t)|^2 \le \|a_t\|^2 = \|a_t^* a_t\| = \|a^2 + t^2\| \le \|a\|^2 + t^2$, from which it follows that $x^2 + y^2 + 2yt \le \|a\|^2$ for every $t \in \mathbb{R}$. But this is possible only if $y = 0$. $\qquad\square$

Theorem 2.6.7 (Gelfand-Naimark). *If $\mathcal{A}$ is a commutative C*-algebra, then the Gelfand transform $a \mapsto \hat{a}$ is an isometric *-isomorphism*

$$\mathcal{A} \xrightarrow{\cong} C_0(\Delta_{\mathcal{A}}).$$

So in particular, one has $\|\hat{a}\|_{\Delta_{\mathcal{A}}} = \|a\|$ and $\widehat{a^*} = \overline{\hat{a}}$ for every $a \in \mathcal{A}$. Finally, the space $\Delta_{\mathcal{A}}$ is compact if and only if $\mathcal{A}$ is unital. In this case the Gelfand transform gives an isomorphism $\mathcal{A} \cong C(\Delta_{\mathcal{A}})$.

Proof We show that in case of a C^*-algebra the Gelfand map is isometric with dense image. This will imply that $\widehat{\mathcal{A}} = \{\hat{a} : a \in \mathcal{A}\}$ is a complete, hence closed, dense subalgebra of $C_0(\Delta_{\mathcal{A}})$, and hence $\widehat{\mathcal{A}} = C_0(\Delta_{\mathcal{A}})$. For $a \in \mathcal{A}$, using Theorem 2.5.2 and Lemma 2.6.4 we get $\|\hat{a}\|_\infty^2 = \|\bar{\hat{a}}\hat{a}\|_\infty = \|\widehat{a^*a}\|_\infty = r(a^*a) = \|a^*a\| = \|a\|^2$, so indeed the Gelfand transform is an isometric map. It remains to show that the image is dense. The image of the Gelfand map $\widehat{\mathcal{A}} = \{\hat{a} : a \in \mathcal{A}\} \subset C_0(\Delta_{\mathcal{A}})$ strictly separates the points of $\Delta_{\mathcal{A}}$: if $m_1, m_2 \in \Delta_{\mathcal{A}}$ such that $\hat{a}(m_1) = \hat{a}(m_2)$ for every $a \in \mathcal{A}$, then we clearly get $m_1 = m_2$, and since we require $m \neq 0$ for the elements $m \in \Delta_{\mathcal{A}}$, we find at least one $a \in \mathcal{A}$ with $\hat{a}(m) = m(a) \neq 0$ for any given $m \in \Delta_{\mathcal{A}}$. Using this, it is then a direct consequence of the Stone-Weierstraß Theorem A.10.1, that $\widehat{\mathcal{A}}$ is dense in $C_0(\Delta_{\mathcal{A}})$ with respect to the supremum-norm whenever $\widehat{\mathcal{A}}$ is invariant under taking complex conjugates, which is the case thanks to Lemma 2.6.6. Finally, if $\Delta_{\mathcal{A}}$ is compact, then $\mathcal{A} \cong C_0(\Delta_{\mathcal{A}})$ is unital. The converse is Theorem 2.4.5(b). □

2.7 The Continuous Functional Calculus

Let a be an element of a commutative unital C^*-algebra $\mathcal{A}$, and let f be a continuous function on its spectrum $\sigma(a) = \mathrm{Im}(\hat{a})$. Then $f \circ \hat{a}$ defines a new element of $\mathcal{A} \cong C(\Delta_{\mathcal{A}})$, which we call $f(a)$. The map $f \mapsto f(a)$ is called the (continuous) *functional calculus* for a.

The functional calculus is one of the most important tools in functional analysis. In case of the algebra $\mathcal{B}(H)$ of bounded operators on a Hilbert space H, it will play a very important role in further sections of this book.

An element a of a C^*-algebra $\mathcal{A}$ is called a *normal element* of $\mathcal{A}$ if a commutes with its adjoint a^*. If $\mathcal{A}$ is unital and $a \in \mathcal{A}$, then there exists a smallest unital C^*-subalgebra $C^*(a, 1)$ containing a, which we call the *unital C^*-algebra generated by a*. In general, we let $C^*(a)$ denote the smallest C^*-subalgebra of $\mathcal{A}$ containing a, which also makes sense if $\mathcal{A}$ has no unit. One notices, that $C^*(a)$ (resp. $C^*(a, 1)$) equals the norm closure in $\mathcal{A}$ of the algebra generated by a and a^* (resp. a, a^* and 1). The element a is normal if and only if the C^*-algebra $C^*(a)$ (resp. $C^*(a, 1)$) is commutative.

In the construction of the functional calculus we will need the following remarkable consequence of the spectral radius formula:

Lemma 2.7.1 *Let $\Phi : \mathcal{A} \to \mathcal{B}$ be a *-homomorphism from the Banach-*-algebra $\mathcal{A}$ into the C^*-algebra $\mathcal{B}$. Then $\|\Phi(a)\| \leq \|a\|$ for every $a \in \mathcal{A}$. In particular, the map Φ is continuous.*

Proof Adjoining units to $\mathcal{A}$ and $\mathcal{B}$ and passing to $\Phi^e : \mathcal{A}^e \to \mathcal{B}^e$ given by $\Phi^e(a, \lambda) = (\Phi(a), \lambda)$ if necessary, we may assume without loss of generality that $\mathcal{A}, \mathcal{B}$ and Φ are unital. Let $a \in \mathcal{A}$. Since $\sigma_{\mathcal{B}}(\Phi(a^*a)) \subseteq \sigma_{\mathcal{A}}(a^*a)$, it follows from the C^*-condition $\|b^*b\| = \|b\|^2$ on $\mathcal{B}$ and Lemma 2.6.4 that $\|\Phi(a)\|^2 = \|\Phi(a^*a)\| = r(\Phi(a^*a)) \leq r(a^*a) \leq \|a^*a\| \leq \|a\|^2$. $\qquad\square$

Lemma 2.7.2 *Let $\mathcal{A} \subset \mathcal{B}$ be unital C^*-algebras and let $a \in \mathcal{A}$ be a normal element. Then $\sigma_{\mathcal{A}}(a) = \sigma_{\mathcal{B}}(a)$.*

Proof One has $C^*(1, a) \subset \mathcal{A} \subset \mathcal{B}$ and therefore $\sigma_{C^*(1,a)}(a) \supset \sigma_{\mathcal{A}}(a) \supset \sigma_{\mathcal{B}}(a)$. So it suffices to show that $\sigma_{C^*(1,a)} \subset \sigma_{\mathcal{B}}(a)$. We can replace the algebra $\mathcal{A}$ with $C^*(1, a)$ and therefore assume that $\mathcal{A}$ is commutative. In a first step we also assume $\mathcal{B}$ to be commutative. Restriction of multiplicative functionals defines a continuous map res : $\Delta_{\mathcal{B}} \to \Delta_{\mathcal{A}}$. Define $\mathrm{res}^* : C(\Delta_{\mathcal{A}}) \to C(\Delta_{\mathcal{B}})$ by $\mathrm{res}^* f(m) = f(\mathrm{res}(m))$. We get a commutative diagram

$$
\begin{array}{ccc}
C(\Delta_{\mathcal{A}}) & \xrightarrow{\;\cong\;} & \mathcal{A} \\
{\scriptstyle \mathrm{res}^*}\big\downarrow & & \big\downarrow \\
C(\Delta_{\mathcal{B}}) & \xrightarrow{\;\cong\;} & \mathcal{B},
\end{array}
$$

whose horizontal arrows are isomorphisms by the Gelfand-Naimark theorem. It follows that res^* is injective, hence res must be surjective. So $\sigma_{\mathcal{A}}(a) = \hat{a}(\Delta_{\mathcal{A}}) = \hat{a}(\mathrm{res}(\Delta_{\mathcal{B}})) = \hat{a}(\Delta_{\mathcal{B}}) = \sigma_{\mathcal{B}}(a)$.

We finally consider the general case, i.e., now $\mathcal{B}$ is no longer restricted to be commutative. Let $\lambda \in \mathbb{C}$ with $\lambda \notin \sigma_{\mathcal{B}}(a)$. We have to show $\lambda \notin \sigma_{\mathcal{A}}(a)$. The element $a - \lambda$ is invertible in $\mathcal{B}$. Put $b = (a - \lambda)^{-1} \in \mathcal{B}$. Then b commutes with a and a^*, so the C^*-algebra $\mathcal{C} = C^*(1, a, b)$ generated by $1, a, b$ is commutative and $\lambda \notin \sigma_{\mathcal{C}}(a)$. By the first part we get $\sigma_{\mathcal{C}}(a) = \sigma_{\mathcal{A}}(a)$. $\qquad\square$

Theorem 2.7.3 *Let $\mathcal{A}$ be a unital C^*-algebra and let $a \in \mathcal{A}$ be a normal element. Then there exists exactly one unital *-homomorphism*

$$\Phi_a : C(\sigma(a)) \to \mathcal{A}$$

with the property that $\Phi_a(\mathrm{Id}) = a$. We write $f(a) = \Phi_a(f)$. Then Φ_a is isometric and the image of Φ_a is $C^*(a, 1)$. If $\mathcal{A}$ is commutative, then $\widehat{\Phi_a(f)} = f \circ \hat{a}$.

If $\mathcal{A} = C^*(1, a)$, then $\hat{a} : \Delta_{\mathcal{A}} \to \sigma(a)$ is a homeomorphism.

If $f : \sigma(a) \to \mathbb{C}$ is given by a power series $f(z) = \sum_{n=0}^{\infty} a_n(z - z_0)^n$ which converges uniformly on $\sigma(a)$, then $f(a) = \sum_{n=0}^{\infty} a_n(a - z_0 1)^n$, and the series converges in norm.

Proof For the uniqueness, let $\Phi, \Psi : C(\sigma(a)) \to \mathcal{A}$ be two unital *-homomorphisms with $\Phi(\mathrm{Id}) = \Psi(\mathrm{Id}) = a$. Then they agree on all polynomials in z and $\bar{z}$, which span a dense subalgebra of $C(\sigma(a))$ by the Theorem of Stone-Weierstraß A.10.1. By Lemma 2.7.1 the maps Φ and Ψ are continuous, hence they agree.

To show existence, let $\mathcal{B} = C^*(a, 1) \subseteq \mathcal{A}$ denote the abelian unital C*-algebra generated by a. Let $\hat{a} : \Delta_\mathcal{B} \to \sigma_\mathcal{B}(a) = \sigma_\mathcal{A}(a)$ denote the Gelfand transform of a. Then $\hat{a}$ is surjective by Theorem 2.5.2 and it is also injective, because if $m_1, m_2 \in \Delta_\mathcal{B}$ with $m_1(a) = \hat{a}(m_1) = \hat{a}(m_2) = m_2(a)$, then both characters will coincide on the linear span of elements of the form $a^l(a^*)^k$, which is dense in $C^*(a, 1)$. Since $\Delta_\mathcal{B}$ is compact, it follows that $\hat{a} : \Delta_\mathcal{B} \to \sigma_\mathcal{B}(a)$ is a homeomorphism. We then get a unital isometric *-isomorphism

$$\Psi : C(\sigma_\mathcal{B}(a)) \to C(\Delta_\mathcal{B}); \Psi(f) = f \circ \hat{a}^{-1}.$$

Composing this with the inverse of the Gelfand transform $\widehat{\quad} : \mathcal{B} \to C(\Delta_\mathcal{B})$ gives us a unital isometric *-isomorphism $\Phi_a : C(\Delta_\mathcal{B}) \to \mathcal{B}$. If we apply Ψ to the identity $\text{Id} : \sigma_\mathcal{B}(a) \to \sigma_\mathcal{B}(a)$ we obtain the function $\hat{a}$, which is mapped to a under the inverse Gelfand transform. Thus we get $\Phi_a(1) = 1$ and $\Phi_a(\text{Id}) = a$.

The assertion for a uniformly convergent power series f follows from applying Φ_a to the polynomials $f_N(z) = \sum_{n=0}^N a_n(z - z_0)^n$ and the fact that Φ_a is isometric. □

Example 2.7.4 If $\mathcal{A} = C(X)$ for some compact Hausdorff space X, the functional calculus sends an element $g \in C(\sigma(f)) = C(f(X))$ to the function $g \circ f \in C(X)$. This follows from the uniqueness assertion in Theorem 2.5.2 and the fact that the homomorphism $\Phi : C(f(X)) \to C(X) : \Phi(g) = g \circ f$ satisfies $\Phi(\text{Id}) = f$.

Corollary 2.7.5 (a) *If $a = a^*$ is a self-adjoint element of the C*-algebra $\mathcal{A}$, then $\sigma_\mathcal{A}(a) \subseteq \mathbb{R}$.*

(b) *Let $\Psi : \mathcal{A} \to \mathcal{B}$ be a unital *-homomorphism between C*-algebras and let $a \in \mathcal{A}$ be a normal element. Then $\Psi(a)$ is normal and $\sigma(\Psi(a)) \subset \sigma(a)$. The diagram*

$$
\begin{array}{ccc}
C(\sigma(a)) & \xrightarrow{\Phi_a} & \mathcal{A} \\
{\scriptstyle \text{res}} \downarrow & & \downarrow {\scriptstyle \Psi} \\
C(\sigma(\Psi(a))) & \xrightarrow{\Phi_{\Psi(a)}} & \mathcal{B}
\end{array}
$$

commutes.

In particular, one has $f(\Psi(a)) = \Psi(f(a))$ for every $f \in C(\sigma(a))$.

(c) *Suppose that $a \in \mathcal{A}$ is a normal element in the unital C*-algebra $\mathcal{A}$ and let $f \in C(\sigma_\mathcal{A}(a))$. Then $f(a)$ is a normal element of $\mathcal{A}$, $\sigma_\mathcal{A}(f(a)) = f(\sigma_\mathcal{A}(a))$ and $g(f(a)) = (g \circ f)(a)$ for all $g \in C(\sigma_\mathcal{A}(f(a)))$.*

Proof (a) We assume $\mathcal{A}$ to be unital. By Theorem 2.7.3 we have $f(\text{Id}) = a = a^* = f(\overline{\text{Id}})$, where $\text{Id} = \text{Id}_{\sigma_\mathcal{A}(a)}$. As the map $C(\sigma(a)) \to \mathcal{A}$, that sends f to $f(a)$, is isometric, hence injective, it follows $\overline{\text{Id}} = \text{Id}$ on $\sigma(a)$ and hence $\sigma(a) \subset \mathbb{R}$.

(b) The two ways through the diagram give us two unital *- homomorphisms $\Pi_1, \Pi_2 : C(\sigma(a)) \to \mathcal{B}$. Both maps send the identity $\mathrm{Id} = \mathrm{Id}_{\sigma_{\mathcal{A}}(a)}$ to $\Phi(a)$, and therefore they agree.

Applying part (b) to the functional calculus homomorphism $\Phi_a : C(\sigma(a)) \to \mathcal{A}$ yields (c) except for $\sigma_{\mathcal{A}}(f(a)) = f(\sigma_{\mathcal{A}}(a))$, where it only gives "$\subset$". If X is a compact Hausdorff space then $\sigma_{C(X)}(f) = f(X)$. If we apply this fact to the algebra $C(\sigma_{\mathcal{A}}(a))$ and observe that the spectrum is preserved under isometric $*$-isomorphisms, we see that $\sigma_{\mathcal{A}}(f(a)) = \sigma_{C^*(a,1)}(f(a)) = \sigma_{C(\sigma_{\mathcal{A}}(a))}(f) = f(\sigma_{\mathcal{A}}(a))$. $\qquad\square$

We now show that injective $*$-homomorphisms between C^*-algebras are automatically isometric.

Corollary 2.7.6 *Suppose that the map $\Psi : \mathcal{A} \to \mathcal{B}$ is an injective $*$-homomorphism from the C^*-algebra $\mathcal{A}$ to the C^*-algebra $\mathcal{B}$. Then $\|\Psi(a)\| = \|a\|$ for every $a \in \mathcal{A}$, i.e., Ψ is isometric.*

Proof By Lemma 2.7.1 we know that $\|\Psi(a)\| \leq \|a\|$ for every $a \in \mathcal{A}$. We want to show equality. As in the proof of Lemma 2.7.1 we may assume without loss of generality that $\mathcal{A}, \mathcal{B}$ and Ψ are unital.

So assume that there is $a \in \mathcal{A}$ with $\|\Phi(a)\| < \|a\|$. By scaling we may assume $\|a\| = 1$. Then $\alpha \overset{\text{def}}{=} \|\Psi(a^*a)\| = \|\Psi(a)\|^2 < \|a\|^2 = \|a^*a\| = 1$. So with $c = a^*a$ we have c self-adjoint with $\sigma(c) \subseteq [-1,1]$ and $\sigma(\Psi(c)) \subseteq [-\alpha, \alpha]$. It follows from Lemma 2.6.4 that $\sigma(c)$ contains either 1 or -1. So we can find a function $0 \neq f \in C(\sigma(c))$ with $f \equiv 0$ on $\sigma(\Psi(c))$. Using the injectivity of Ψ it follows that $0 \neq \Psi(f(c)) = f(\Psi(c)) = 0$, a contradiction. $\qquad\square$

2.8 Exercises and Notes

Exercise 2.1 Let $(V, \|\cdot\|)$ be a Banach space. Let $\mathcal{A} = \mathcal{B}(V)$ be the set of all bounded operators on V, i.e., the set of all linear operators $T : \mathcal{A} \to \mathcal{A}$, such that

$$\|T\|_{op} \overset{\text{def}}{=} \sup_{v \neq 0} \frac{\|Tv\|}{\|v\|} < \infty.$$

Show: $\|\cdot\|_{op}$ is a norm, it is called the *operator norm*. Show that $\mathcal{A}$ is a unital Banach algebra with this norm.

Exercise 2.2 Let V be a Banach space, $D \subset \mathbb{C}$ open and $f : D \to V$ holomorphic. Show that f is continuous.

Exercise 2.3 Suppose that for an element a of a unital Banach-*-algebra one has $\sigma_{\mathcal{A}}(a^*) = \overline{\sigma_{\mathcal{A}}(a)}$.

Exercise 2.4 Suppose that $\mathcal{A}$ is a Banach-*-algebra. Show that any left unit is a unit. In other words, suppose that $a = ea$ holds for every a in the Banach-*-algebra $\mathcal{A}$. Show that e is a unit.

Exercise 2.5 Give an example of a unital Banach algebra $\mathcal{A}$ and two elements $x, y \in \mathcal{A}$ with $xy = 1$, but $yx \neq 1$.

Exercise 2.6 Consider the Banach algebra $l^1(\mathbb{Z})$ with the usual convolution product. Show that $f^*(n) \overset{\text{def}}{=} \overline{f(n)}$ defines an involution on $l^1(\mathbb{Z})$ making $l^1(\mathbb{Z})$ into a Banach *-algebra, which is not symmetric, i.e., there exists a self-adjoint element $f \in l^1(\mathbb{Z})$ such that $\sigma(f) \not\subset \mathbb{R}$.

Exercise 2.7 Let $(V, \|\cdot\|)$ be a Banach space, and let $(a_n)_{n \in \mathbb{N}}$ be a sequence in $\mathcal{A}$. Show that, if $\sum_n \|a_n\| < \infty$, then the series $\sum_n a_n$ converges in $\mathcal{A}$.

Exercise 2.8 Let $(\mathcal{A}, \|\cdot\|)$ be a Banach algebra. Show that for every $a \in \mathcal{A}$ the series

$$\exp(a) \overset{\text{def}}{=} \sum_{n=0}^{\infty} \frac{a^n}{n!}$$

converges and that, for $a, b \in \mathcal{A}$ with $ab = ba$, one has $\exp(a+b) = \exp(a)\exp(b)$.

Exercise 2.9 Let $\mathcal{B} \subset \mathcal{A}$ a closed sub algebra of the Banach algebra $\mathcal{A}$. Suppose that $\mathcal{A}$ is unital and $1 \in \mathcal{B}$. Show that for $x \in \mathcal{B}$,

$$r_{\mathcal{A}}(x) = r_{\mathcal{B}}(x),$$

where $r_{\mathcal{A}}$ is the spectral radius with respect to the algebra $\mathcal{A}$ and likewise for $\mathcal{B}$.

Exercise 2.10 Let $\mathcal{A}$ be a unital Banach algebra. Let $x, y \in \mathcal{A}$ with $xy = yx$. Show that the spectral radius satisfies

$$r(xy) \leq r(x)r(y), \qquad r(x + y) \leq r(x) + r(y).$$

Exercise 2.11 Show that commuting idempotents repel each other. More precisely, let $\mathcal{A}$ be a Banach algebra, and let $e \neq f$ in $\mathcal{A}$ with $e^2 = e$, $f^2 = f$, and $ef = fe$. Show that $\|e - f\| \geq 1$.

Exercise 2.12 Let $\mathcal{A}$ be the disk-algebra of Example 2.2.8.. Show that the algebra of polynomial functions is dense in $\mathcal{A}$.

(Hint: For $f \in \mathcal{A}$ and $0 < r < 1$ consider $f_r(z) = f(rz)$.)

Exercise 2.13 Show that the disk-algebra, equipped with the involution

$$f^*(z) \overset{\text{def}}{=} \overline{f(\bar{z})}$$

is a Banach-*-algebra, but not a C^*-algebra.

Exercise 2.14 Let $\mathcal{A} = C_0(X)$ for a locally compact Hausdorff space X. For $x \in X$ let $m_x : \mathcal{A} \to \mathbb{C}$ be defined by $m_x(f) = f(x)$. Show that the map $x \mapsto m_x$ is a homeomorphism from X to the structure space $\Delta_\mathcal{A}$.

(Hint: By passing to the one-point compactification reduce to the case when X is compact. For a given ideal I in $\mathcal{A}$ consider the set V of all $x \in X$ such that $f(x) = 0$ for every $f \in I$. Show that V is non-empty for a proper ideal I.)

Exercise 2.15 Let A be a commutative C^*-algebra. A linear functional L is called *positive* if $L(aa^*) \geq 0$ for every $a \in A$. Show that every positive functional is continuous.

Exercise 2.16 (Wiener's Lemma) Suppose that $f : \mathbb{R} \to \mathbb{C}$ is a 2π-periodic function such that

$$f(x) = \sum_{n\in\mathbb{Z}} a_n e^{inx} \quad \text{with} \quad \sum_{n\in\mathbb{Z}} |a_n| < \infty.$$

Show that if $f(x) \neq 0$ for every $x \in \mathbb{R}$, then there exist $b_n \in \mathbb{C}$ such that

$$\frac{1}{f(x)} = \sum_{n\in\mathbb{Z}} b_n e^{inx} \quad \text{with} \quad \sum_{n\in\mathbb{Z}} |b_n| < \infty.$$

Exercise 2.17 Let $\mathcal{A}$ and $\mathcal{B}$ be commutative C^*-algebras, and let $\phi : \mathcal{A} \to \mathcal{B}$ be a linear map with $\phi(aa') = \phi(a)\phi(a')$ for any $a, a' \in \mathcal{A}$. Show that ϕ is a continuous $*$-homomorphism.

(Hint: Consider the map $\phi^* : \Delta_\mathcal{B} \to \Delta_\mathcal{A}$ given by $\phi^*(m) = m \circ \phi$.)

Exercise 2.18 Assume that a is a normal element in the non-unital C^*-algebra $\mathcal{A}$. Let $C_0(\sigma_\mathcal{A}(a)) = \{f \in C(\sigma_\mathcal{A}(a)) : f(0) = 0\}$.

(a) Show that there exists a unique $*$-homomorphism $\Phi : C_0(\sigma_\mathcal{A}(a)) \to \mathcal{A}$ such that $\Phi(\text{Id}) = a$ where $\text{Id} = \text{Id}_{\sigma_\mathcal{A}(a)}$.

(b) Show that the $*$-homomorphism Φ in (a) is isometric and satisfies $\Phi(C_0(\sigma_\mathcal{A}(a))) = C^*(a)$.

(Hint: Apply Theorem 2.7.3 to the unitization $\mathcal{A}^e$ of $\mathcal{A}$.)

Exercise 2.19 Let a be a self-adjoint element in the C^*-algebra $\mathcal{A}$. Then a is called *positive* if $\sigma_\mathcal{A}(a) \subseteq [0, \infty)$. Show that for each positive element $a \in \mathcal{A}$ and for each $n \in \mathbb{N}$ there exists a unique positive element $b \in \mathcal{A}$ with $b^n = a$.

Exercise 2.20 Let a be any self-adjoint element in the C^*-algebra $\mathcal{A}$. Then there exist unique positive elements a^+ and a^- such that $a = a^+ - a^-$ and $a^+a^- = a^-a^+ = 0$.

Exercise 2.21 Suppose that a is a self-adjoint element in the unital C^*-algebra $\mathcal{A}$, and let $f(z) = \sum_{n=0}^{\infty} \beta_n z^n$ be a power series, which converges absolutely for every $z \in \sigma_{\mathcal{A}}(a)$. Then $f(a) = \sum_{n=0}^{\infty} \beta_n a^n$.

Notes

It is a pleasure for us to recommend Kaniuth's book [Kan09] for further reading on the theory of commutative Banach Algebras.

Chapter 3
Duality for Abelian Groups

In this chapter we are mainly interested in the study of *abelian* locally compact groups A, their dual groups $\widehat{A}$ together with various associated group algebras. Using the Gelfand-Naimark Theorem as a tool, we shall then give a proof of the Plancherel Theorem, which asserts that the Fourier transform extends to a unitary equivalence of the Hilbert spaces $L^2(A)$ and $L^2(\widehat{A})$. We also prove the Pontryagin Duality Theorem that gives a canonical isomorphism between A and its bidual $\widehat{\widehat{A}}$.

3.1 The Dual Group

A locally compact abelian group will be called an *LCA-group* for short. A *character* of an LCA-group A is a continuous group homomorphism

$$\chi : A \rightarrow \mathbb{T},$$

where $\mathbb{T}$ is the *circle group*, i.e., the multiplicative group of all complex numbers of absolute value one. The set $\widehat{A}$ of all characters on A forms a group under point-wise multiplication

$$(\chi \cdot \mu)(x) = \chi(x) \cdot \mu(x), \qquad x \in A.$$

The inverse element to a given $\chi \in \widehat{A}$ is given by $\chi^{-1}(x) = \frac{1}{\chi(x)} = \overline{\chi(x)}$. The group $\widehat{A}$ is called the *dual group* of A.

Examples 3.1.1

- As explained in Example 1.7.1, the dual group of $\mathbb{Z}$ is $\mathbb{R}/\mathbb{Z}$ and vice versa.

- The characters of the additive group $\mathbb{R}$ are the maps $\chi_t : x \mapsto e^{2\pi i x t}$, where t varies in $\mathbb{R}$. We then get an isomorphism of groups $\mathbb{R} \cong \widehat{\mathbb{R}}$ mapping t to χ_t.

Definition In what follows next we want to show that $\widehat{A}$ carries a natural topology that makes it a topological group. For a given topological space X let $C(X)$ be the

A. Deitmar, S. Echterhoff, *Principles of Harmonic Analysis*, Universitext,
DOI 10.1007/978-3-319-05792-7_3, © Springer International Publishing Switzerland 2014

complex vector space of all continuous maps from X to $\mathbb{C}$. For a compact set $K \subset X$ and an open set $U \subset \mathbb{C}$ define the set

$$L(K, U) \stackrel{\text{def}}{=} \{f \in C(X) : f(K) \subset U\}.$$

This is the set of all f that map a given compact set into a given open set. The topology generated by the sets $L(K, U)$ as K and U vary, is called the *compact-open topology*.

Lemma 3.1.2 (a) *Let X be a topological space. With the compact-open topology, $C(X)$ is a Hausdorff space.*

(b) *A net (f_i) in $C(X)$ converges in the compact-open topology if and only if it converges uniformly on every compact subset of X.*

(c) *If X is locally compact, then a net (f_i) converges in the compact-open topology if and only if it converges locally uniformly.*

(d) *If X is compact, the compact-open topology on $C(X)$ coincides with the topology given by the sup-norm.*

(e) *If $C(X)$ is endowed with the compact-open topology, then each point evaluation map $\delta_x : C(X) \to \mathbb{C}; f \mapsto f(x)$ is continuous.*

Proof (a) Let $f \neq g$ in $C(X)$. Choose $x \in X$ such that $f(x) \neq g(x)$, and choose disjoint open neighborhoods S, T in $\mathbb{C}$ of $f(x)$ and $g(x)$. Then the sets $L(\{x\}, S)$ and $L(\{x\}, T)$ are disjoint open neighborhoods of f and g, so $C(X)$ is a Hausdorff space.

(b) Fix $\varepsilon > 0$, let $f_i \to f$ be a net converging in the compact-open topology and let $K \subset X$ be a compact subset. For $z \in \mathbb{C}$ and $r > 0$ let $B_r(z)$ be the open ball of radius r around z and let $\bar{B}_r(z)$ be its closure. For $x \in X$ let U_x be the inverse image under f of the open ball $B_{\epsilon/3}(f(x))$. Then U_x is an open neighborhood of x and f maps its closure $\bar{U}_x$ into the closed ball $\bar{B}_{\epsilon/3}(f(x))$. As K is compact, there are $x_1, \ldots x_n \in K$ such that K is a subset of the union $U_{x_1} \cup \cdots \cup U_{x_n}$. Since closed subsets of compact sets are compact, the set $\bar{U}_{x_i} \cap K$ is compact. Let L be the intersection of the sets $L(\bar{U}_{x_i} \cap K, B_{2\epsilon/3}(f(x_i)))$. Then L is an open neighborhood of f in the compact-open topology. Therefore, there exists an index j_0 such that for $j \geq j_0$ each f_j lies in L. Let $j \geq j_0$ and $x \in K$. Then there exists i such that $x \in U_{x_i}$. Therefore,

$$|f_j(x) - f(x)| \leq |f_j(x) - f(x_i)| + |f(x_i) - f(x)|$$

$$< \frac{2\varepsilon}{3} + \frac{\varepsilon}{3} = \varepsilon.$$

It follows that the net converges uniformly on K. The converse direction is trivial.

(c) Let X be a locally compact space and let (f_j) be a net in $C(X)$ which converges to $f \in C(X)$ in the compact-open topology, i.e., it converges uniformly on compact sets. As every $x \in X$ has a compact neighborhood, (f_j) converges uniformly on a neighborhood of a given x, hence it converges locally uniformly. Conversely, assume that (f_j) converges locally uniformly and let $K \subset X$ be compact. For each $x \in K$ there exists an open neighborhood U_x on which the net (f_j) converges uniformly. These U_x form an open covering of K, hence finitely many suffice, i.e., $K \subset U_{x_1} \cup \cdots \cup U_{x_n}$ for some $x_1, \ldots, x_n \in K$. As (f_j) converges uniformly on each U_{x_i}, in converges uniformly on K.

(d) If X is compact, the compact-open topology and the sup-norm topology generate the same set of convergent nets. Therefore they have the same closed sets, so they are equal. For the last point, (e), let (f_i) be a net in $C(X)$ convergent to f. Then $\delta_x(f_i) = f_i(x)$ converges to $f(x) = \delta_x(f)$, so the evaluation map is continuous. $\square$

By definition, the dual group $\widehat{A}$ is a subset of the set $C(A)$ of all continuous maps from A to $\mathbb{C}$. It is a consequence of Lemma 3.1.2 (e) that $\widehat{A}$ is closed in the compact-open topology of $C(A)$.

Examples 3.1.3.

- The compact-open topology on the dual $\widehat{\mathbb{Z}} \cong \mathbb{T}$ of $\mathbb{Z}$ is the natural topology of the circle group $\mathbb{T}$.

- The compact-open topology on the dual $\widehat{\mathbb{T}} \cong \mathbb{Z}$ of $\mathbb{T}$ is the discrete topology.

- The compact-open topology on the dual $\widehat{\mathbb{R}} \cong \mathbb{R}$ of $\mathbb{R}$ is the usual topology of $\mathbb{R}$.

Proposition 3.1.4 *With the compact-open topology, $\widehat{A}$ is a topological group that is Hausdorff.*

Later we will see that $\widehat{A}$ is also locally compact, i.e., an LCA-group.

Proof We have to show that the map $\alpha : \widehat{A} \times \widehat{A} \to \widehat{A}$, that sends a pair (χ, η) to $\chi \eta^{-1}$, is continuous. For two pairs $(\chi, \eta), (\chi' \eta')$ and $x \in A$ we have

$$|\chi(x)\eta^{-1}(x) - \chi'(x)\eta'^{-1}(x)| \leq |\chi(x)\eta^{-1}(x) - \chi(x)\eta'^{-1}(x)|$$
$$+ |\chi(x)\eta'^{-1}(x) - \chi'(x)\eta'^{-1}(x)|$$
$$= |\eta^{-1}(x) - \eta'^{-1}(x)| + |\chi(x) - \chi'(x)|,$$

Let $K \subset A$ be compact and let $\varepsilon > 0$. Then

$$B_{K,\varepsilon}(\chi\eta^{-1}) = \{\gamma \in \widehat{A} : \|\gamma - \chi\eta^{-1}\|_K < \varepsilon\}$$

is an open neighborhood of $\chi\eta^{-1}$ and sets of this form are a neighborhood base. The estimate above shows that the open neighborhood $B_{K,\varepsilon/2}(\chi) \times B_{K,\varepsilon/2}(\eta)$ of (χ, η) is mapped to $B_{K,\varepsilon}(\chi\eta^{-1})$, so α is continuous. $\square$

The observation, that the dual group of the compact group $\mathbb{T}$ is the discrete group $\mathbb{Z}$ and vice versa, is an example of the following general principle:

Proposition 3.1.5

(a) *If A is compact, then $\widehat{A}$ is discrete.*

(b) *If A is discrete, then $\widehat{A}$ is compact.*

Proof Let A be compact, and let L be the set of all $\eta \in \widehat{A}$ such that $\eta(A)$ lies in the open set $\{\mathrm{Re}(\cdot) > 0\}$. As A is compact, L is an open unit-neighborhood in $\widehat{A}$. For every $\eta \in \widehat{A}$, the image $\eta(A)$ is a subgroup of $\mathbb{T}$. The only subgroup of $\mathbb{T}$, however, that is contained in $\{\mathrm{Re}(\cdot) > 0\}$, is the trivial group. Therefore $L = \{1\}$, and so $\widehat{A}$ is discrete.

For the second part, assume that A is discrete. Then $\widehat{A} = \mathrm{Hom}(A, \mathbb{T})$ is a subset of the set $\mathrm{Map}(A, \mathbb{T})$ of all maps from A to $\mathbb{T}$. The set $\mathrm{Map}(A, \mathbb{T})$ can be identified with the product $\prod_{a \in A} \mathbb{T}$. By Tychonov's Theorem, the latter is a compact Hausdorff space in the product topology and $\widehat{A}$ forms a closed subspace. As A is discrete, the inclusion $\widehat{A} \hookrightarrow \prod_{a \in A} \mathbb{T}$ induces a homeomorphism of $\widehat{A}$ onto its image in the product space. Hence $\widehat{A}$ is compact. □

3.2   The Fourier Transform

Let A be an LCA-group and consider its convolution algebra $L^1(A)$. In this section we want to show that the topological space $\widehat{A}$ is canonically homeomorphic to the structure space $\Delta_{L^1(A)}$ of the commutative Banach algebra $L^1(A)$. Since this structure space is locally compact, this will show that the dual group $\widehat{A}$ is an LCA-group. Recall that the Fourier transform $\hat{f} : \widehat{A} \to \mathbb{C}$ of a function $f \in L^1(A)$ is defined as

$$\hat{f}(\chi) = \int_A f(x)\overline{\chi(x)}\,dx.$$

Theorem 3.2.1 *The map $\chi \mapsto d_\chi$ from the dual group $\widehat{A}$ to the structure space $\Delta_{L^1(A)}$ defined by*

$$d_\chi(f) = \hat{f}(\chi)$$

is a homeomorphism. In particular, $\widehat{A}$ is a locally compact Hausdorff space, so $\widehat{A}$ is an LCA-group.

It follows that for every $f \in L^1(A)$ the Fourier transform $\hat{f}$ is a continuous function on the dual group $\widehat{A}$, which vanishes at infinity.

Proof By Lemma 1.7.2 it follows that d_χ indeed lies in the structure space of the Banach algebra $L^1(A)$.

Injectivity Assume $d_\chi = d_{\chi'}$, then $\int_A f(x)\overline{(\chi(x) - \chi'(x))}\, dx = 0$ for every $f \in C_c(A)$. This implies that the continuous functions χ and χ' coincide.

Surjectivity Let $m \in \Delta_{L^1(A)}$. As $C_c(A)$ is dense in $L^1(A)$, there exists an element $g \in C_c(A)$ with $m(g) \neq 0$. For $x \in A$ define $\chi(x) = \overline{m(L_x g)/m(g)}$. The continuity of m and Lemma 1.4.2 implies that χ is a continuous function on A. One computes,

$$m(L_x g)m(L_y g) = m(L_x g * L_y g) = m(L_{xy} g * g) = m(L_{xy} g)m(g).$$

Dividing by $m(g)^2$ and taking complex conjugates, one gets the identity $\chi(x)\chi(y) = \chi(xy)$, so χ is a multiplicative map from A to $\mathbb{C}^\times$. Let $f \in C_c(A)$. Then one can write the convolution $f * g$ as $\int_A f(x)L_x g\, dx$, and this integral may be viewed as a vector-valued integral with values in the Banach space $L^1(A)$ as in Sect. B.6. One uses the continuity of the linear functional m and Lemma B.6.5 to get

$$\int_A f(x)\overline{\chi(x)}\, dx = \frac{1}{m(g)} \int_A f(x)m(L_x g)\, dx = \frac{1}{m(g)} m\left(\int_A f(x)L_x g\, dx \right)$$

$$= \frac{1}{m(g)} m(f * g) = \frac{m(f)m(g)}{m(g)} = m(f).$$

Let (ϕ_U) be a Dirac net in $C_c(A)$. Then $\phi_U * \overline{\chi}$ converges point-wise to $\overline{\chi}$ and so for $x \in A$ and $\varepsilon > 0$ there exists a unit-neighborhood U, such that

$$|\chi(x)| \leq \left| \phi_U * \overline{\chi}(x) \right| + \varepsilon = \left| \int_A L_x \phi_U(y)\overline{\chi(y)}\, dy \right| + \varepsilon$$

$$= \left| m(L_x \phi_U) \right| \leq \lim_U \|L_x \phi_U\|_1 + \varepsilon = 1 + \varepsilon.$$

As ε is arbitrary, we get $|\chi(x)| \leq 1$ for every $x \in A$. By $\chi(x^{-1}) = \chi(x)^{-1}$ we infer $|\chi(x)| = 1$ for every $x \in A$. So the map χ lies in $\widehat{A}$, and the map d is surjective.

Continuity Let $\chi_j \to \chi$ be a net in $\widehat{A}$ which converges locally uniformly on A. Let $f \in L^1(A)$ and choose $\varepsilon > 0$. We have to show that there exists j_0 such that for $j \geq j_0$ one has $|\widehat{f}(\chi_j) - \widehat{f}(\chi)| < \varepsilon$. Let $g \in C_c(A)$ with $\|f - g\|_1 < \varepsilon/3$. Since $\chi_j \to \chi$ uniformly on $\mathrm{supp}(g)$, there exists j_0 such that for $j \geq j_0$ it holds $|\widehat{g}(\chi_j) - \widehat{g}(\chi)| < \varepsilon/3$. For $j \geq j_0$ one has

$$|\widehat{f}(\chi_j) - \widehat{f}(\chi)| \leq |\widehat{f}(\chi_j) - \widehat{g}(\chi_j)| + |\widehat{g}(\chi_j) - \widehat{g}(\chi)| + |\widehat{g}(\chi) - \widehat{f}(\chi)|$$

$$< \frac{\varepsilon}{3} + \frac{\varepsilon}{3} + \frac{\varepsilon}{3} = \varepsilon.$$

The *continuity of the inverse map d^{-1}* is a direct consequence of the following lemma.

Lemma 3.2.2 *Let $\chi_0 \in \widehat{A}$. Let K be a compact subset of A, and let $\varepsilon > 0$. Then there exist $l \in \mathbb{N}$, functions $f_0, f_1, \ldots, f_l \in L^1(A)$, and $\delta > 0$ such that for $\chi \in \widehat{A}$ the condition $|\hat{f}_j(\chi) - \hat{f}_j(\chi_0)| < \delta$ for every $j = 0, \ldots, l$ implies $|\chi(x) - \chi_0(x)| < \varepsilon$ for every $x \in K$.*

Proof For $f \in L^1(A)$ we have

$$\hat{f}(\chi) - \hat{f}(\chi_0) = \int_A f(x)\overline{(\chi(x) - \chi_0(x))}\,dx$$

$$= \int_A f(x)\overline{\chi_0(x)}\,(\overline{\chi(x)}\chi_0(x) - 1)\,dx$$

$$= \widehat{f\bar{\chi}_0}(\chi\bar{\chi}_0) - \widehat{f\bar{\chi}_0}(1).$$

So without loss of generality we can assume $\chi_0 = 1$.

Let $f \in L^1(A)$ with $\hat{f}(1) = \int_A f(x)\,dx = 1$. Then there is a unit-neighborhood U in A with $\|L_u f - f\|_1 < \varepsilon/3$ for every $u \in U$. As K is compact, there are finitely many $x_1, \ldots, x_l \in A$ such that K is a subset of $x_1 U \cup \cdots \cup x_l U$. Set $f_j = L_{x_j} f$ as well as $f_0 = f$ and let $\delta = \varepsilon/3$. Let $\chi \in \widehat{A}$ with $|\hat{f}_j(\chi) - 1| < \varepsilon/3$ for every $j = 0, \ldots, l$. Now let $x \in K$. Then there exists $1 \le j \le l$ and $u \in U$ such that $x = x_j u \in x_j U$. One gets

$$|\chi(x) - 1| = |\overline{\chi(x)} - 1|$$

$$\le |\overline{\chi(x)} - \overline{\chi(x)}\hat{f}(\chi)| + |\hat{f}(\chi)\overline{\chi(x)} - \hat{f}_j(\chi)| + |\hat{f}_j(\chi) - 1|$$

$$= |1 - \hat{f}(\chi)| + |\widehat{L_x f}(\chi) - \widehat{L_{x_j} f}(\chi)| + |\hat{f}_j(\chi) - 1|$$

$$< \frac{\varepsilon}{3} + \frac{\varepsilon}{3} + \frac{\varepsilon}{3} = \varepsilon,$$

where the last inequality uses

$$|\widehat{L_x f}(\chi) - \widehat{L_{x_j} f}(\chi)| \le \|L_x f - L_{x_j} f\|_1 = \|L_{x_j}(L_u f - f)\|_1$$

$$= \|L_u f - f\|_1 < \varepsilon/3.$$

The lemma and the theorem are proven. $\square$

3.3 The C^*-Algebra of an LCA-Group

In this section we introduce the C^*-algebra $C^*(A)$ of the LCA-group A as a certain completion of the convolution algebra $L^1(A)$. We show that restriction of multiplicative functionals from $C^*(A)$ to the dense subalgebra $L^1(A)$ defines a homeomorphism between $\Delta_{C^*(A)}$ and $\Delta_{L^1(A)}$. Hence by the results of the previous section, $\Delta_{C^*(A)}$ is canonically homeomorphic to the dual group $\widehat{A}$. The Gelfand-Naimark Theorem then implies that the Fourier transform on $L^1(A)$ extends to an isometric *-isomorphism between $C^*(A)$ and $C_0(\widehat{A})$. These results will play an important role in the proof of the Plancherel Theorem in the following section.

Let $f \in L^1(A)$ and $\phi, \psi \in L^2(A)$. For every $y \in A$ one has

$$|\langle L_y\phi, \psi\rangle| \leq \|L_y\phi\|_2\|\psi\|_2 = \|\phi\|_2\|\psi\|_2.$$

This implies that the integral $\int_A f(y)\langle L_y\phi, \psi\rangle\, dy$ exists, and one has the estimate

$$\left|\int_A f(y)\langle L_y\phi, \psi\rangle dy\right| \leq \|f\|_1\|\phi\|_2\|\psi\|_2.$$

In other words, the anti-linear map that sends ψ to the integral $\int_A f(y)\langle L_y\phi, \psi\rangle\, dy$ is bounded, hence continuous by Lemma C.1.2. As every continuous anti-linear map on a Hilbert space is represented by a unique vector, there exists a unique element $L(f)\phi$ in $L^2(A)$ such that $\langle L(f)\phi, \psi\rangle = \int_A f(y)\langle L_y\phi, \psi\rangle\, dy$ for every $\psi \in L^2(A)$. The above estimate gives $|\langle L(f)\phi, \psi\rangle| \leq \|f\|_1\|\phi\|_2\|\psi\|_2$. In particular, for $\psi = L(f)\phi$ one concludes $\|L(f)\phi\|_2^2 \leq \|f\|_1\|\phi\|_2\|L(f)\phi\|_2$, hence $\|L(f)\phi\|_2 \leq \|f\|_1\|\phi\|_2$, which implies that the linear map $\phi \mapsto L(f)\phi$ is bounded, hence continuous. Note that for $\phi \in C_c(G)$ one has $L(f)\phi = f * \phi$ by Lemma 3.3.1 below.

Lemma 3.3.1 *If $f \in L^1(A)$ and $\phi \in L^1(A)\cap L^2(A)$, then $L(f)\phi = f * \phi = \phi * f$.*

Proof Let $\psi \in C_c(A)$. Then the inner product $\langle L(f)\phi, \psi\rangle$ equals $\int_A f(y) \int_A \phi(y^{-1}x)\overline{\psi(x)}\, dx\, dy$. This integral exists if f, ϕ, ψ are replaced with their absolute values. Therefore we can apply Fubini's Theorem to get $\langle L(f)\phi, \psi\rangle = \langle f * \phi, \psi\rangle$, whence the claim.

Lemma 3.3.2 *The map L from $L^1(A)$ to the space $\mathcal{B}(L^2(A))$ is an injective, continuous homomorphism of Banach-*-algebras.*

Proof The map is linear and satisfies $\|L(f)\|_{op} \leq \|f\|_1$, therefore is continuous. For $f, g \in L^1(A)$ and ϕ in the dense subspace $C_c(A)$ of $L^2(A)$ the above lemma and the associativity of convolution implies

$$L(f * g)\phi = (f * g) * \phi = f * (g * \phi) = L(f)L(g)\phi,$$

so L is multiplicative. For $\phi, \psi \in C_c(A)$ we get

$$\langle f * \phi, \psi\rangle = \int_A \int_A f(y)\phi(y^{-1}x)\overline{\psi(x)}\, dy\, dx$$

$$= \int_A \int_A f(y)\phi(x)\overline{\psi(yx)}\, dx\, dy$$

$$= \int_A \int_A \phi(x)\Delta(y^{-1})\overline{f(y^{-1})}\psi(y^{-1}x)\, dy\, dx$$

$$= \langle \phi, f^* * \psi\rangle,$$

where we used the transformation $x \mapsto yx$ followed by the transformation $y \mapsto y^{-1}$. This shows $L(f^*) = L(f)^*$.

For the injectivity, let $f \in L^1(G)$ with $L(f) = 0$. Then in particular $f * \phi = 0$ for every $\phi \in C_c(A)$. Using Lemma 1.6.6 this implies $f = 0$.

Definition We define the *group C^*-algebra* $C^*(A)$ of the LCA-group A to be the norm-closure of $(L^1(A))$ in the C^*-algebra $\mathcal{B}(L^2(A))$. As $L^1(A)$ is a commutative Banach algebra, $C^*(A)$ is a commutative C^*-algebra.

Theorem 3.3.3 *The map $L^* : \Delta_{C^*(A)} \rightarrow \Delta_{L^1(A)}$ given by $m \mapsto m \circ L$ is a homeomorphism. It follows $\Delta_{C^*(A)} \cong \widehat{A}$ and $C^*(A) \cong C_0(\widehat{A})$.*

Proof As the image of L is dense in $C^*(A)$, it follows that $m \circ L \neq 0$ for every $m \in \Delta_{C^*(A)}$ and that L^* is injective. Therefore by Lemma 2.4.7 it suffices to show that L^* is surjective.

To prove this, let $m \in \Delta_{L^1(A)}$ and $\chi \in \widehat{A}$ such that $m(f) = \widehat{f}(\chi)$ for every $f \in L^1(A)$. We have to show that m is continuous in the C^*-norm, because then it has a unique extension to $C^*(A)$. For this let $\mu_0 \in \Delta_{C^*(A)}$ be fixed. Then there is $\chi_0 \in \widehat{A}$ such that for $f \in L^1(A)$ the identity $\widehat{f}(\chi_0) = \mu_0(f)$ holds, where we have written $\mu_0(L(f)) = \mu_0(f)$. For $f \in L^1(A)$, one has

$$m(f) = \int_A f(x)\overline{\chi(x)}\,dx = \int_A f(x)\overline{\chi(x)}\chi_0(x)\overline{\chi_0(x)}\,dx = \mu_0(f\bar{\chi}\chi_0).$$

It follows that $|m(f)| = |\mu_0(f\bar{\chi}\chi_0)| \leq \|f\bar{\chi}\chi_0\|_{C^*(A)}$. So we have to show that for $f \in L^1(A)$ the C^*-norm of f equals the C^*-norm of $f\eta$ for any $\eta \in \widehat{A}$. As the C^*-norm is the operator norm in $\mathcal{B}(L^2(A))$, we consider $\phi, \psi \in L^2(A)$, and we compute

$$\langle L(\eta f)\phi, \psi \rangle = \int_A \eta(x)f(x)\langle L_x\phi, \psi\rangle\,dx$$

$$= \int_A \eta(x)f(x)\int_A \phi(x^{-1}y)\overline{\psi(y)}\,dy\,dx$$

$$= \int_A f(x)\int_A (\bar{\eta}\phi)(x^{-1}y)\overline{(\bar{\eta}\psi)(y)}\,dy\,dx$$

$$= \langle L(f)(\bar{\eta}\phi), \bar{\eta}\psi\rangle.$$

Putting $\psi = L(\eta f)\phi$, we get

$$\|L(\eta f)\phi\|_2^2 = \langle L(f)(\bar{\eta}\phi), \bar{\eta}L(\eta f)\phi\rangle \leq \|L(f)(\bar{\eta}\phi)\|_2\|\bar{\eta}L(\eta f)\phi\|_2.$$

Since $\|\bar{\eta}L(\eta f)\phi\|_2 = \|L(\eta f)\phi\|_2$ it follows $\|L(\eta f)\phi\|_2 \leq \|L(f)(\bar{\eta}\phi)\|_2$ and so the operator norm of $L(\eta f)$ is less than or equal to the operator norm of $L(f)$. By symmetry we get equality and the theorem follows. $\square$

Corollary 3.3.4 *Let A be an LCA-group. Then the Fourier transform $L^1(A) \to C_0(\widehat{A})$, mapping f to $\hat{f}$, is injective.*

Proof Let $\mathcal{A} = L^1(A)$. As $\widehat{A} \cong \Delta_{\mathcal{A}} \cong \Delta_{C^*(A)}$, the Fourier transform is the composition of the injective maps $\mathcal{A} \to C^*(A) \to C_0(\widehat{A})$. ☐

3.4 The Plancherel Theorem

In this section we will construct the Plancherel measure on the dual group $\widehat{A}$ relative to a given Haar measure on the LCA group A and we will state the Plancherel Theorem, which says that the Fourier transform extends to a unitary equivalence

$$L^2(A) \cong L^2(\widehat{A}).$$

The proof of the Plancherel theorem will be postponed to the following section, where it will be shown as a consequence of Pontryagin duality.

Lemma 3.4.1 *Let $\phi, \psi \in L^2(A)$. Then the convolution integral $\phi * \psi(x) = \int_A \phi(y)\psi(y^{-1}x)\,dy$ exists for every $x \in A$ and defines a continuous function in x. The convolution product $\phi * \psi$ lies in $C_0(A)$ and its sup-norm satisfies $\|\phi * \psi\|_A \leq \|\phi\|_2\|\psi\|_2$. Finally one has $\phi * \phi^*(1) = \|\phi\|_2^2$.*

Proof With ψ, also the function $L_x\psi^*$ lies in $L^2(A)$, as A is abelian, hence unimodular. The convolution integral is the same as the inner product $\langle \phi, L_x\psi^* \rangle$, hence the integral exists for every $x \in A$. The continuity follows from Lemma 1.4.2 and the fact that the map $L^2(A) \to \mathbb{C}$, given by $\psi \mapsto \langle \phi, \psi \rangle$ is continuous. Next use the Cauchy-Schwarz inequality to get

$$\|\phi * \psi\|_A = \sup_{x \in A} |\langle \phi, L_x\psi^* \rangle| \leq \|\phi\|_2\|\psi\|_2.$$

Choose sequences (ϕ_n) and (ψ_n) in $C_c(A)$ with $\|\phi_n - \phi\|_2, \|\psi_n - \psi\|_2 \to 0$. Then it follows from the above inequality that $\phi_n * \psi_n \in C_c(A)$ converges uniformly to $\phi * \psi$. It follows that $\phi * \psi \in C_0(A)$ since $C_0(A)$ is complete. The final assertion $\phi * \phi^*(1) = \|\phi\|_2^2$ is clear by definition. ☐

The space $\mathcal{C} = C_0(A) \times C_0(\widehat{A})$ is a Banach space with the norm

$$\|(f, \eta)\|_0^* = \max\left(\|f\|_A, \|\eta\|_{\widehat{A}}\right).$$

We embed $C_0(A) \cap L^1(A)$ into this product space by mapping f to $(f, \hat{f})$ and we denote the closure of $C_0(A) \cap L^1(A)$ inside $\mathcal{C}$ by

$$C_0^*(A).$$

This is a Banach space the norm of which we write as $\|f\|_0^*$.

Lemma 3.4.2 *Let p_0 and p_* be the projections from C to $C_0(A)$ and $C_0(\widehat{A})$, respectively. Then the restrictions of p_0 and p_* to $C_0^*(A)$ are both injective. Hence we can consider $C_0^*(A)$ as a subspace of $C_0(A)$ as well as of $C_0(\widehat{A})$.*

Proof Let $f \in C_0^*(A)$ and write f_* for $p_*(f)$ and f_0 for $p_0(f)$. We have to show that if one of these two is zero, then so is the other. Let (f_n) be a sequence in $C_0(A) \cap L^1(A)$ converging to f in $C_0^*(A)$. Then f_n converges to f_* in $C^*(A)$ and to f_0 uniformly on A. So for $\psi \in L^2(A)$ the sequence $f_n * \psi$ converges to $f_*(\psi)$ in $L^2(A)$. If ψ is in $C_c(A)$, then $f_n * \psi$ also converges uniformly to $f_0 * \psi$. So for every $\phi \in C_c(A)$, the sequence $\langle f_n * \psi, \phi \rangle$ converges to $\langle f^*(\psi), \phi \rangle$ and by uniform convergence also to $\langle f_0 * \psi, \phi \rangle$, i.e., we have $\langle f_*(\psi), \phi \rangle = \langle f_0 * \psi, \phi \rangle$. As this holds for all $\psi, \phi \in C_c(G)$, we conclude $f_* = 0 \Leftrightarrow f_0 = 0$ as claimed.   □

A given element f of $C_0^*(A)$ can be viewed as an element of $C_0(A)$, or of $C^*(A) \cong C_0(\widehat{A})$. We will freely switch between these two viewpoints in the sequel. If we want to emphasize the distinction, we write f for the function on A and $\hat{f}$ for its Fourier transform, the function on $\widehat{A}$.

For $g \in C^*(A)$ and $\phi \in L^2(A)$ we from now on write $L(g)\phi$ for the element $g(\phi)$ of $L^2(A)$.

Lemma 3.4.3 *Let $f \in C_0^*(A)$. If the Fourier transform $\hat{f}$ is real-valued, then $f(1)$ is real. If $\hat{f} \geq 0$, then $f(1) \geq 0$. Here 1 denotes the unit element of A.*

Proof Suppose that $\hat{f}$ is real-valued. Then $\hat{f} = \overline{\hat{f}} = \widehat{f^*}$, so we get $f = f^*$, and therefore $f(1) = f^*(1) = \overline{f(1)}$. Now suppose $\hat{f} \geq 0$. Then there exists $g \in C_0(\widehat{A}) \cong C^*(A)$ with $g \geq 0$ and $\hat{f} = g^2$. Let $\phi = \phi^* \in C_c(A)$. Then $L(g)\phi \in L^2(A)$, so $(L(g)\phi) * (L(g)\phi)^*(1) = \|L(g)\phi\|_2^2 \geq 0$. Now g is a limit in $C^*(A)$ of a sequence (g_n) in $L^1(A)$. We can assume $g_n = g_n^*$ for every $n \in \mathbb{N}$. Using Lemma 3.3.1 we have

$$(L(g)\phi) * (L(g)\phi)^* = \lim_n (L(g_n)\phi) * (L(g_n)\phi)^* = \lim_n (g_n * \phi) * (g_n * \phi)^*$$

$$= \lim_n g_n * \phi * \phi * g_n = \lim_n g_n * g_n * \phi * \phi$$

$$= \lim_n L(g_n * g_n)(\phi * \phi) = L(f)(\phi * \phi) = f * \phi * \phi.$$

We get $f * \phi * \phi(1) \geq 0$, and therefore $f(1) \geq 0$ by Lemma 1.6.6. since we can let $\phi * \phi$ run through a Dirac net.   □

Lemma 3.4.4 (a) *The space $L^1(A) * C_c(A)$ is a subspace of $C_0(A)$.*

(b) *Let $f \in C^*(A)$, and let $\phi, \psi \in C_c(A)$. Then $L(f)(\phi * \psi)$ lies in $C_0^*(A) \cap L^2(A)$, viewed as a subspace of $C_0(A)$. One has $\widehat{L(f)(\phi * \psi)} = \hat{f}\hat{\phi}\hat{\psi}$.*

Proof (a) Let $f \in L^1(A)$ and $\phi \in C_c(A)$. Choose a sequence $f_n \in C_c(A)$ such that $\|f_n - f\|_1 \to 0$. Then $f_n * \phi \in C_c(A)$, and for every $x \in A$ we have $|f * \phi(x) - f_n * \phi(x)| \le \|f - f_n\|_1 \|\phi\|_\infty$. This shows that $f * \phi$ is a uniform limit of functions in $C_0(A)$. Since $C_0(A)$ is complete with respect to $\| \cdot \|_A$, the result follows.

For (b) let now $f \in C^*(A)$. There is a sequence $f_n \in L^1(A)$ converging to f in $C^*(A)$. Then $L(f_n)(\phi * \psi) = f_n * \phi * \psi$ lies in $C_0(A) \cap L^1(A)$. We have to show that the ensuing sequence $f_n * \phi * \psi$ is a Cauchy sequence in $C_0^*(A)$. This means that the sequence, as well as its Fourier transform, are both Cauchy sequences in $C_0(A)$ and $C_0(\widehat{A})$, respectively. Observe first that $\widehat{(f_n * \phi * \psi)} = \hat{f}_n \hat{\phi} \hat{\psi}$. Now $\hat{f}_n$ converges uniformly on $\widehat{A}$, so $\widehat{(f_n * \phi * \psi)}$ converges uniformly to $\hat{f} \hat{\phi} \hat{\psi}$, hence is Cauchy in $C_0(\widehat{A})$. By Lemma 3.4.1 we conclude that for $m, n \in \mathbb{N}$ one has $\|(f_m - f_n) * \phi * \psi\|_A \le \|(f_m - f_n) * \phi\|_2 \|\psi\|_2$. The right hand side tends to zero as m, n grow large, so $f_n * \phi * \psi$ is a Cauchy sequence in $C_0(A)$. Since $L(f)(\phi * \psi) \in L^2(A)$, the result follows. $\qquad\square$

Lemma 3.4.5 *Let (ϕ_U) be a Dirac net in $C_c(A)$. Then*

(a) *$(f * \phi_U)$ converges to f in $C^*(A)$ for every $f \in C^*(A)$,*

(b) *$(f * \phi_U)$ converges uniformly to f for every $f \in C_0(A)$,*

(c) *$(f * \phi_U)$ converges to f in $C_0^*(A)$ for every $f \in C_0^*(A)$,*

(d) *$(\widehat{\phi_U})$ converges locally uniformly to 1 on $\widehat{A}$.*

Proof For (a) observe that the result holds for the dense subspace $L^1(A)$ by Lemma 1.6.6. Then a standard $\varepsilon/3$-argument extends it to all of $C^*(A)$. For (b) we can use the same argument with $L^1(A)$ replaced by the dense subspace $C_c(A)$ of $C_0(A)$. Then (c) is a consequence of (a) and (b). For the proof of (d) let $C \subseteq \widehat{A}$ be a compact set. Choose a positive $\psi \in C_c(\widehat{A})$ with $\psi \equiv 1$ on C and let $f \in C^*(A)$ with $\hat{f} = \psi$. Then $\|\widehat{\phi_U} \psi - \psi\|_{\widehat{A}} = \|\phi_U * f - f\|_{op} \to 0$ by (a) and the result follows. $\qquad\square$

Lemma 3.4.6 *Let $\eta \in C_c(\widehat{A})$ be real-valued, and let $\varepsilon > 0$. Then there are $f_1, f_2 \in C_0^*(A) \cap L^2(A)$, considered as subspace of $C_0(A)$, such that*

- *the Fourier transforms $\hat{f}_1, \hat{f}_2$ lie in $C_c(\widehat{A})$,*

- *they satisfy $\hat{f}_1 \le \eta \le \hat{f}_2$, further $\|\hat{f}_1 - \hat{f}_2\|_{\widehat{A}} < \varepsilon$, and $\mathrm{supp}(\hat{f}_i) \subset \mathrm{supp}(\eta)$ for $i = 1, 2$,*

- *as well as $0 \le f_2(1) - f_1(1) < \varepsilon$.*

In particular, every $\eta \in C_c(\widehat{A})$ is the uniform limit of functions of the form $\hat{f}$ with $f \in C_0^*(A)$ of support contained in $\mathrm{supp}(\eta)$.

Proof For any Dirac function ϕ in $C_c(A)$ one has $\hat{\phi} \in C_0(\widehat{A})$ by Theorem 3.2.1 and by Lemma 3.4.5 the ensuing function $\hat{\phi}$ can be chosen to approximate the constant 1 arbitrarily close on any compact set. Note that the Fourier transform of a function of the form $h * h^*$ is ≥ 0. Let $K \subset \widehat{A}$ be the support of η. As $C_c(A)$ contains Dirac functions of arbitrary small support, we conclude that for every $\delta > 0$ there exists a function $\phi_\delta \in C_c^+(A)$ such that the function $\psi_\delta = \phi_\delta * \phi_\delta^*$ satisfies

$$1 - \delta \leq \hat{\psi}_\delta(\chi) \leq 1 + \delta \qquad \text{for every } \chi \in K.$$

Fix $\phi \in C_c^+(A)$ such that $\psi = \phi * \phi^*$ satisfies $\hat{\psi}(\chi) \geq 1$ for every $\chi \in K$. Let $f \in C^*(A)$ with $\hat{f} = \eta$ and set

$$f_1 = f * (\psi_\delta - \delta\psi), \qquad f_2 = f * (\psi_\delta + \delta\psi).$$

According to Lemma 3.4.4, the functions f_1 and f_2 lie in the space $C_0(A) \cap L^2(A)$. For every $\chi \in \widehat{A}$ it holds,

$$\hat{f}_1(\chi) = \hat{f}(\chi)(\hat{\psi}_\delta(\chi) - \delta\hat{\psi}(\chi)) \leq \eta(\chi) \leq \hat{f}_2(\chi).$$

Further, as $\hat{f}(\chi) = \eta(\chi)$, one has $\text{supp}(\hat{f}_i) \subset \text{supp}(\eta)$. The other properties follow by choosing δ small enough.   □

Proposition 3.4.7 *Let $\psi \in C_c(\widehat{A})$ be real-valued. Then the supremum of the set*

$$\{f(1) : f \in C_0^*(A),\ \hat{f} \leq \psi\}$$

equals the infimum of the set

$$\{f(1) : f \in C_0^*(A),\ \hat{f} \geq \psi\}.$$

We denote this common value by $I(\psi)$. We extend I to all of $C_c(\widehat{A})$ by setting $I(u + iv) = I(u) + iI(v)$, where u and v are real-valued. Then I is a Haar integral on $C_c(\widehat{A})$.

We write this integral as

$$I(\psi) = \int_{\widehat{A}} \psi(\chi)\,d\chi.$$

Proof It follows from Lemma 3.4.3 that the supremum is less or equal to the infimum and Lemma 3.4.6 implies that they coincide. Thus I exists. It is clearly linear and it is positive by Lemma 3.4.3. For the invariance let $\psi \in C_c(\widehat{A})$ be real-valued, and let $f \in C_0^*(A)$ with $\hat{f} \leq \psi$. For $\chi \in \widehat{A}$ we then have $L_\chi \hat{f} \leq L_\chi \psi$. Further, $L_\chi \hat{f} = \widehat{\chi f}$ as well as $\chi f(1) = f(1)$. This implies the invariance of I. The proof of the proposition is finished.   □

We close this section with formulating the Plancherel theorem for LCA groups. The proof will be given as a consequence of the Pontryagin Duality Theorem in the following section.

Theorem 3.4.8 (Plancherel Theorem). *For a given Haar measure on A there exists a uniquely determined Haar measure on $\widehat{A}$, called the Plancherel measure, such that for $f \in L^1(A) \cap L^2(A)$ one has*

$$\|f\|_2 = \|\hat{f}\|_2.$$

This implies that the Fourier transform extends to an isometry from $L^2(A)$ to $L^2(\widehat{A})$. Indeed, it is also surjective, so the Fourier transform extends to a canonical unitary equivalence $L^2(A) \cong L^2(\widehat{A})$.

In the special case of a compact group we derive from this, that the characters form an orthonormal basis of $L^2(A)$.

Corollary 3.4.9 *Let A be a compact LCA-group. Then the elements of the dual group $\widehat{A}$ form an orthonormal basis of $L^2(A)$.*

Proof According to our conventions, we assume the Haar measure of A to be normalized in a way that the total volume is one. As A is compact, any continuous function on A, in particular every character, lies in $L^2(A)$. We show that the characters of A form an orthonormal system, i.e., that for $\chi, \eta \in \widehat{A}$ we have

$$\langle \chi, \eta \rangle = \delta_{\chi,\eta} = \begin{cases} 1 & \chi = \eta, \\ 0 & \chi \neq \eta. \end{cases}$$

If $\chi = \eta$, then

$$\langle \chi, \eta \rangle = \int_A \underbrace{\chi(x)\overline{\chi(x)}}_{=1}\, dx = \int_A dx = 1.$$

If $\chi \neq \eta$, then pick $x_0 \in A$ with $\chi(x_0) \neq \eta(x_0)$. We obtain

$$\chi(x_0)\langle \chi, \eta \rangle = \int_A \chi(x_0 x)\overline{\eta(x)}\, dx = \int_A \chi(x)\overline{\eta(x_0^{-1}x)}\, dx = \eta(x_0)\langle \chi, \eta \rangle,$$

which implies $\langle \chi, \eta \rangle = 0$ as claimed. It follows that the Fourier transform of a character χ is the map δ_χ with $\delta_\chi(\eta) = \delta_{\chi,\eta}$. These maps form an orthonormal basis of the Hilbert space $L^2(\widehat{A})$ for the discrete group $\widehat{A}$. Since the Fourier transform is a unitary equivalence, the characters form an orthonormal basis of $L^2(A)$. $\qquad\square$

3.5 Pontryagin Duality

In the previous sections we saw that the dual group $\widehat{A}$ of an LCA group A, which consists of all continuous homomorphisms of A into the circle group $\mathbb{T}$, is again an LCA group. So we can also consider the dual group $\widehat{\widehat{A}}$ of $\widehat{A}$. There is a canonical homomorphism $\delta : A \to \widehat{\widehat{A}}$, which we write as $x \mapsto \delta_x$, and which is given by

$$\delta_x(\chi) = \chi(x).$$

We call δ the *Pontryagin map*. To see that for each $x \in A$ the map $\delta_x : \widehat{\widehat{A}} \to \mathbb{T}$ is indeed a continuous group homomorphism, and hence an element of $\widehat{\widehat{A}}$, we first observe that

$$\delta_x(\chi\mu) = \chi\mu(x) = \chi(x)\mu(x) = \delta_x(\chi)\delta_x(\mu)$$

for all $\chi, \mu \in \widehat{A}$, which implies that δ_x is a homomorphism. Since convergence in $\widehat{A}$ with respect to the compact open topology implies point-wise convergence we see that if a net $\chi_j \to \chi$ converges in $\widehat{A}$, then the net $\delta_x(\chi_j) = \chi_j(x)$ converges to $\chi(x) = \delta_x(\chi)$, which proves continuity of δ_x for each $x \in A$.

Examples 3.5.1.

- If $A = \mathbb{R}$ we know that $\mathbb{R} \cong \widehat{\mathbb{R}}$ via $t \mapsto \chi_t$ with $\chi_t(s) = e^{2\pi i s t}$. Thus we can also identify $\mathbb{R}$ with its bidual by mapping $s \in \mathbb{R}$ to a character $\mu_s : \widehat{\mathbb{R}} \to \mathbb{T}$, $\mu_s(\chi_t) = e^{2\pi i t s}$. It is easy to check that the map $\mu_s = \delta_s$ with $\delta : \mathbb{R} \to \widehat{\widehat{\mathbb{R}}}$ coincides with the above defined Pontryagin map. So we see in particular that the Pontryagin map is an isomorphism of groups in the case $A = \mathbb{R}$.

- Very similarly, we see that the Pontryagin maps $\delta : \mathbb{T} \to \widehat{\widehat{\mathbb{T}}}$ and $\delta : \mathbb{Z} \to \widehat{\widehat{\mathbb{Z}}}$ transform to the identity maps under the identifications $\mathbb{Z} \cong \widehat{\mathbb{T}}$ and $\mathbb{T} \cong \widehat{\mathbb{Z}}$ as explained in Example 1.7.1.

Proposition 3.5.2 *Let A be an LCA-group. The Pontryagin map is an injective continuous group homomorphism from A to $\widehat{\widehat{A}}$. In particular, if $1 \neq x \in A$ there exists some $\chi \in \widehat{A}$ such that $\chi(x) \neq 1$.*

Proof Note first that the Pontryagin map δ is a group homomorphism, since $\delta_{xy}(\chi) = \chi(xy) = \chi(x)\chi(y) = \delta_x(\chi)\delta_y(\chi)$. It suffices to show continuity at the unit element 1. So let V be an open unit-neighborhood in $\widehat{\widehat{A}}$. Then there exists a compact set $K^* \subset \widehat{A}$ and an $\varepsilon > 0$, such that V contains the open unit-neighborhood

$$B_{K^*,\varepsilon} = \left\{ \alpha \in \widehat{\widehat{A}} : |\alpha(\chi) - 1| < \varepsilon \ \forall_{\chi \in K^*} \right\}.$$

Let $L \subset A$ be a compact unit-neighborhood. As K^* is compact, there are $\chi_1 \ldots, \chi_n \in K^*$ such that $K^* \subset B_{L,\varepsilon/2}(\chi_1) \cup \cdots \cup B_{L,\varepsilon/2}(\chi_n)$, where

$$B_{L,\varepsilon}(\chi) = \left\{ \chi' \in \widehat{A} : \|\chi' - \chi\|_L < \varepsilon \right\}.$$

For $j = 1, \ldots, n$ let $U_j = \{x \in A : |\chi_j(x) - 1| < \varepsilon/2\}$. Let $U = \overset{\circ}{L} \cap U_1 \cap \cdots \cap U_n$. Then U is a unit-neighborhood for which we have $x \in U \Rightarrow |\chi(x) - 1| < \varepsilon \ \forall_{\chi \in K^*}$. So $\delta(U) \subset V$ and δ is continuous.

We still have to show that $\delta : A \to \widehat{\widehat{A}}$ is injective. So assume that $1 \neq x \in A$ with $\delta_x = 1_{\widehat{A}}$. Then $\chi(x) = 1$ for every $\chi \in \widehat{A}$. Choose $g \in C_c(A)$ with $g(1) = 1$ and $g(x^{-1}) = 0$. Then $L_x(g) \neq g$, but by Lemma 1.7.2 we have $\widehat{L_x(g)}(\chi) = \bar{\chi}(x)\hat{g}(\chi) = \hat{g}(\chi)$ for every $\chi \in \widehat{A}$. This contradicts the fact that the Fourier transform is injective by Corollary 3.3.4. □

Lemma 3.5.3 *Let $f \in C_0^*(A)$ be such that its Fourier transform lies in $C_c(\widehat{A})$. Then for every $x \in A$ one has $f(x) = \hat{\hat{f}}(\delta_{x^{-1}})$.*

Proof One has for $x \in A$,

$$f(x) = L_{x^{-1}} f(1) = \int_{\widehat{A}} \widehat{L_{x^{-1}} f}(\chi)\, d\chi = \int_{\widehat{A}} \hat{f}(\chi)\delta_x(\chi)\, d\chi = \hat{\hat{f}}(\delta_{x^{-1}}). \qquad \square$$

Lemma 3.5.4 *For an LCA-group A the following hold.*

(a) $C_c(A)$ is dense in $C_0^*(A)$.

(b) $C_c(\widehat{A}) \cap \{\hat{f} : f \in C_0^*(A) \cap L^2(A)\}$ is dense in $C_0^*(\widehat{A})$.

(c) $C_c(\widehat{A}) \cap \{\hat{f} : f \in C_0^*(A) \cap L^2(A)\}$ is dense in $L^2(\widehat{A})$.

Proof (a) As $C_0(A) \cap L^1(A)$ is dense in $C_0^*(A)$ by definition, it suffices to show that for a given f in this space there exists a sequence in $C_c(A)$ converging to f in the norms $\|\cdot\|_A$ and $\|\cdot\|_1$ simultaneously. Let $n \in \mathbb{N}$, and let $K_n \subset A$ be a compact set with $|f| < 1/n$ outside K_n. Choose a function χ_n in $C_c(A)$ with $0 \leq \chi_n \leq 1$, which is constantly equal to 1 on K_n. Set $f_n = \chi_n f$. Then the sequence f_n converges to f in both norms. Parts (b) and (c) follow from part (a) and Lemma 3.4.6. □

Theorem 3.5.5 (Pontryagin Duality). *The Pontryagin map $\delta : A \to \widehat{\widehat{A}}$ is an isomorphism of LCA groups.*

Proof We already know that δ is an injective continuous group homomorphism. We will demonstrate that it has a dense image. Assume this is not the case. Then there is an open subset U of $\widehat{\widehat{A}}$, which is disjoint from $\delta(A)$. By Lemma 3.4.6 applied to $\widehat{A}$, there exists $\psi \in C_0^*(\widehat{A})$, which is non-zero such that $\hat{\psi}$ is supported in U, i.e., it satisfies $\hat{\psi}(\delta(A)) = 0$. By Lemma 3.5.4, there exists a sequence (f_n) in $C_0^*(A)$ such that $\psi_n \overset{\text{def}}{=} \hat{f}_n$ lies in $C_c(\widehat{A})$ and converges to ψ in $C_0^*(\widehat{A})$. The inversion formula of Lemma 3.5.3 shows that $f_n(x) = \hat{\psi}_n(\delta_{x^{-1}})$ for every $x \in A$. This implies that the sequence f_n tends to zero uniformly on A. On the other hand $\hat{f}_n$ converges to ψ uniformly on $\widehat{A}$. This implies that (f_n) is a Cauchy sequence in $C_0^*(A)$ so it converges in this space. As the limit is unique, it follows from Lemma 3.4.2 that $\psi = 0$ in contradiction to our assumption. So the image of δ is indeed dense in $\widehat{\widehat{A}}$.

We next show that δ is a proper map, i.e., that the inverse image of a compact set is compact. For this let $K \subset \widehat{\widehat{A}}$ be compact. It suffices to show that the function $\check{\delta}(x) = \delta(x^{-1})$ is proper. By Lemma 3.4.6, there exists $\psi \in C_0^*(\widehat{A})$ such that $\hat{\psi}$ has compact support, is ≥ 0 on $\widehat{\widehat{A}}$ and ≥ 1 on K. As above, there is a sequence (f_n) in $C_0^*(A)$ such that $\psi_n \overset{\text{def}}{=} \hat{f}_n \geq 0$ lies in $C_c(\widehat{A})$ and converges to ψ in $C_0^*(\widehat{A})$. Fix n with $\|\hat{\psi}_n - \hat{\psi}\|_{\widehat{\widehat{A}}} < 1/2$. We also have $f_n(x) = \hat{\psi}_n(\delta_{x^{-1}})$ for every $x \in A$ again and, as f_n is in $C_0(A)$, there exists a compact set $C \subset A$ such that $|f_n| < 1/2$ outside C. As $\hat{\psi}_n$ is $\geq 1/2$ on K, it follows that the pre-image of K under $\check{\delta}$ is contained in C. As δ is continuous, this pre-image is closed, hence compact, so δ is proper.

It remains to show that δ is a *closed map*, i.e., that it maps closed sets to closed sets. Then δ is a homeomorphism, i.e., the theorem follows. So we finish our proof with the following lemma.

Lemma 3.5.6 *Let $\phi : X \to Y$ be a continuous map between locally compact Hausdorff spaces. If ϕ is proper, then it is closed.*

Proof Let T be a closed subset of X. We show first that

(∗) For every compact set $L \subset Y$ the intersection $\phi(T) \cap L$ is closed.

For this recall that $\phi^{-1}(L)$ is compact and therefore $T \cap \phi^{-1}(L)$ is compact and so $\phi(T) \cap L = \phi(T \cap \phi^{-1}(L))$ is compact and therefore closed.

Now we use (∗) to deduce that $\phi(T)$ is closed. Let y be in the closure of $\phi(T)$. Let L be a compact neighborhood of y. For every neighborhood U of y one has $U \cap (L \cap \phi(T)) \neq \emptyset$, so y is in $\overline{L \cap \phi(T)} = L \cap \phi(T) \subset \phi(T)$. This means that $\phi(T)$ is closed.   □

Proposition 3.5.7 *The Fourier transform induces an isometric isomorphism of Banach spaces $\mathcal{F} : C_0^*(A) \to C_0^*(\widehat{A})$ with inverse map given by the dual Fourier transform $\widehat{\mathcal{F}} : C_0^*(\widehat{A}) \to C_0^*(A);$ $\widehat{\mathcal{F}}(\psi)(x) \overset{\text{def}}{=} \hat{\psi}(\delta_{x^{-1}}).$*

Proof Let B be the space of all $f \in C_0^*(A)$ such that $\hat{f}$ lies in $C_c(\widehat{A})$. For $f \in B$ we have $\widehat{\mathcal{F}} \circ \mathcal{F}(f) = f$ by Lemma 3.5.3. Further, one has

$$\|f\|_0^* = \max(\|\hat{f}\|_{\widehat{A}}, \|f\|_A) = \max(\|\hat{f}\|_{\widehat{A}}, \|\widehat{\mathcal{F}} \circ \mathcal{F}(f)\|_A)$$

$$= \max(\|\hat{f}\|_{\widehat{A}}, \|\hat{f}\|_{\widehat{\widehat{A}}}) = \|\mathcal{F}(f)\|_0^*.$$

As the set $\mathcal{F}(B)$ is dense in $C_0^*(\widehat{A})$ by Lemma 3.5.4, the Fourier transform defines a surjective isometry from the closure of B to $C_0^*(\widehat{A})$. Conversely, this means that $\widehat{\mathcal{F}}$ is an isometry from $C_0^*(\widehat{A})$ to $C_0^*(A)$. Since $\widehat{\mathcal{F}} = \mathcal{F}_{\widehat{A}} \circ \delta^{-1}$, where $\mathcal{F}_{\widehat{A}}$ denotes the Fourier transform on $\widehat{A}$ and since $\mathcal{F}_{\widehat{A}}(C_0^*(\widehat{A}))$ contains a subset of $C_c(\widehat{\widehat{A}})$ that is dense in $C_0^*(\widehat{\widehat{A}})$ by Lemma 3.5.4, it follows from Pontryagin duality that $\widehat{\mathcal{F}}(C_0^*(\widehat{A}))$ is dense in $C_0^*(A)$. Since it is isometric it must be an isomorphism of Banach spaces as claimed.   □

Theorem 3.5.8 (Inversion Formula). *Let $f \in L^1(A)$ be such that its Fourier transform $\hat{f}$ lies in $L^1(\widehat{A})$. Then f is a continuous function, and for every $x \in A$ one has*

$$f(x) = \hat{\hat{f}}(\delta_{x^{-1}}).$$

Proof Let $f \in L^1(A)$ with $\hat{f} \in L^1(\widehat{A})$. Then $\hat{f}$ lies in $C_0(\widehat{A}) \cap L^1(\widehat{A})$, which is a subspace of $C_0^*(\widehat{A})$. By Proposition 3.5.7, there exists $g \in C_0^*(A)$ with $\hat{g} = \hat{f}$ and $g(x) = \hat{\hat{f}}(\delta_{x^{-1}})$ for every $x \in A$. Since the Fourier transform is injective on $C^*(A)$, we have $f = g$. □

We are now ready for the proof of the Plancherel Theorem.

Proof of Theorem 3.4.8 Let $f \in L^1(A) \cap L^2(A)$. By Lemma 3.4.1 one has $f * f^* \in L^1(A) \cap C_0(A)$. The continuous function $h = f * f^* = |\hat{f}|^2$ is positive. Let $\phi \in C_c(\widehat{A})$ satisfy $0 \le \phi \le h$. Then

$$\int_{\widehat{A}} \phi(\chi) \, d\chi \le f * f^*(1) = \|f\|_2^2 < \infty.$$

Therefore h is integrable, so $\widehat{f * f^*} \in L^1(A)$. By Theorem 3.5.8 it follows that $\|f\|_2^2 = f * f^*(1) = \widehat{\widehat{f * f^*}}(1) = \widehat{|\hat{f}|^2}(1) = \|\hat{f}\|_2^2$. As $L^1(A) \cap L^2(A)$ is dense in $L^2(A)$, the Fourier-transform $f \mapsto \hat{f}$ extends uniquely to an isometric linear map $L^2(A) \to L^2(\widehat{A})$. By Lemma 3.4.6 the image in $L^2(\widehat{A})$ is dense, so the map is surjective. □

With the help of the Plancherel theorem, we can see that there are indeed many functions f, to which the inversion formula applies.

Proposition 3.5.9 *Let $\phi, \psi \in L^1(A) \cap L^2(A)$, and let $f = \phi * \psi$. Then $f \in L^1(A)$ and $\hat{f} \in L^1(\widehat{A})$, so the inversion formula applies to f.*

Proof We have $\hat{f} = \widehat{\phi * \psi} = \hat{\phi}\hat{\psi}$ is the point-wise product of L^2-functions on $\widehat{A}$, hence $\hat{f} \in L^1(\widehat{A})$. □

3.6 The Poisson Summation Formula

Let A be an LCA group, and let B be a closed subgroup of A. We want to study the relations between the dual group $\widehat{A}$ of A and the dual groups $\widehat{B}$ and $\widehat{A/B}$ of the subgroup B and the quotient group A/B. The Poisson Summation Formula relates the Fourier transform of A to the transforms on B and A/B.

We introduce some further notation: If E is a subset of A we denote by $E^{\perp}$ the *annihilator* of E in $\widehat{A}$, i.e., the set of all characters $\chi \in \widehat{A}$ with $\chi(E) = 1$. Similarly, if $L \subset \widehat{A}$, we denote by $L^{\perp}$ the *annihilator of L in A*, i.e., the set of all $x \in A$ such that $\chi(x) = 1$ for every $\chi \in L$. In short, we have

$$E^{\perp} = \{\chi \in \widehat{A} : \chi(x) = 1 \; \forall x \in E\}$$
$$L^{\perp} = \{x \in A : \chi(x) = 1 \; \forall \chi \in L\}.$$

It is easy to see that $E^{\perp}$ is a closed subgroup of $\widehat{A}$, and $L^{\perp}$ is a closed subgroup of A. Recall that the Pontryagin isomorphism $\delta : A \to \widehat{\widehat{A}}$ is defined by putting $\delta_x(\chi) = \chi(x)$ for every $x \in A$.

Proposition 3.6.1 *Let A be an LCA group, and let B be a closed subgroup of A. Then the following are true:*

(a) $B^{\perp}$ *is isomorphic to* $\widehat{A/B}$ *via* $\chi \mapsto \widetilde{\chi}$ *with* $\widetilde{\chi} \in \widehat{A/B}$ *defined by* $\widetilde{\chi}(xB) \overset{\text{def}}{=} \chi(x)$.

(b) $(B^{\perp})^{\perp} = B$.

(c) $\widehat{A}/B^{\perp}$ *is isomorphic to* $\widehat{B}$ *via* $\chi \cdot B^{\perp} \mapsto \chi|_B$.

Proof This is a straightforward verification (See Exercise 3.10). □

As a direct corollary we get

Corollary 3.6.2 *Let B be a closed subgroup of the LCA-group A. Then the restriction map* $\operatorname{res}_B^A : \widehat{A} \to \widehat{B}$ *defined by* $\chi \mapsto \chi|_B$ *is surjective with kernel* $\widehat{A/B}$.

Note that one could formulate the above result in more fancy language as follows: If B is a closed subgroup of A, then we get the short exact sequence

$$1 \longrightarrow B \overset{\iota}{\longrightarrow} A \overset{q}{\longrightarrow} A/B \longrightarrow 1$$

of LCA groups. The above result then says that the dual sequence

$$1 \longrightarrow \widehat{A/B} \overset{\hat{q}}{\longrightarrow} \widehat{A} \overset{\hat{\iota}}{\longrightarrow} \widehat{B} \longrightarrow 1$$

is also an exact sequence of LCA groups, where for any continuous homomorphism $\psi : A_1 \to A_2$ between two LCA groups A_1, A_2, we denote by $\hat{\psi} : \widehat{A_2} \to \widehat{A_1}$ the homomorphism defined by $\hat{\psi}(\chi) = \chi \circ \psi$ for $\chi \in \widehat{A_2}$. One should not mistake this notation with the notion of the Fourier transform of a function. Note that if $\iota : B \to A$ is the inclusion map, then $\hat{\iota}(\chi) = \chi \circ \iota = \chi|_B$, so $\hat{\iota} = \operatorname{res}_B^A$.

We now come to Poisson's summation formula. Recall from Theorem 1.5.3 together with Corollary 1.5.5 that for any closed subgroup B of the LCA group A we can choose Haar measures on A, B and A/B in such a way that for every $f \in C_c(A)$ we get the quotient integral formula

$$\int_{A/B} \int_B f(xb)\, db\, dx\, B = \int_A f(x)\, dx.$$

In what follows we shall always assume that the Haar measures are chosen this way.

Theorem 3.6.3 (Poisson's Summation Formula). *Let B be a closed subgroup of the LCA group A. For $f \in L^1(A)$ define $f^B \in L^1(A/B)$ as $f^B(xB) = \int_B f(xb)\, db$. Then, if we identify $\widehat{A/B}$ with $B^{\perp}$ as in Proposition 3.6.1, we get $\widehat{f^B} = \hat{f}\,|_{B^{\perp}}$. If, in addition, $\hat{f}\,|_{B^{\perp}} \in L^1(B^{\perp})$, then we get*

$$\int_B f(xb)\, db = \int_{B^{\perp}} \hat{f}(\chi)\chi(x)\, d\chi,$$

for almost all $x \in A$, where Haar measure on $\widehat{B^{\perp}} \cong \widehat{A/B}$ is the Plancherel measure with respect to the chosen Haar measure on A/B. If f^B is everywhere defined and continuous, the above equation holds for all $x \in A$.

Proof For $\chi \in B^{\perp}$ we have $\chi(xb) = \chi(x)$ for every $x \in A$ and $b \in B$. We therefore get from Theorem 1.5.3,

$$\widehat{f^B}(\chi) = \int_{A/B} f^B(xB)\bar{\chi}(x)\, dx\, B = \int_{A/B} \int_B f(xb)\bar{\chi}(xb)\, db\, dx\, B$$

$$= \int_A f(x)\bar{\chi}(x)\, dx = \hat{f}(\chi)$$

for every $\chi \in B^{\perp}$. Moreover, if $\hat{f}\,|_{B^{\perp}} \in L^1(B^{\perp}) = L^1(\widehat{A/B})$, then the inversion formula of Theorem 1.5.3 implies that

$$\int_B f(xb)\, db = f^B(xB) = \widehat{\widehat{f^B}}\,(\delta_{x^{-1}B})$$

$$= \widehat{\hat{f}\,|_{B^{\perp}}}(\delta_{x^{-1}B}) = \int_{B^{\perp}} \hat{f}(\chi)\overline{\chi(x)}\, d\chi.$$

almost everywhere. It holds everywhere if, in addition, the defining integral for f^B exists everywhere and f^B is continuous.   □

Example 3.6.4. (The Poisson Summation formula for $\mathbb{R}$) Let A be the group $(\mathbb{R}, +)$ with the usual topology. Then $A \cong \hat{A}$ via the map $y \mapsto \chi_y$ where $\chi_y(x) = e^{2\pi i x y}$. Let B be the closed subgroup $\mathbb{Z}$. Then the above identification maps B bijectively to $B^{\perp}$. For $f \in L^1(\mathbb{R})$ such that $\hat{f}\,|_{\mathbb{Z}} \in L^1(\mathbb{Z})$, the equality

$$\sum_{k \in \mathbb{Z}} f(x + k) = \sum_{k \in \mathbb{Z}} \hat{f}(k) e^{2\pi i k x}$$

holds almost everywhere in x, where $\hat{f}(x) = \int_{\mathbb{R}} f(y) e^{-2\pi i x y}\, dy$. In particular, define the *Schwartz space* $\mathcal{S}(\mathbb{R})$ as the space of all C^{∞}- functions $f : \mathbb{R} \to \mathbb{C}$ such that for any two integers $m, n \geq 0$ the function $x^n f^{(m)}(x)$ is bounded. Then the Fourier transform maps $\mathcal{S}(\mathbb{R})$ bijectively to itself (Exercise 3.14). For $f \in \mathcal{S}(\mathbb{R})$, both sums in the Poisson summation formula converge uniformly and define continuous functions, which then must be equal in every point. For $x = 0$ we get the elegant formula

$$\sum_{k \in \mathbb{Z}} f(k) = \sum_{k \in \mathbb{Z}} \hat{f}(k).$$

For applications of this formula to theta series and the Riemann zeta function, see [Dei05].

3.7   Exercises and Notes

Exercise 3.1. Let $\mathcal{U}$ be a basis for the topology on the LCA-group A. Let $\mathcal{U}_c$ denote the set of all $U \in \mathcal{U}$ that are relatively compact. Show that the set $\mathcal{B}$ of all $L(\bar{U}, V)$, where $U \in \mathcal{U}_c$ and V is open in $\mathbb{T}$, generates the topology of $\widehat{A}$.

Exercise 3.2. Show that if an LCA-group A is second countable, then so is its dual $\widehat{A}$.

Exercise 3.3. Let b be the map $b : \mathbb{R} \rightarrow \prod_{t \in \mathbb{R}} \mathbb{T}$ sending $x \in \mathbb{R}$ to the element $b(x)$ with coordinates $b(x)_t = e^{2\pi i t x}$. Let B denote the closure of $b(\mathbb{R})$ in the product space. By Tychonov's Theorem the product is compact; therefore B is a compact group called the *Bohr compactification of* $\mathbb{R}$. Show that B is separable but not second countable.

(Hint: Use the fact that $\mathbb{Q}$ is dense in $\mathbb{R}$. Show that B is isomorphic to the dual group of $\mathbb{R}_{\mathrm{disc}}$, which is the group $(\mathbb{R}, +)$ with the discrete topology. Then use Exercise 3.2)

Exercise 3.4. Verify the statements in Example 3.1.3.

Exercise 3.5. Let A and B be two LCA groups. Show that $\widehat{A \times B} = \widehat{A} \times \widehat{B}$.

Exercise 3.6. Show that the multiplicative group $\mathbb{C}^\times$ is locally compact with the topology of $\mathbb{C}$ and that
$$\widehat{\mathbb{C}^\times} \cong \mathbb{Z} \times \mathbb{R}.$$

Exercise 3.7. Let $(A_j)_{j \in J}$ be a family of discrete groups. Show that there is a canonical isomorphism
$$\widehat{\bigoplus_{j \in J} A_j} \cong \prod_{j \in J} \widehat{A_j}.$$

Exercise 3.8.

(a) Let (A_j, p_i^j) be a projective system of compact groups. Show that there is a canonical isomorphism of topological groups,

$$\widehat{\varprojlim A_j} \cong \varinjlim \widehat{A_j}.$$

(b) Let (B_j, d_i^j) be a direct system of discrete groups satisfying the Mittag-Leffler condition. Show that there is a canonical isomorphism

$$\widehat{\varinjlim B_j} \cong \varprojlim \widehat{B_j}.$$

Exercise 3.9. Let A be an LCA group, and let $f \in L^1(A)$ such that $\widehat{f} \in L^1(\widehat{A})$. Show $f \in L^2(A)$.

Exercise 3.10. For a closed subgroup B of the LCA-group A and a closed subgroup L of $\widehat{A}$ let

$$B^\perp = \{\chi \in \widehat{A} : \chi(B) = 1\}$$
$$L^\perp = \{x \in A : \delta_x(L) = 1\}.$$

Show that $B^\perp$ is canonically isomorphic to $\widehat{A/B}$, $(B^\perp)^\perp = B$, and $\widehat{A}/B^\perp$ is canonically isomorphic to $\widehat{B}$.

Exercise 3.11. For a continuous group homomorphism $\phi : A \to B$ between LCA groups, define $\hat{\phi} : \widehat{B} \to \widehat{A}$ by

$$\hat{\phi}(\chi) \stackrel{\text{def}}{=} \chi \circ \phi.$$

Show that for any two composable homomorphisms ϕ and ψ one has $\widehat{\phi \circ \psi} = \hat{\psi} \circ \hat{\phi}$. This means that $A \mapsto \widehat{A}$ defines a contravariant functor on the category of LCA groups and continuous group homomorphisms.

Exercise 3.12. A *short exact sequence* of LCA groups is a sequence of continuous group homomorphisms

$$A \stackrel{\alpha}{\hookrightarrow} B \stackrel{\beta}{\twoheadrightarrow} C$$

such that α is injective, β is surjective, the image of α is the kernel of β, the group A carries the subspace topology and C carries the quotient topology. Show that a short exact sequence like this induces a short exact sequence of the dual groups

$$\widehat{C} \stackrel{\hat{\beta}}{\hookrightarrow} \widehat{B} \stackrel{\hat{\alpha}}{\twoheadrightarrow} \widehat{A}.$$

Exercise 3.13 Let $A = \mathbb{R}$, and choose the Lebesgue measure as Haar measure. Identify $\widehat{A}$ with $\mathbb{R}$ via $x \mapsto \chi_x$ with $\chi_x(y) = e^{2\pi i xy}$. Show that via this identification, the Lebesgue measure is the Plancherel measure on $\widehat{A}$.

(Hint: Use the fact that $\int_{\mathbb{R}} e^{-\pi x^2} dx = 1$ and compute the Fourier transform of $f(x) = e^{-\pi x^2}$.)

Exercise 3.14. Show that $\hat{f} \in \mathcal{S}(\mathbb{R})$ for every $f \in \mathcal{S}(\mathbb{R})$ and that the map $\mathcal{F} : \mathcal{S}(\mathbb{R}) \to \mathcal{S}(\mathbb{R})$ defined by $\mathcal{F}(f) = \hat{f}$ is a bijective linear map with

$$\mathcal{F}^{-1}(g)(y) = \int_{\mathbb{R}} g(x) e^{2\pi i xy} dx.$$

Exercise 3.15. Let $f \in \mathcal{S}(\mathbb{R})$ and set $g(x) = \sum_{k \in \mathbb{Z}} f(x + k)$. Show that g is a smooth function on $\mathbb{R}$.

(Hint: The estimate $|f(x)| \leq C/(1 + x^2)$ for a constant C shows point-wise convergence. The same holds for the n-th derivative $f^{(n)}$ instead of f. Now integrate n times.)

Exercise 3.16. As an application of Theorem 3.6.3, show that for every Schwartz function $f \in \mathcal{S}(\mathbb{R})$,

$$\sum_{k \in \mathbb{Z}} f(k) = \sum_{k \in \mathbb{Z}} \hat{f}(k)$$

holds, where $\hat{f}(x) = \int_{\mathbb{R}} f(y) e^{-2\pi i xy} dy$.

Exercise 3.17. Let $A = \mathbb{R}^n$ with Lebesgue measure as Haar measure and identify $\mathbb{R}^n$ with $\widehat{\mathbb{R}}^n$ via $x \mapsto \chi_x$ with $\chi_x(y) = e^{-2\pi i \langle x, y \rangle}$, where $\langle x, y \rangle$ denotes the standard inner product on $\mathbb{R}^n$. Let $\mathcal{S}(\mathbb{R}^n)$ denote the space of all C^∞- functions $f : \mathbb{R}^n \to \mathbb{C}$ such that for any two multi-indices $\alpha, \beta \in \mathbb{N}_0^n$ the function

$$x^\alpha \partial^\beta (f) \stackrel{\text{def}}{=} x_1^{\alpha_1} \ldots x_n^{\alpha_n} \frac{\partial^{|\beta|} f}{x_1^{\beta_1} \ldots x_n^{\beta_n}}$$

is bounded, where $|\beta| = \beta_1 + \cdots + \beta_n$. Formulate and prove the analogues of Exercise 3.14 and Exercise 3.16 in this setting.

Exercise 3.18. (Parseval's equation) Let A be an LCA group, and let $\widehat{A}$ be equipped with the Plancherel measure with respect to a given Haar measure on A. Show that the equation

$$\langle f, g \rangle = \int_A f(x) \bar{g}(x) dx = \int_{\widehat{A}} \hat{f}(\chi) \bar{\hat{g}}(\chi) d\chi = \langle \hat{f}, \hat{g} \rangle$$

holds for all $f, g \in L^1(A) \cap L^2(A)$.

Exercise 3.19. A finite abelian group A can be equipped either with the counting measure or with the normalized Haar measure that gives A the volume 1. What is the Plancherel measure in either case?

Exercise 3.20. For a finite abelian group A, let $C(A)$ be the space of all function from A to $\mathbb{C}$. For a group homomorphism $\phi : A \to B$ between finite abelian groups let $\phi^* : C(B) \to C(A)$ be defined by $\phi^* f = f \circ \phi$, and let $\phi_* : C(A) \to C(B)$ be defined by

$$\phi_* g(b) \overset{\text{def}}{=} \sum_{a : \phi(a) = b} g(a),$$

where the empty sum is interpreted as zero. Show that for composable homomorphisms one has $(\phi\psi)_* = \phi_* \psi_*$ and $(\phi\psi)^* = \psi^* \phi^*$.

Exercise 3.21. For a finite abelian group A let $\mathbb{F} : C(A) \to C(\widehat{A})$ be the Fourier transform. Show that for every group homomorphism $\phi : A \to B$ between finite abelian groups the diagram

$$
\begin{array}{ccc}
C(A) & \xrightarrow{\ \mathbb{F}\ } & C(\widehat{A}) \\
\uparrow{\scriptstyle \phi^*} & & \uparrow{\scriptstyle \widehat{\phi}_*} \\
C(B) & \xrightarrow{\ \mathbb{F}\ } & C(\widehat{B})
\end{array}
$$

Exercise 3.22. An LCA-group A is called *monothetic*, if it contains a dense cyclic subgroup. Show that a compact LCA-group A is monothetic if and only if its dual $\widehat{A}$ is isomorphic to a subgroup of $\mathbb{T}_d$, where $\mathbb{T}_d$ is the circle group with the discrete topology.

Notes

In principle, the ideas for the proofs of the Plancherel Theorem and the Pontryagin Duality Theorem given in this chapter goes back to the paper [Wil62] of J.H. Williamson. However, to our knowledge, this book is the first that exploits the very natural isomorphism $C_0^*(A) \cong C_0^*(\widehat{A})$.

Chapter 4
The Structure of LCA-Groups

In this chapter we will apply the duality theorem for proving structure theorems for LCA groups. As main result we will show that all such groups are isomorphic to groups of the form $\mathbb{R}^n \times H$ for some $n \in \mathbb{N}_0$, such that H is a locally compact abelian group that contains an open compact subgroup K. This theorem will imply better structure theorems if more information on the group is available. For instance it will follow that every compactly generated LCA group is isomorphic to a group of the form $\mathbb{R}^n \times \mathbb{Z}^m \times K$ for some compact group K and some $n, m \in \mathbb{N}_0$, and every compactly generated locally euclidean group is isomorphic to one of the form $\mathbb{R}^n \times \mathbb{Z}^m \times \mathbb{T}^l \times F$, for some finite group F and some nonnegative integers n, m and l. To prepare the proofs of these theorems, we start with a section on connectedness in locally compact groups. The main result in that section shows that every totally disconnected locally compact group G has a unit-neighborhood base consisting of compact open subgroups of G. The structure theorems will be shown in the second section of this chapter.

4.1 Connectedness

Recall that a topological space X is called *connected* if it cannot be written as the disjoint union of two nonempty open subsets, i.e., if $U, V \subset X$ are two open (or closed) subsets of X such that $X = U \cup V$ and $U \cap V = \emptyset$, then $X = U$ or $X = V$. Another way to formulate connectedness is to say that the space X is connected if and only if any set $U \subset X$, which is open and closed at the same time, must be equal to $\emptyset$ or X.

A subset $A \subset X$ is called connected if it is connected in the subspace topology, i.e., a subset $A \subset X$ is connected if for all open sets $U, V \subset X$ with $A \subset U \cup V$ and $A \cap U \cap V = \emptyset$ it follows that $A \subset U$ or $A \subset V$. If A and B are two connected subsets of X with non-empty intersection, then the union $A \cup B$ is also connected. Also, the closure of a connected set is always connected, but intersections of connected sets are not necessarily connected, for this see the examples below. If $f : X \to Y$ is a continuous map, then the image of any connected subset of X is a connected subset of Y (See Exercise 4.1).

A. Deitmar, S. Echterhoff, *Principles of Harmonic Analysis,* Universitext,
DOI 10.1007/978-3-319-05792-7_4, © Springer International Publishing Switzerland 2014

If $x \in X$, then the *connected component* $C(x)$ of x is the union of all connected subsets containing x. It is the biggest connected subset of X that contains x. The set $C(x)$ is closed in X, and the space X can be written as the disjoint union of its connected components. A topological space X is called *totally disconnected* if $C(x) = \{x\}$ for every $x \in X$.

Examples 4.1.1

- Any interval in the real line is connected, and so is any ball in $\mathbb{R}^n$ (See Exercise 4.1).

- A circle in $\mathbb{R}^2$ is connected, since it can be written as the image of an interval under a continuous function. But the intersections of two circles may consist of two distinct points, in which case it is not connected. This shows that intersections of connected sets might not be connected.

- Every discrete topological space X is totally disconnected.

- The Cantor set is an example of a non-discrete totally disconnected set (See Exercise 4.2).

Proposition 4.1.2 *Let G be a topological group, and let G_0 be the connected component of the unit in G. Then G_0 is a closed normal subgroup of G. Moreover, the coset xG_0 is the connected component of x for every $x \in G$ and the quotient group G/G_0 is totally disconnected with respect to the quotient topology.*

Proof The set xG_0 is the image of G_0 under the homeomorphism $l_x : G \to G$, which maps y to xy; therefore the set xG_0 is the connected component of x. If $x \in G_0$, then $xG_0 \cap G_0 \neq \emptyset$, and hence $xG_0 = G_0$. Since $y \mapsto y^{-1}$ and $y \mapsto xyx^{-1}$ are also homeomorphisms from G to G, it also follows that $G_0^{-1} = G_0$ and $xG_0x^{-1} = G_0$ for every $x \in G$. This shows that G_0 is a closed normal subgroup of G.

To see that G/G_0 is totally disconnected, we have to show that every subset $A \subset G/G_0$ that contains more then one element, is not connected. Take the inverse image $B = q^{-1}(A) \subset G$. Then B contains at least two disjoint cosets xG_0, yG_0 with $x, y \in G$. Hence, B is not connected. Thus there exist open subsets $W_1, W_2 \subset G$ with $B \cap W_1 \cap W_2 = \emptyset$, $B \cap W_i \neq \emptyset$ for $i = 1, 2$, and $B \subset W_1 \cup W_2$. Then $xG_0 \cap W_1 \cap W_2 = \emptyset$, and $xG_0 \subset W_1 \cup W_2$ for every $x \in B$. Since xG_0 is connected, it follows that for each $x \in B$ there exists exactly one $i \in \{1, 2\}$ with $xG_0 \subset W_i$. Thus, for $V_i = q(W_i)$ it follows that V_1, V_2 are non-empty open subsets of G/G_0 with $A \cap V_1 \cap V_2 = \emptyset$, $A \cap V_i \neq \emptyset$ for $i = 1, 2$, and $A \subset V_1 \cup V_2$. . $\qquad\square$

Definition A topological space X is called *locally connected*, if each point has a connected neighborhood.

Lemma 4.1.3 (a) *If X is a locally connected space, then every connected component $C(x)$ is open.*

(b) *If the topological group G possesses a connected unit-neighborhood, then its connected component G_0 is open.*

Proof (a) Let $x \in X$ and let $y \in C(x)$. Let U denote a connected neighborhood of y. As $C(x) = C(y)$, it follows $U \subset C(x)$ and therefore $C(x)$ is open.

(b) If G has a connected unit-neighborhood U, then every point x has a connected neighborhood, for example xU. Therefore G is locally connected and the claim follows from part (a). □

In what follows next, we shall prepare a few structural facts for groups with compact open neighborhoods of the identity that will be used later.

Proposition 4.1.4 *Let G be a topological group, and let U be a compact and open neighborhood of the identity e in G. Then U contains a compact and open subgroup K of G.*

Proof By Lemma 1.1.6, we can find an open neighborhood $V = V^{-1}$ of e such that $UV = VU = U$. Since $e \in U$, it follows that $V \subset U$, and then $V^2 \subset VU \subset U$. By induction we then see that $V^n \subset U$ for every $n \in \mathbb{N}$, from which it follows that the group $K = \bigcup_{n \in \mathbb{N}} V^n$ generated by V lies in U. By Lemma 1.1.7 (b) and (c) we know that K is an open and closed subgroup of G. Thus, since it is contained in the compact set U, it is also compact. □

Definition Let X be a topological space. Recall that a system $\mathcal{B}$ of open subsets of X is called *base of the topology* if every open set is a union of sets in $\mathcal{B}$. For example, in a metric space, the open balls form a base of the topology.

Proposition 4.1.5 *Every totally disconnected locally compact Hausdorff space X has a base for its topology consisting of open and compact subsets of X.*

Proof It is enough to show that for each $x \in X$ and each compact neighborhood U of x there exists an open and closed subset V of X such that $x \in V \subset U$. For this let M denote the set of all $y \in U$ such that there exists a relatively open and closed subset $C_y \subset U$ with $y \in C_y$ and $x \notin C_y$. Then M is the union of all such sets. This implies in particular, that M is relatively open in U. Let $A = U \setminus M$. Then $x \in A$ and A is closed. We shall show

(a) If $W \subset U$ is relatively open in U and $\overline{W} \setminus W$ is a subset of M, then there exists $\tilde{W} \supset W$ open and closed in U, such that $A \cap \tilde{W} = A \cap W$.

(b) The set A is connected.

Once this is shown, we conclude that $A = \{x\}$ since X is totally disconnected. It then follows that the compact boundary $\partial U = U \smallsetminus \mathring{U}$ of U lies in M and lies in the union of finitely many sets $C_{y_1}, \dots, C_{y_n}$. The set $V = U \smallsetminus \left(\bigcup_{i=1}^n C_{y_i} \right)$ is then a relatively open and closed subset of U that contains x. Since it lies in the interior $\mathring{U}$ of U, it is also open in X. Thus it satisfies the above stated requirements.

To show (a), let W be as assumed. Then the compact set $\overline{W} \smallsetminus W$ can be covered by finitely many of the open sets C_y, hence we get $\left(\overline{W} \smallsetminus W \right) \subset C$ for a set of the form $C = C_{y_1} \cup \dots \cup C_{y_n} \subset M$. Then C is closed and open in U. We set $\tilde{W} = W \cup C$. Since C contains $\overline{W} \smallsetminus W$ it follows $W \cup C = \overline{W} \cup C$, so $\tilde{W}$ is closed and open in U. Finally, as C is contained in M, we have $\tilde{W} \cap A = W \cap A$.

To end the proof, we conclude (b) from (a). Let B_1, B_2 be closed (hence compact) subsets of A such that $B_1 \cap B_2 = \emptyset$ and $A = B_1 \cup B_2$. We have to show that one of these sets is empty. Assume that $x \in B_1$. Since B_1, B_2 are compact with $B_1 \cap B_2 = \emptyset$ we may choose a relatively open subset $W \subset U$ such that $B_2 \subset W$ and $\overline{W} \cap B_1 = \emptyset$ by the Lemma of Urysohn (A.8.1). By (a) there exists $\tilde{W}$ open and closed in U with $B_2 = \tilde{W} \cap A$. Then $x \notin \tilde{W}$ and so $\tilde{W} \subset M$, which implies $B_2 = \emptyset$. $\square$

Theorem 4.1.6 (a) *Let G be a locally compact group that is totally disconnected. Then every unit-neighborhood U in G contains an open and compact subgroup of G.*

(b) *A locally compact group is profinite if and only if it is compact and totally disconnected.*

Proof (a) follows from the Propositions 4.1.4 and 4.1.5.

(b) Let $G = \lim_{\leftarrow} G_i$ be profinite, then $G \subset \prod_i G_i$ and the product is compact and totally disconnected, hence the closed subset G is compact and totally disconnected. For the converse, let G be compact and totally disconnected. By part (a), G possesses a unit-neighborhood base of open subgroups. By compactness, every open subgroup U has finite index. Let U be an open subgroup and let N be the kernel of the group homomorphism $G \to \mathrm{Per}(G/U)$, where $\mathrm{Per}(G/U)$ is the permutation group of the finite set G/U. As U is open, this homomorphism is continuous, so N is an open normal subgroup, which is contained in U, since the coset eU is stable under $n \in N$. Therefore, G has a unit-neighborhood base consisting of open normal subgroups. Consider the natural map

$$\phi : G \to \lim_{\substack{\leftarrow \\ N}} G/N,$$

where the limit is extended over all open normal subgroups. We claim that ϕ is an isomorphism of locally compact groups.

Injectivity Let $x \in G \setminus \{1\}$. There exists an open normal subgroup N with $x \notin N$. So x is not equal to 1 in G/N and ϕ is injective.

Surjectivity Let $y \in \varprojlim_N G/N$. For each N, let A_N denote the inverse image in G of $y_N \in G/N$. The family of closed sets (A_N) has the finite intersection property, so there exists an element x in their intersection. This is a pre-image for y.

Continuity Every projection $G \to G/N$ is continuous and therefore so is ϕ.

Continuity of the Inverse A continuous surjective map $\phi : X \to Y$ between compact Hausdorff spaces is automatically closed, for the image of compact sets are compact.

$\square$

We will also need the concept of path-connectedness, which we present next.

Definition A topological space X is called *path-connected* if any two points can be joined by a continuous path, i.e., if for any two $x, y \in X$ there exists a continuous map $p : [0,1] \to X$ with $p(0) = x$ and $p(1) = y$.

Lemma 4.1.7 (a) *Any path-connected space is connected.*

(b) *A connected topological group having a path-connected neighborhood of the unit element, is path-connected.*

(c) *If K is a compact LCA-group which is connected, then the discrete group $\widehat{K}$ has no non-trivial elements of finite order.*

Proof (a) Let X be a path-connected space and let $X = U \cup V$ where U and V are open subsets with $U \cap U = \emptyset$. Suppose U is non-empty and fix a point $x \in U$. Let $y \in X$ be any other point and let $p : [0,1] \to X$ be a path with $p(0) = x$ and $p(1) = y$. Then $[0,1] = p^{-1}(U) \cup p^{-1}(V)$ is a disjoint decomposition of the unit interval into open subsets with $p^{-1}(U) \neq \emptyset$. The unit interval being connected, it follows that $p^{-1}(V) = \emptyset$ and so $y \in U$, i.e., $U = X$, so X is connected.

(b) Let G be a connected topological group and $U \subset G$ a path-connected unit-neighborhood. The subgroup H generated by U is an open subgroup and as G is connected, $H = G$. So any $x \in G$ can be written in the form $x = x_1^{\varepsilon_1} \cdots x_n^{\varepsilon_n}$, where $x_1, \ldots, x_n \in U$ and $\varepsilon_j \in \{\pm 1\}$. Let $p_1, \ldots, p_n : [0,1] \to G$ be paths connecting the unit $1 \in G$ to the points $x_1, \ldots, x_n$, i.e., $p_j(0) = 1$ and $p_j(1) = x_j$. Then $p(t) = p_1(t)^{\varepsilon_1} \cdots p_n(t)^{\varepsilon_n}$ is a path connecting $1 \in G$ to x, which means that G is path-connected.

(c) Let K be a compact LCA-group and let $D = \widehat{K}$ be its dual group. Suppose that D has a non-trivial element χ of finite order, say $\chi^n = 1$. We have to show that K is non-connected. Let $\mu_n \subset \mathbb{C}$ be the finite group of n-th root of unity, then the continuous group homomorphism $K \to \mu_n$ mapping $k \in K$ to $\chi(k)$, is non-trivial and so $\chi^{-1}(1)$ is an open subgroup different from K, so K is non-connected. $\square$

In the proof of the next theorem we shall need the following path lifting lemma.

Lemma 4.1.8 *Suppose that $\Lambda \subseteq G$ is a discrete subgroup of the topological group G and let $q : G \to G/\Lambda$ denote the quotient map. Suppose that $\sigma : [0,1] \to G/\Lambda$ is a path, connecting $0 = q(0)$ to $q(x)$ for some given $x \in G$. Then there exists a path $\tilde{\sigma} : [0,1] \to G$ connecting 0 to x in G such that $\sigma = q \circ \tilde{\sigma}$.*

Proof First choose an open unit neighborhood U in G such $U \cap \Lambda = \emptyset$. Then $V = q(U)$ is an open neighborhood of 0 in G/Λ and for each $y \in G$ the map $q : yU \to q(y)V$ is a homeomorphism. Let $\sigma : [0,1] \to G/\Lambda$ be as in the lemma. Then there exists a partition $0 = t_0 < t_1 < \cdots < t_l = 1$ of $[0,1]$ such that $\sigma([t_i, t_{i+1}]) \subseteq \sigma(t_i)V$ for every $0 \le i \le l - 1$. We show by induction on $j \in \{0, \dots, l\}$ that there exists a continuous path $\tilde{\sigma} : [0, t_j] \to G/\Lambda$ with $q \circ \tilde{\sigma} = \sigma$ on $[0, t_j]$. This is clear for $j = 0$, so assume it is true for $0 \le j < l$. Then $\sigma([t_j, t_{j+1}]) \subseteq \sigma(t_j)V = q(\tilde{\sigma}(t_j)U)$. Since $q : \tilde{\sigma}(t_j)U \to \sigma(t_j)V$ is a homeomorphism, we find a continuous map $\tilde{\sigma}_j : [t_j, t_{j+1}] \to \tilde{\sigma}(t_j)U$ such that $q \circ \tilde{\sigma}_j = \sigma$ on $[t_j, t_{j+1}]$ and $\tilde{\sigma}_j(t_j) = \tilde{\sigma}(t_j)$. Glueing both maps at t_j gives the desired path $\tilde{\sigma} : [0, t_{j+1}] \to G$. $\square$

Theorem 4.1.9 *Let the LCA-group K be compact, path-connected and second countable. Then K is isomorphic to a product of countably many circle groups, i.e., $K \cong \prod_{i \in I} T_i$, where $T_i \cong \mathbb{T}$ for every $i \in I$ and the index set I is countable.*

Remark The restriction of second countability in this theorem is required, as Shelah has shown in 1974 that the question, whether a path-connected compact LCA group is a product of circle groups, is non-decidable in the context of the usual Zermelo-Frenkel set theory plus axiom of choice. For details see Theorem 8.48 in [HM06].

The proof of this theorem requires the concept of a divisible hull of a torsion-free abelian group which we shall introduce first. For this it is best to write abelian groups *additively*, i.e., use the symbol $+$ to denote the group law. For $a \in A$ and $n \in \mathbb{N}$ we then write

$$na = a + \cdots + a \quad (n\text{-times}).$$

We further write $(-n)a = -na$ for $n \in \mathbb{N}$.

Definition A group G is called *torsion-free* if it has no elements of finite order. If the group is abelian and written additively, this means that $na = 0$ implies $a = 0$ for $n \in \mathbb{N}$ and $a \in A$.

An abelian group A is called *divisible*, if for every $a \in A$ and every $n \in \mathbb{N}$ there exists $b \in A$ such that $a = nb$. The additive group of rationals, $(\mathbb{Q}, +)$, is divisible, the group $(\mathbb{Z}, +)$ is not.

Lemma 4.1.10 *Let the abelian group A be divisible and torsion-free. Then A is a* $\mathbb{Q}$*-vector space. More precisely, there exists a unique map* $\mathbb{Q} \times A \to A$ *making A a vector space over the field of rationals.*

Proof Let A be divisible and torsion-free. We first show that for $a \in A$ and $n \in \mathbb{N}$ the element $b \in A$ with $nb = a$ is uniquely determined. For this assume that b' also satisfies $nb' = a$. Then $n(b-b') = a - a = 0$ and so $b - b' = 0$ which means $b = b'$. We then write this element b as $b = \frac{1}{n}a$. We define a map $\mathbb{Q} \times A \to A$ by sending $\left(\frac{k}{n}, a\right)$ to $k\left(\frac{1}{n}a\right)$. The axioms of a $\mathbb{Q}$-vector space are verified in a straightforward manner.

For the uniqueness, assume there is a second map $(r, a) \mapsto r \circ a$ making A a vector space over $\mathbb{Q}$. For $k \in \mathbb{N}$ we get

$$k \circ a = (1 + \cdots + 1) \circ a = a + \cdots + a = ka.$$

Next for $k = -1$ we have $a + (k \circ a) = 1 \circ a + (-1) \circ a = (1-1) \circ a = 0 \circ a = 0$ and so $(-1) \circ a = -a = (-1)a$, so in total we have $k \circ a = ka$ for every $k \in \mathbb{Z}$. For $r = \frac{k}{n} \in \mathbb{Q}$ we finally have

$$n(r \circ a) = n \circ r \circ a = (nr) \circ a = k \circ a = ka = nra,$$

so $r \circ a = ra$ and the uniqueness is proven. $\square$

Lemma 4.1.11 *Let A be a torsion-free abelian group. Then there exists a* $\mathbb{Q}$*-vector space* $A_\mathbb{Q}$*, called the* divisible hull *and an injective group homomorphism* $\phi : A \hookrightarrow A_\mathbb{Q}$ *such that every group homomorphism* $A \to V$ *to a* $\mathbb{Q}$*-vector space V factors uniquely over* ϕ*. The vector space* $A_\mathbb{Q}$ *is generated by the image of A and is unique up to isomorphism.*

For the reader familiar with the notion of tensor products over rings, the group $A_\mathbb{Q}$ can also be defined as the tensor product $A \otimes \mathbb{Q}$ over the ring $\mathbb{Z}$.

Definition We define the *rank* of a torsion-free abelian group A as the dimension of the vector space $A_\mathbb{Q}$. It may be infinite. If the group is finitely generated, say with r generators, then $A \cong \mathbb{Z}^r$ by Theorem I.8.4 of [Lan02]. Therefore the rank is finite and equals the smallest number of generators. On the other hand, there are groups of finite rank which are not finitely generated as for example the group $\mathbb{Q}$ itself has rank one but is not finitely generated as an additive group.

Proof of Lemma 4.1.11 We define $A_\mathbb{Q}$ to be the quotient of $A \times \mathbb{N}$ by the following equivalence relation. We say $(a, m) \sim (b, n)$ if and only if $na = mb$. It is easy to see that this establishes an equivalence relation. We write the class of (a, m) as $\frac{a}{m}$ and we turn $A_\mathbb{Q}$ into an abelian group by setting

$$\frac{a}{m} + \frac{b}{n} = \frac{na + mb}{mn}.$$

The axioms of abelian groups are readily verified, and so is the fact that the map $a \mapsto \frac{a}{1}$ injects A into $A_{\mathbb{Q}}$. We note that $A_{\mathbb{Q}}$ is still torsion-free as $n\frac{a}{m} = 0$ implies $0 = mn\frac{a}{m} = na$ and hence $a = 0$ and so $\frac{a}{m} = 0$. The equation $n\frac{a}{mn} = \frac{a}{m}$ implies that $A_{\mathbb{Q}}$ is divisible, hence a $\mathbb{Q}$-vector space, which is clearly generated by the image of A. Finally, let $\psi : A \rightarrow V$ be a group homomorphism to a $\mathbb{Q}$-vector space V. Define $\psi_{\mathbb{Q}} : A_{\mathbb{Q}} \rightarrow V$ by $\psi_{\mathbb{Q}}\left(\frac{a}{m}\right) = \frac{1}{m}\psi(a)$. Then $\psi_{\mathbb{Q}}$ is the unique $\mathbb{Q}$-linear map such that $\psi = \psi_{\mathbb{Q}} \circ \phi$. $\qquad\square$

Proof of Theorem 4.1.9   Let $D = \widehat{K}$ be the dual group. Then D is discrete and, as K is connected, Lemma 4.1.7 implies that D is *torsion-free*. So D may be considered a subgroup of the $\mathbb{Q}$-vector space $D_{\mathbb{Q}}$ of Lemma 4.1.11.

We claim that D is countable. Recall that the elements of D form an orthonormal basis of the Hilbert space $L^2(K)$ by Corollary 3.4.9. This Hilbert space is separable by Lemma B.4.9, hence any orthornormal basis is countable.

We next show that every finite rank subgroup of the group D is finitely generated. Let F be a finite rank subgroup of D and $L = \widehat{F}$, then the inclusion of F into D dualizes to a surjection $K \twoheadrightarrow L$, so L is compact and path-connected. Let r be the rank of F. Then F is the union of all its finitely generated subgroups, where it suffices to consider those subgroups of full rank. This means that $F \cong \varinjlim F_j$, where $(F_j)_{j \in J}$ is the directed family of all finitely generated subgroups of full rank. Our goal is to show that this limit stops, i.e., that the map $F_j \rightarrow F$ is an isomorphism for some j. By Exercise 3.8, dualizing gives

$$L \cong \varprojlim T_j,$$

where $T_j = \widehat{F_j}$ is isomorphic to the torus group $\mathbb{T}^r$ and each projection in the projective system is surjective. As $\mathbb{T}^r \cong \mathbb{R}^r/\mathbb{Z}^r$ we can fix some index $v \in J$ and an isomorphism $T_v \cong \mathbb{R}^r/\mathbb{Z}^r$. Then for each $j \geq v$ there is a subgroup $\Lambda_j \subset \mathbb{Z}^r$ of full rank and an isomorphism $\psi_j : T_j \rightarrow \mathbb{R}^r/\Lambda_j$ such that the diagram

$$
\begin{array}{ccc}
T_j & \xrightarrow{\;\;\pi_v^j\;\;} & T_v \\[4pt]
{\scriptstyle \psi_j}\Big\downarrow & & \Big\downarrow{\scriptstyle \cong} \\[4pt]
\mathbb{R}^r/\Lambda_j & \xrightarrow[\;\;p_j\;\;]{} & \mathbb{R}^r/\mathbb{Z}^r
\end{array}
$$

commutes, where p_j is the natural projection. Let Λ be the intersection of all Λ_j for $j \geq v$. The group $G = \mathbb{R}^r/\Lambda$ injects to the projective limit and we claim that G actually equals the path-connected group L. For this let $x \in L$ and let p be a path joining it to the neutral element. Lemma 4.1.8 implies that for every projection $\pi_j : L \rightarrow T_j$ with $j \geq v$ the projected path $\pi_j \circ p$ lifts to a path in G, and therefore the whole path lies in G, which implies that $x \in G$. We get $L \cong \mathbb{R}^r/\Lambda$ and as L is compact, Λ has full rank. So the limit stops and hence every finite rank subgroup of D is finitely generated.

According to Exercise 3.7, the assertion of the theorem is equivalent to saying that D is a direct sum of cyclic groups. Hence the theorem will follow from the next lemma.

Lemma 4.1.12 *Let D be a countable torsion-free abelian group such that every finite rank subgroup of D is finitely generated. Then D is a direct sum of cyclic groups.*

Proof The $\mathbb{Q}$-vector space $D_{\mathbb{Q}}$ is generated by D, hence it contains a basis consisting of elements of D. So let $v_1, v_2, \ldots$ be a basis of $D_{\mathbb{Q}}$ with $v_j \in D$ for each j. As an application of the theory of elementary divisors, we shall inductively construct a basis $w_1, w_2, \ldots$ of D_Q such that

$$\mathbb{Q}w_1 \oplus \cdots \oplus \mathbb{Q}w_n = \mathbb{Q}v_1 \oplus \cdots \oplus \mathbb{Q}v_n$$

and

$$(\mathbb{Q}w_1 \oplus \cdots \oplus \mathbb{Q}w_n) \cap D = \mathbb{Z}w_1 \oplus \cdots \oplus \mathbb{Z}w_n$$

holds for every $n \in \mathbb{N}$. To start, let $F = \mathbb{Q}v_1 \cap D$. Then $\mathbb{Q}F = \mathbb{Q}v_1$ is one-dimensional, so F has rank one. By the assumption, F is finitely generated, being of rank one, it has one generator w_1. This concludes the construction of w_1. Now assume $w_1, \ldots, w_n$ have been constructed. The group $G = (\mathbb{Q}w_1 \oplus \cdots \oplus \mathbb{Q}w_n \oplus \mathbb{Q}v_{n+1}) \cap D$ has rank $n + 1$ and is a subgroup of D, hence is isomorphic to $\mathbb{Z}u_1 \oplus \cdots \oplus \mathbb{Z}u_{n+1}$ for some $u_1, \ldots, u_{n+1}$. We claim that the u_j can be chosen such that $u_j = w_j$ for $1 \leq j \leq n$. Using the basis (u_j) this is equivalent to the following. Suppose that $w_1, \ldots, w_n \in \mathbb{Z}^{n+1}$ are linearly independent over $\mathbb{Q}$ and that $(\mathbb{Q}w_1 \oplus \cdots \oplus \mathbb{Q}w_n) \cap \mathbb{Z}^{n+1} = \mathbb{Z}w_1 \oplus \cdots \oplus \mathbb{Z}w_n$, then there exists $w_{n+1} \in \mathbb{Z}^{n+1}$ such that $\mathbb{Z}^{n+1} = \mathbb{Z}w_1 \oplus \cdots \oplus \mathbb{Z}w_{n+1}$. Consider the integral $(n + 1) \times n$ matrix B with columns $w_1, \ldots, w_n$. By Theorem III.7.9 of [Lan02], there exist invertible integral matrices S, T such that the matrix SBT has non-zero entries only on the main diagonal. The property $(\mathbb{Q}w_1 \oplus \cdots \oplus \mathbb{Q}w_n) \cap \mathbb{Z}^{n+1} = \mathbb{Z}w_1 \oplus \cdots \oplus \mathbb{Z}w_n$ prevails for the columns $w_1', \ldots, w_n'$ of the matrix SBT, which implies that each diagonal entry of SBT can be chosen equal to 1, so the $w_1', \ldots, w_n'$ are the first n standard basis vectors. Let w_{n+1}' be the last standard basis vector, then the vector $w_{n+1} = S^{-1}w_{n+1}'$ will do the job. This finishes the construction of the sequence (w_n). From the properties of this sequence it follows that

$$D = \bigoplus_{n=1}^{\infty} \mathbb{Z}w_n$$

and so the Lemma holds. The theorem also follows.   □

4.2   The Structure Theorems

The main results of this section are the following three Structure Theorems.

Theorem 4.2.1 (First Structure Theorem). *Let A be an LCA group. Then there exists $n \in \mathbb{N}_0$ and an LCA group H such that*

(a) *A is isomorphic to $\mathbb{R}^n \times H$.*

(b) *H contains an open compact subgroup K.*

A topological group G is called *compactly generated* if there exists a compact set $V \subset G$ generating G as a group. In this case the set $W = V \cup V^{-1}$ is compact as well and that $G = \bigcup_{n \in \mathbb{N}} W^n$. Note that every connected locally compact group is automatically compactly generated. Indeed, if V is any compact unit-neighborhood in G, then $\langle V \rangle$ is an open and closed subgroup of G and hence contains the connected component G_0. But $G = G_0$ if G is connected. For compactly generated LCA groups we shall prove

Theorem 4.2.2 (Second Structure Theorem). *Let A be a compactly generated LCA group. Then there exist $n, m \in \mathbb{N}_0$ and a compact group K such that A is isomorphic to $\mathbb{R}^n \times \mathbb{Z}^m \times K$.*

For the proof of this theorem we shall use without proof the well known structure theorem for finitely generated abelian groups, as can be found in many text books on algebra. It says that every finitely generated abelian group B is isomorphic to a group of the form $\mathbb{Z}^m \times F$ for some nonnegative integer $m \in \mathbb{N}_0$ and some finite abelian group F. Moreover, recall that every finite abelian group F is known to be a direct product of finite cyclic group.

A third structure theorem, which will be proved in this section, deals with LCA groups which are locally euclidean.

Definition A topological space is called *locally euclidean* of *dimension n* if every point has an open neighborhood homeomorphic to $\mathbb{R}^n$.

Remark The notion of a locally euclidean group is closely related to the notion of an abelian *Lie group*, which is by definition a differentiable manifold with a smooth abelian group structure [War83]. The connection is this: For a given abelian Lie group, the underlying topological group is a locally euclidean group. The other way round, every locally euclidean group with countably generated topology allows for a unique differentiable structure that makes it an abelian Lie group. The latter is a deep theorem, known as the *Montgomery-Zippin Theorem*, see [MZ55]. The proof of the existence part of this theorem in the special case of LCA groups follows from our third structure theorem stated below. In this case, the proof was first obtained by Pontryagin in [Pon34].

Two abelian Lie groups are isomorphic if and only if their underlying LCA groups are isomorphic. This rounds up to saying that the notion of an abelian Lie group essentially coincides with the notion of a locally euclidean LCA-group with countably many connected components.

Examples 4.2.3

- Every discrete abelian group is a locally euclidean LCA-group since it is locally homeomorphic to $\mathbb{R}^0 = \{0\}$.

- $\mathbb{R}^n$ and $\mathbb{T}^n$ are locally euclidean LCA groups of dimension n for every $n \in \mathbb{N}_0$. There is nothing to show for $\mathbb{R}^n$ and for $n = 0$. For $\mathbb{T}^n$ with $n > 0$ consider the set $V = \left(-\frac{1}{2}, \frac{1}{2}\right)^n \subset \mathbb{R}^n$ and the map $\phi : V \to \mathbb{T}^n$ given by

$$\phi(x_1, \ldots, , x_n) = \left(e^{2\pi i x_1}, \ldots, e^{2\pi i x_n}\right).$$

  Then ϕ is a homeomorphism of $V \cong \mathbb{R}^n$ onto a unit neighborhood of $\mathbb{T}^n$.

- If the groups $A_1, \ldots, A_l$ are locally euclidean LCA groups of dimensions $n_1, \ldots, n_l$, then the direct product $A_1 \times \cdots \times A_l$ is a locally euclidean LCA-group of dimension $n_1 + n_2 + \cdots + n_l$.

Theorem 4.2.4 (Third Structure Theorem). *Suppose that A is a locally euclidean LCA-group. Then A is isomorphic to $\mathbb{R}^n \times \mathbb{T}^m \times D$ for some nonnegative integers $n, m \in \mathbb{N}_0$ and some discrete abelian group D.*

Combining Theorems 4.2.2 and 4.2.4 together with the isomorphisms $\widehat{\mathbb{R}^n} \cong \mathbb{R}^n$, $\widehat{\mathbb{Z}^m} \cong \mathbb{T}^m$, $\widehat{\mathbb{T}^l} \cong \mathbb{Z}^l$ and the fact that the dual of a discrete group is compact and the dual of a compact group is discrete, we get the following corollary as consequence of the Pontryagin Duality Theorem.

Corollary 4.2.5 *Let A be an LCA-group. Then the following are equivalent:*

(a) *A is compactly generated.*

(b) *$\widehat{A}$ is locally euclidean.*

Using the easy fact that quotients of compactly generated groups are compactly generated, Theorem 4.2.4 together with the structure theorem for finitely generated abelian groups immediately implies

Corollary 4.2.6 *Let A be a locally compact abelian group. Then the following are equivalent:*

(a) *A is a compactly generated locally euclidean LCA-group.*

(b) *There exist $n, m, l \in \mathbb{N}_0$ and some finite abelian group F such that $A \cong \mathbb{R}^n \times \mathbb{T}^m \times \mathbb{Z}^l \times F$.*

Using Theorem 4.1.9, we derive the following classification of second countable path-connected LCA groups.

Corollary 4.2.7 *Assume that the LCA-group A is second countable and path-connected. Then A is isomorphic to $\mathbb{R}^n \times \prod_{i \in I} T_i$ for some $n \geq 0$, where each T_i is a circle group, i.e., $T_i \cong \mathbb{T}$ and the set I is countable.*

Proof Being path-connected, A is connected, so by Theorem 4.2.2 the group A is isomorphic to $\mathbb{R}^n \times K$ for a compact group K. Then $K \cong A/\mathbb{R}^n$ is path-connected and by Theorem 4.1.9, K is isomorphic to a product of circle groups.   □

The proofs of the structure theorems require several technical lemmas. We start with

Lemma 4.2.8 *Let A and B be abelian topological groups, and let $q : A \to B$ be a continuous and surjective homomorphism. Suppose further that there exists a continuous section $s : B \to A$ for q, i.e., s is a continuous homomorphism such that $q \circ s = \mathrm{id}_B$. Then A is isomorphic to $C \times B$ with $C = \ker(q) \subset A$.*

Proof Just check that the map $\phi : A \to C \times B$ given by $\phi(x) = \left(x \cdot (s \circ q)(x)^{-1}, q(x)\right)$ is a continuous homomorphism with inverse given by $\phi^{-1}(c, b) = c \cdot s(b)$.   □

Example 4.2.9 If A is an abelian topological group such that there exists a surjective continuous homomorphism $q : A \to \mathbb{Z}^l$, then A is isomorphic to $N \times \mathbb{Z}^l$ for $N = \ker(q)$. Indeed, if we choose $x_i \in A$ such that $q(x_i) = e_i$, where e_i denotes the ith unit vector in $\mathbb{Z}^l$, then $s : \mathbb{Z}^l \to A$ given by $s(n_1, \dots, n_l) = x_1^{n_1} \cdots x_l^{n_l}$ is a (continuous) section for q and the lemma applies.

In the course of this section we shall need other criteria that imply that a given group can be decomposed as a direct product of two subgroups. The main ingredient for this is the Open Mapping Theorem for σ-compact locally compact groups that we will formulate below. Recall that a topological space X is called σ-*compact* if it is the union of countably many compact subsets of X. Every compactly generated group G is σ-compact, since then $G = \bigcup_{n \in \mathbb{Z}} V^n$ for some compact subset $V \subset G$. Since every connected locally compact group is compactly generated (see the discussion preceding Theorem 4.2.2) it follows that such groups are also σ-compact.

Theorem 4.2.10 (Open Mapping Theorem). *Suppose that G and H are locally compact groups such that G is σ-compact. Suppose further that $\phi : G \to H$ is a continuous and surjective homomorphism. Then ϕ is open. In particular, if ϕ is a continuous bijective homomorphism from G to H where G is σ-compact, then ϕ is a topological isomorphism.*

Proof Let 1_G and 1_H be the units in G and H, respectively. It is enough to show that $\phi(U)$ is a neighborhood of 1_H whenever U is a neighborhood of 1_G. To see this, choose a compact neighborhood V of 1_G such that $V = V^{-1}$ and $V^2 \subset U$. Since G is σ-compact, we can choose a countable family $\{K_n : n \in \mathbb{N}\}$ of compact subsets of G such that $G = \bigcup_{n \in \mathbb{N}} K_n$. For each $n \in \mathbb{N}$ we may then find a finite subset $F_n \subset K_n$

such that $K_n \subset \bigcup_{x \in F_n} xV$. Put $F = \bigcup_{n \in \mathbb{N}} F_n$. Then F is a countable subset of G such that $G = \bigcup_{x \in F} xV$. Since ϕ is onto, it follows that $H = \bigcup_{x \in F} \phi(xV)$. Since $\phi(xV)$ is compact, hence closed for every $x \in F$ and since H is a Baire space by Proposition A.9.1, it follows that there exists at least one $x \in F$ such that the interior of $\phi(xV) = \phi(x)\phi(V)$ is nonempty. But this implies that $\phi(V)$ has nonempty interior as well. So choose a nonempty open set W in H with $W \subset \phi(V)$. Then $W^{-1}W$ is an open set in H with

$$1_H \in W^{-1}W \subset \phi(V)^{-1}\phi(V) \subset \phi\left(V^{-1}V\right) = \phi\left(V^2\right) \subset \phi(U). \qquad \square$$

Example 4.2.11 The condition of σ-compactness in the open mapping theorem is indeed necessary, as the following example shows. Let $H = \mathbb{R}$ with the usual topology, and let $G = \mathbb{R}$ with the discrete topology. Then the identity map $\phi : G \to H$ is a surjective continuous homomorphism, which is not open.

Corollary 4.2.12 *Suppose that G is a locally compact group and that N, H are closed subgroups of G such that*

(a) *N and H are σ-compact,*

(b) *$N \cap H = \{1_G\}$ and $N \cdot H = G$,*

(c) *$n \cdot h = h \cdot n$ for all $n \in N, h \in H$.*

Then the map $\phi : N \times H \to G$ given by $\phi(n, h) = n \cdot h$ is an isomorphism of locally compact groups.

Proof Conditions (b) and (c) imply that Φ is a continuous bijective homomorphism and (a) implies that $N \times H$ is σ-compact. The result then follows from the Open Mapping Theorem. $\qquad \square$

Example 4.2.13 The condition of σ-compactness in the corollary is necessary. As an example let $G = \mathbb{R} \times \mathbb{R}_d$, where $\mathbb{R}_d$ denotes the real line equipped with the discrete topology. Let $H = \mathbb{R}(1, 1)$ and $N = \mathbb{R}(1, -1)$. As groups, one has $G \cong H \times N$, but as H and N both have the discrete topology, the underlying map form G to $H \times N$ is not continuous.

Remark Recall that an abelian group D is called *divisible* if for every $x \in D$ and for every $n \in \mathbb{N}$ there exists an element $y \in D$ such that $y^n = x$. It is easy to check that $\mathbb{R}^r \times \mathbb{T}^s$ is divisible for all $r, s \in \mathbb{N}_0$, but $\mathbb{Z}^l$ is certainly not divisible if $l > 0$. Divisible groups have the following remarkable extension property for homomorphisms to D:

Lemma 4.2.14 *Let B be a subgroup of the abelian group A, and let $\psi : B \to D$ be a homomorphism to a divisible group D. Then there exists a homomorphism $\tilde{\psi} : A \to D$ with $\tilde{\psi}|_B = \psi$.*

Proof This is a straightforward application of Zorn's Lemma as soon as we can show that the lemma applies to the case where A is generated by $B \cup \{x\}$ for some element $x \in A \setminus B$. If $x^n \notin B$ for every $0 \neq n \in \mathbb{Z}$, we may define $\tilde{\psi}(bx^n) = \psi(b)$ for all $b \in B$, $n \in \mathbb{Z}$. Suppose now that there exists $m \in \mathbb{N}$ with $x^m \in B$ and suppose that m is minimal with this property. Since D is divisible, we can choose an element $d \in D$ with $d^m = \psi(x^m)$. Define $\tilde{\psi}(bx^n) = \psi(b)d^n$. To check that this gives a well-defined homomorphism we have to show that if $bx^n = b'x^l$, for some $b, b' \in B$ and $n, l \in \mathbb{Z}$, then $\psi(b)d^n = \psi(b')d^l$. Assume that $n > l$. Since $x^{n-l} = b^{-1}b' \in B$ there exists some $q \in \mathbb{N}$ with $n - l = qm$. Then $b^{-1}b' = (x^m)^q$, which implies that $\psi\left(b^{-1}b'\right) = \psi(x^m)^q = (d^m)^q = d^{n-l}$. But this is equivalent to $\psi(b)d^n = \psi(b')d^l$.   □

Lemma 4.2.15 *Let A be an LCA group and let B and D be closed subgroups of A such that*

- *B and D are σ-compact.*
- *$B \cdot D$ is open in A and $B \cap D = \{1\}$.*
- *D is divisible.*

Then there exists a closed subgroup C of A with $B \subset C$, B is open in C, such that $\phi : C \times D \to A$, $\phi(c, d) = c \cdot d$ is an isomorphism of locally compact groups.

Proof Note first that $\phi : B \times D \to B \cdot D$, $\phi(b, d) = bd$ is an isomorphism of LCA groups by Corollary 4.2.12. Let $\psi : B \cdot D \to D$ be the projection onto the second factor. Since D is divisible, it extends to some homomorphism $\tilde{\psi} : A \to D$. Since $B \cdot D$ is open in A, this extension is automatically continuous (since it is continuous at the unit of A). Since the inclusion $\iota : D \to A$ is a section for $\tilde{\psi}$ it follows from Lemma 4.2.8 that $A \cong \ker(\tilde{\psi}) \times D$. Put $C = \ker(\tilde{\psi})$ and observe that $B = C \cap (B \cdot D)$. Hence, since $B \cdot D$ is open in A we have B open in C.   □

Example 4.2.16 As a special case of the above corollary it follows that every divisible open subgroup D of an abelian topological group A splits off as a direct factor of A. To be more precise, there exists a discrete subgroup C of A such that $A \cong C \times D$ via the obvious map.

In this special case one can omit the conditions that A is locally compact and D is σ-compact, since the only place where these conditions were used in the above lemma was to ensure that $B \times D \cong B \cdot D$ as topological groups, which is clear in case of $B = \{1\}$.

Definition We say that a topological group A is *topologically generated* by some subset $U \subset A$ if the group $\langle U \rangle$ generated by U is dense in A. In particular, A is topologically generated by a single element $x \in A$ if $\{x^n : n \in \mathbb{Z}\}$ is dense in A. This is clearly equivalent to the existence of a homomorphism $\phi : \mathbb{Z} \to A$ with dense image and $\phi(1) = x$. A group which is topologically generated by a single element x is called monothetic (see also Exercise 3.22).

Lemma 4.2.17 (a) *Let A be an LCA-group and let $\phi : A \to \mathbb{T}$ be an injective continuous group homomorphism. Then ϕ is either an isomorphism or A is discrete.*

(b) *Every monothetic LCA group A is either compact or topologically isomorphic to $\mathbb{Z}$.*

Proof (a) As a first step we assume A is totally disconnected and show that A is discrete. By Theorem 4.1.6, A contains a compact open subgroup K. Then $\phi(K)$ is a compact subgroup of $\mathbb{T}$, hence finite or equal to $\mathbb{T}$. If $\phi(K)$ equals $\mathbb{T}$, then the open mapping theorem says that ϕ maps K isomorphically to $\mathbb{T}$, contradicting the fact that A is totally disconnected. Hence K is finite, so K is a finite unit-neighborhood. Being Hausdorff, A is discrete.

Next we assume that ϕ is not surjective and show that A is discrete. If $x \in \mathbb{T} \setminus \phi(A)$ and $y \in \mathbb{T}$ with $y^k = x$ for some $k \in \mathbb{Z}$, then $y \notin \phi(A)$ as well. Therefore, there exists a sequence $x_k^{\pm 1} \notin \phi(A)$, tending to 1. Hence the connected component of $\phi(A)$ is trivial, and the same holds for A, i.e., A is totally disconnected, hence discrete.

Finally suppose ϕ is surjective and that A is not totally disconnected, i.e., $A_0 \neq \{1\}$, where A_0 is the connected component of A. If $A_0 \neq A$, then the second observation, applied to A_0, shows that A_0 is discrete, hence trivial, contradicting our assumption. Therefore A is connected, hence σ-compact. By The Open Mapping Theorem 4.2.10, ϕ is an isomorphism.

Now for (b) let A be as in the assumption, then there exists a group homomorphism $\psi : \mathbb{Z} \to A$ with dense image. The dual $\hat{\psi} : \widehat{A} \to \widehat{\mathbb{Z}} \cong \mathbb{T}$ must therefore be injective, hence by part (a) the group $\widehat{A}$ is either discrete or isomorphic to $\mathbb{T}$ and so A is either compact or isomorphic to $\mathbb{Z}$. $\square$

Lemma 4.2.18 *Suppose that A is an LCA group such that there exists a compact unit-neighborhood V and $x_1, \dots, x_l \in A$ with*

$$A = V \cdot \overline{\langle x_1, \dots, x_l \rangle}.$$

Then either A is compact or there exists an $i \in \{1, \dots, l\}$ with $\langle x_i \rangle = \{x_i^n : n \in \mathbb{Z}\}$ closed in A and isomorphic to $\mathbb{Z}$.

Proof For each $i \in \{1, \dots, l\}$ it follows from Lemma 4.2.17 that $\langle x_i \rangle$ is closed and isomorphic to $\mathbb{Z}$ or that $\overline{\langle x_i \rangle}$ is compact. If $\overline{\langle x_i \rangle}$ is compact for every $1 \leq i \leq n$, it follows that

$$\overline{\langle x_1, \dots, x_l \rangle} = \overline{\langle x_1 \rangle} \cdots \overline{\langle x_l \rangle},$$

since products of compact sets are compact and hence closed. It then follows that $A = V \cdot \overline{\langle x_1, \dots, x_l \rangle}$ is compact, too. $\square$

Lemma 4.2.19 *Let A be a compactly generated LCA group. Then there exists a closed subgroup L of A such that $L \cong \mathbb{Z}^l$ for some $l \in \mathbb{N}_0$, and A/L is compact.*

Proof Choose a compact unit-neighborhood $V = V^{-1}$ such that $A = \langle V \rangle = \bigcup_{n \in \mathbb{N}} V^n$. Since V^2 is compact, we find $x_1, \dots, x_m \in A$ such that $V^2 \subset \bigcup_{i=1}^{m} V x_i$.

We first claim that $A = V \cdot H$ for $H = \langle x_1, \dots, x_m \rangle$. To see this, we show by induction that $V^n \subset V \cdot H$ for every $n \in \mathbb{N}$. By definition of H, this is clear for $n = 2$, and if it is known for $n \in \mathbb{N}$, then

$$V^{n+1} = V \cdot V^n \subset V \cdot (V \cdot H) = V^2 \cdot H \subset (V \cdot H) \cdot H = V \cdot H.$$

By Lemma 4.2.18 either A is compact and the proof is finished, or there is j such that $T = \langle x_j \rangle$ is discrete and infinite. We assume $j = m$. The LCA-group $\bar{A} = A/T$ is compactly generated and with $\bar{V} = VT/T$ we have $\bar{A} = \bar{V} \langle x_1, \dots, x_{m-1} \rangle$. By induction on m we can assume that there is a subgroup $\bar{L}$ in $\bar{A}$ with $\bar{L} \cong \mathbb{Z}^l$ and $\bar{A}/\bar{L}$ is compact. Let $\tau_1, \dots, \tau_l \in A$ be any pre-images of the generators of $\bar{L}$. We claim that the group $L = \langle \tau_1, \dots, \tau_l, x_m \rangle$ is discrete and isomorphic to $\mathbb{Z}^{l+1}$. Only discreteness is non-trivial. Let $\bar{V}$ be a unit-neighborhood in $\bar{A}$ with $\bar{V} \cap \bar{L} = \{1\}$ and let V be its pre-image in A. Further let $U \subset A$ be a unit-neighborhood in A with $U \cap \langle x_m \rangle = \{1\}$. Set $W = V \cap U$, then one has $W \cap L = \{1\}$. $\square$

Definition We say that two locally compact groups G and G' are *locally isomorphic* if there exist open unit neighborhoods V and V' in G and G', respectively, together with a homeomorphism $\phi : V \to V'$ such that $\phi(xy) = \phi(x)\phi(y)$ and $\phi(x^{-1}) = \phi(x)^{-1}$ for all $x, y \in V$ such that $xy \in V$ (resp. $x^{-1} \in V$).

Lemma 4.2.20 *Suppose that A is a connected LCA-group locally isomorphic to $\mathbb{R}^n$. Then A is isomorphic to $\mathbb{R}^n/L$ for some discrete subgroup L of $\mathbb{R}^n$. If, in addition, A does not contain any infinite compact subgroup, then A is isomorphic to $\mathbb{R}^n$.*

Proof Assume that $A \neq \{e\}$, i.e., $n > 0$. Then there exists a neighborhood V of $\{e\}$ in A, an $\varepsilon > 0$, and a homeomorphism $\phi : U_\varepsilon(0) := \{x \in \mathbb{R}^n : \|x\|_2 < \varepsilon\} \to V$ such that $\phi(x + y) = \phi(x)\phi(y)$ for all $x, y \in U_\varepsilon(0)$ such that $x + y \in U_\varepsilon(0)$. Define $\Phi : \mathbb{R}^n \to A$ as follows: given $x \in \mathbb{R}^n$ choose $m \in \mathbb{N}$ such that $\|\frac{1}{m}x\|_2 < \varepsilon$ and then put $\Phi(x) = \phi(\frac{1}{m}x)^m$. Φ is well-defined, because if $m, k \in \mathbb{N}$ such that $\|\frac{1}{m}x\|_2, \|\frac{1}{k}x\|_2 < \varepsilon$, then we also have $\|\frac{1}{mk}x\|_2 < \varepsilon$ and it follows from the local additivity of ϕ that

$$\phi\left(\frac{1}{m}x\right)^m = \phi\left(k \cdot \frac{1}{km}x\right)^m = \phi\left(\frac{1}{km}x\right)^{km} = \phi\left(m\frac{1}{km}x\right)^k = \phi\left(\frac{1}{k}x\right)^k.$$

A similar computation shows that the group homomorphism Φ is continuous, since it is continuous at 0. As A is connected, we have $A = \bigcup_{m=1}^{\infty} V^m$, and since $\Phi(\mathbb{R}^n) \supseteq \phi(U) = V$, it follows that Φ is onto. The space $\mathbb{R}^n$ is σ-compact, so it follows from the

Open Mapping Theorem that Φ is open, i.e., that $A \cong \mathbb{R}^n/L$ with $L = \ker\Phi \subset \mathbb{R}^n$. L is discrete since $L \cap U_\varepsilon(0) = \{0\}$.

If A does not contain nontrivial infinite compact subgroups, then Φ is also injective, because if there would exist a $0 \neq x \in \mathbb{R}^n$ with $\Phi(x) = e$, then $\Phi(\mathbb{R} \cdot x) = \Phi([0,1] \cdot x)$ would be a nontrivial infinite compact subgroup of A. $\square$

Lemma 4.2.21 *Suppose that A is a locally euclidean LCA-group. Then the connected component A_0 of A is open in A.*

Proof This follows from Lemma 4.1.3. $\square$

Definition If A, B are LCA groups and if $\phi : A \to B$ is any continuous group homomorphism, then there exists a *dual homomorphism* $\widehat{\phi} : \widehat{B} \to \widehat{A}$ given by $\widehat{\phi}(\chi) = \chi \circ \phi$. Recall from Corollary 3.6.2 (see also Exercise 3.12) that any short exact sequence of locally compact abelian groups

$$0 \to A \xrightarrow{\iota} B \xrightarrow{q} C \to 0$$

(where we assume that ι is a topological embedding and q is open) dualizes to give a short exact sequence

$$0 \to \widehat{C} \xrightarrow{\widehat{q}} \widehat{B} \xrightarrow{\widehat{\iota}} \widehat{A} \to 0.$$

We shall freely use this fact in the following lemma, which gives the proof of the general structure theorem in the special case where A has co-compact connected component A_0.

Lemma 4.2.22 *Suppose that A is a non-compact LCA group such that A/A_0 is compact. Then $A \cong \mathbb{R}^n \times K$ for some $n \in \mathbb{N}$ and some compact group K.*

Proof Let $q : A \to A/A_0$ denote the quotient map and choose a compact symmetric unit-neighborhood V such that $q(V) = A/A_0$. Then A coincides with the subgroup $\langle V \rangle = \bigcup_{n \in \mathbb{N}} V^n$ generated by V because the open subgroup $\langle V \rangle$ of A contains the connected component A_0 and $q(\langle V \rangle) \supset q(V) = A/A_0$. By Lemma 4.2.19 there exists a closed subgroup $L \subset A$ isomorphic to $\mathbb{Z}^n$ for some $n \in \mathbb{N}$ such that $C = A/L$ is compact (we have $n > 0$ since A is not compact). We therefore get a short exact sequence

$$0 \to \mathbb{Z}^n \to A \to C \to 0,$$

which dualizes to a short exact sequence

$$0 \to \widehat{C} \to \widehat{A} \to \mathbb{T}^n \to 0,$$

since $\widehat{\mathbb{Z}^n} \cong \mathbb{T}^n$. Since C is compact, $\widehat{C}$ is a discrete subgroup of $\widehat{A}$. This implies that the map $\widehat{A} \to \mathbb{T}^n$ is locally a homeomorphism, so $\widehat{A}$ is locally isomorphic to $\mathbb{R}^n$. But then the connected component $(\widehat{A})_0$ of $\widehat{A}$ is an open subgroup of $\widehat{A}$.

Note that $\widehat{A}$ (and hence also $(\widehat{A})_0$) does not contain any infinite compact subgroup, because if $E \subset \widehat{A}$ would be such group, then it would follow from dualizing the sequence $0 \to E \to \widehat{A} \to \widehat{A}/E \to 0$ and the duality theorem that there is a short exact sequence $0 \to B \to A \to \widehat{E} \to 0$ with $\widehat{E}$ an infinite discrete group. But this would contradict the assumption that A/A_0 is compact.

We can now apply Lemma 4.2.20 to see that $(\widehat{A})_0$ is isomorphic to $\mathbb{R}^n$ for some $n \in \mathbb{N}_0$. Since $\mathbb{R}^n$ is divisible, it follows from Example 4.2.16 that $\widehat{A}$ is isomorphic to $\mathbb{R}^n \times H$ for some discrete abelian group H. The Pontryagin duality theorem then implies that $A \cong \widehat{\mathbb{R}^n} \times \widehat{H} \cong \mathbb{R}^n \times K$, if K denotes the compact group $\widehat{H}$.   □

Collecting all the previous information, the proof of the general structure theorem is now very easy:

Proof of Theorem 4.2.1   Let A be any LCA group. We have to show that A is isomorphic to some direct product $\mathbb{R}^n \times H$ for some $n \in \mathbb{N}_0$ and some LCA group H such that H contains a compact open subgroup.

To show this, let A_0 be the connected component of A. Then A/A_0 is totally disconnected and therefore contains an open compact subgroup E by Theorem 4.1.6. The inverse image M of E in A is then open in A with M/A_0 compact. Since $M_0 = A_0$ we can apply Lemma 4.2.22 to M to see that $M \cong \mathbb{R}^n \times K$ for some $n \in \mathbb{N}_0$ and some compact group K. Since M is open in A and $\mathbb{R}^n$ is divisible, we can apply Lemma 4.2.15 (to $\mathbb{R}^n$ and K) to see that there exists a closed subgroup H of A containing K as an open subgroup such that $A \cong \mathbb{R}^n \times H$.   □

From this we easily obtain the structure theorem for compactly generated LCA groups:

Proof of Theorem 4.2.2   Let A be a compactly generated LCA group. By the general structure theorem we know that $A \cong \mathbb{R}^n \times H$ for some $n \in \mathbb{N}_0$ and some LCA group H such that H has a compact open subgroup C. Since quotients of compactly generated groups are compactly generated it follows that H/C is a finitely generated discrete group, and hence H/C is isomorphic to $\mathbb{Z}^l \times F$ for some $l \in \mathbb{N}_0$ and some finite group F. Let K be the inverse image of F in H under the quotient map. Then K is compact and it follows then from Example 4.2.9 that $H \cong \mathbb{Z}^l \times K$. Thus $A \cong \mathbb{R}^n \times \mathbb{Z}^l \times K$.   □

We finally want to prove the structure theorem for locally euclidean LCA groups. For this we need

Lemma 4.2.23 *Let L be a discrete subgroup of $\mathbb{R}^n$. Then $L \cong \mathbb{Z}^l$ for some $0 \le l \le n$ and $\mathbb{R}^n/L \cong \mathbb{R}^{n-l} \times \mathbb{T}^l$.*

Proof We first show that $L \cong \mathbb{Z}^l$ for some $0 \le l \le n$. We give the proof by induction on n. The lemma is clearly true for $n = 0$ and also for $L = \{0\}$ (with $l = 0$). So suppose that $n > 0$ and $L \neq \{0\}$. Since L is discrete, we can find some x_1 in L such that $\|x_1\|_2$ is minimal among all elements $0 \neq x \in L$. Put $R = \mathbb{R} \cdot x_1 \cong \mathbb{R}$. Then $R \cap L = \mathbb{Z} \cdot x_1 \cong \mathbb{Z}$ because if there were an element $y \in (R \cap L) \setminus \mathbb{Z} \cdot x_1$, then there would exist (after passing to $-x_1$ if necessary) an element $n \in \mathbb{N}_0$ with $n x_1 < y < (n+1) x_1$ and then $y - n x_1 \in L \setminus \{0\}$ with $\|y - n \cdot x_1\|_2 < \|x_1\|_2$, which would contradict the minimality condition for x_1.

Let $\varepsilon = \frac{1}{2} \|x_1\|_2$, and let $q : \mathbb{R}^n \to \mathbb{R}^n / R$ denote the quotient map. We claim that $q(U_\varepsilon(0)) \cap q(L) = \{0\}$. To see this, suppose that there exists $x \in \mathbb{R}^n$ with $\|x\|_2 < \frac{1}{2} \|x_1\|_2$ such that $0 \neq q(x) \in q(L)$. Then there exists a $t \in \mathbb{R}$ with $x + t \cdot x_1 \in L$. If $x + t \cdot x_1 = 0$ we have $x \in R$ and $q(x) = 0$, which is impossible. If $x + t \cdot x_1 \neq 0$ the minimality condition on x_1 implies that $\|x_1\|_2 \leq \|x + t \cdot x_1\|_2 < (\frac{1}{2} + |t|) \|x_1\|_2$, from which it follows that $|t| > \frac{1}{2}$. Choose $n \in \mathbb{Z}$ with $|t + n| < \frac{1}{2}$. Then $x + (t + n) \cdot x_1$ is also in L, which then implies that $x \in R$ or that $|t + n| > \frac{1}{2}$. Both are impossible.

Since $q(U_\varepsilon(0)) \cap q(L) = \{0\}$ it follows that $q(L)$ is a discrete subgroup of $\mathbb{R}^n / R \cong \mathbb{R}^{n-1}$ and by the induction assumption we know that $q(L) \cong \mathbb{Z}^k$ for some $0 \leq k \leq n - 1$. Since $R \cap L = \mathbb{Z} \cdot x_1 \cong \mathbb{Z}$ we obtain a short exact sequence

$$\{0\} \to \mathbb{Z} \to L \to \mathbb{Z}^k \to \{0\}$$

of abelian groups. But then it follow from Example 4.2.9 that $L \cong \mathbb{Z}^{k+1}$.

If $l > 0$ let $\{x_1, \dots, x_l\}$ be a (minimal) set of generators for L. Then $\{x_1, \dots, x_l\}$ is a set of linearly independent vectors in $\mathbb{R}^m$, since otherwise $L \cong \mathbb{Z}^l$ could be realized as a discrete subgroup of some $\mathbb{R}^k$ with $k < l$, which would contradict the first part of the proof. Choose vectors $y_1, \dots, y_m \in \mathbb{R}^n$ with $m = n - l$ such that $\{x_1, \dots, x_l, y_1, \dots, y_m\}$ is a linear basis of $\mathbb{R}^n$. Then one easily checks that

$$\mathbb{R}^n / L \cong \mathbb{R} x_1 / \mathbb{Z} x_1 \times \cdots \times \mathbb{R} x_l / \mathbb{Z} x_l \times \mathbb{R} y_1 \times \cdots \times \mathbb{R} y_m \cong \mathbb{T}^l \times \mathbb{R}^m. \qquad \square$$

For the proof of the structure theorem for locally euclidean LCA groups we need the following lemma.

Lemma 4.2.24 *Suppose that K is a compact normal subgroup of the locally compact group G such that G/K is compactly generated. Then G is also compactly generated.*

Proof Let $C \subset G/K$ be a compact generating set for G/K. By Remark 1.5.2 we find a compact set $L \subset G$ with $\pi(L) = C$, where $\pi : G \to G/K$ denotes the quotient map. But then $L \cdot K$ is a compact generating set for G. $\qquad \square$

Proof of Theorem 4.2.4 We first reduce the proof to the case of a connected group. So assume we have shown the theorem for connected groups and let A be an arbitrary locally euclidean LCA-group. Let A_0 denote the connected component and let D denote the quotient group A/A_0, which by Lemma 4.1.3 is discrete. The exact sequence $1 \to A_0 \to A \to D \to 1$ dualizes to $1 \to L \to \widehat{A} \to \widehat{A_0} \to 1$, where L is the compact group $\widehat{D}$. If we can show that $A_0 \cong \mathbb{R}^n \times \mathbb{T}^l$ it follows that $\widehat{A_0} \cong \mathbb{R}^n \times \mathbb{Z}^l$ is compactly generated, and as L is compact, it follows that A is compactly generated, too. Hence, by the Second Structure Theorem the group $\widehat{A}$ is isomorphic to $\mathbb{R}^n \times \mathbb{Z}^m \times K$, so that A is isomorphic to $\mathbb{R}^n \times \mathbb{T}^m \times \widehat{K}$. As $\widehat{K}$ is discrete, the general version of the Third Structure Theorem follows from the version for connected groups.

So we now assume the group A to be connected. In this case, the First Structure Theorem tells us that A is isomorphic to $\mathbb{R}^m \times K$ for a compact group K. The locally euclidean group A is path-connected by Lemma 4.1.7. As $A \cong \mathbb{R}^m \times K$ we have $K \cong A/\mathbb{R}^m$ and therefore the compact group K is path-connected. As $\mathbb{R}^n$ is second countable and A is generated by a set homeomorphic to $\mathbb{R}^n$, the space A is second countable, and so is K. By Theorem 4.1.9 we have $K \cong \prod_{i \in I} T_i$ where $T_i \cong \mathbb{T}$ for each $i \in I$.

So far we have shown that the connected locally euclidean group A is isomorphic to $\mathbb{R}^m \times \prod_{i \in I} \mathbb{T}$. It only remains to show that the product is finite. For this we need the concept of contractible spaces.

Definition Let X be a topological space. A subset C of X is called *contractible in X* or *X-contractible* if there exists a point $x_0 \in X$ and a *zero-homotopy*, i.e., a continuous map $h : [0, 1] \times C \to X$ such that

$$h(0, x) = x_0 \qquad \text{and} \qquad h(1, x) = x$$

for every $x \in C$. If C is contractible in X, then every subset of C is contractible in X.

Examples 4.2.25

- The space $X = \mathbb{R}^n$ is contractible in itself, as the map

$$h(t, x) = tx$$

  is a zero-homotopy.

- Complex analysis tells us that the circle group $\mathbb{T}$ is not contractible in itself, as otherwise the path $\gamma(t) = e^{2\pi i t}$ would be homotopic to a constant path, contradicting the fact that $\int_\gamma \frac{1}{z} dz = 2\pi i$ and the homotopy-invariance of path integrals.

- We can elaborate the last example a bit further. Let Y be an arbitrary topological space and let $X = Y \times \mathbb{T}$. Then for a given point $y_0 \in Y$ the set $\{y_0\} \times \mathbb{T}$ is not contractible in X, as any zero-homotopy would, after projection to $\mathbb{T}$, give a zero-homotopy of $\mathbb{T}$.

Now the group A possesses a unit-neighbourhood U which is homeomorphic to $\mathbb{R}^n$, hence contractible in A. Assume that the product in $A \cong \mathbb{R}^m \times \prod_{i \in I} \mathbb{T}$ is infinite, then every unit-neighborhood contains a non-empty set of the form $V \times \prod_{i \in E} V_i \times \prod_{i \notin E} \mathbb{T}$ for some finite set $E \subset I$. Therefore, U contains a subset of the form $\{y_0\} \times \mathbb{T}$ in A, which cannot be contractible, a contradiction. Therefore the product must be finite. $\qquad\square$

Remark 4.2.26 The proof of the third structure theorem for locally euclidean groups becomes much easier under the stronger assumption that the LCA-group A is locally isomorphic to $\mathbb{R}^n$. In this case the connected component A_0 is open in A and locally isomorphic to $\mathbb{R}^n$. Thus, by Lemma 4.2.20 and Lemma 4.2.23, A_0 is isomorphic to $\mathbb{R}^k \times \mathbb{T}^l$ for some $k, l \in \mathbb{N}_0$ with $k + l = n$. Hence A_0 is divisible and it follows from Lemma 4.2.23 (applied to $B = \{e\}$ and $D = A_0$) that $A \cong \mathbb{R}^k \times \mathbb{T}^l \times D$ with $D \cong A/A_0$ discrete. In particular, the proof does not use the quite intricate Theorem 4.1.9.

4.3 Exercises

Exercise 4.1 (a) Show that any interval in $\mathbb{R}$ is connected.

(b) Let A be a connected subset of the topological space X. Show that the closure $\overline{A}$ is connected, too.

(c) Let $f : X \to Y$ be a continuous map. Show that if $A \subset X$ is connected, then so is $f(A) \subset Y$.

Exercise 4.2 Recall that the Cantor set $C \subset [0, 1]$ is defined as the complement $[0, 1] \smallsetminus U$, where U is the union of all intervals $\left(\frac{3n+1}{3^l}, \frac{3n+2}{3^l}\right)$ with $l \in \mathbb{N}$ and $n \in \mathbb{N}_0$ with $0 \le n < 3^{l-1}$. Show that C is totally disconnected.

Exercise 4.3 Let X_i be a topological space for every i in the nonempty index set I. Show that the direct product $\prod_{i \in I} X_i$ is connected (resp. totally disconnected) if and only if all X_i are connected (resp. totally disconnected).

Exercise 4.4 Let G be a locally compact group. Show that the connected component G_0 of G is the intersection of all open subgroups of G.

Exercise 4.5 Show that every projective limit of finite groups is totally disconnected (compare with Exercise 3.8). Give concrete examples of infinite compact abelian groups that are totally disconnected.

Exercise 4.6 Suppose that G is a compact totally disconnected group.

(a) Show that G has a unit-neighborhood base consisting of open normal subgroups of G.

(b) Show that G is a projective limit of finite groups.

Exercise 4.7 Let G be a topological group, and let H be a closed subgroup of G such that H and G/H are connected. Show that also G is connected. Show that the groups $U(n)$ and $SO(n)$ are connected for every $n \in \mathbb{N}$. Is $O(n)$ connected?

Exercise 4.8 Show that an LCA-group, which is countable, must be discrete.

Exercise 4.9 (a) Describe all closed subgroups of $\mathbb{R}^n$.

(b) Show that every closed subgroup and every quotient of a locally euclidean LCA-group is again a locally euclidean LCA-group.

(c) Show that every closed subgroup and every quotient of a compactly generated LCA-group is again a compactly generated LCA-group.

Chapter 5
Operators on Hilbert Spaces

In this chapter, we will apply the results of Chap. 2 on C^*-algebras to operators on Hilbert spaces. In particular, we will discuss the continuous functional calculus for normal bounded operators on Hilbert space, which turns out to be a powerful tool.

The space $\mathcal{B}(H)$ of all bounded linear operators on a Hilbert space H is a Banach algebra with the operator norm (Example 2.1.1), and, as we have seen in Example 2.6.1, even a C^*-algebra. We will write

$$\sigma(T) = \sigma_{\mathcal{B}(H)}(T)$$

for the spectrum of T with respect to the C^*-algebra $\mathcal{B}(H)$ and call it simply the *spectrum of the operator T*.

5.1 Functional Calculus

Let H be a Hilbert space, and let T be a bounded *normal operator* on H, this means that T commutes with its adjoint T^*, i.e., T is normal as an element of the C^*-algebra $\mathcal{B}(H)$. We then can apply the results of Sect. 2.7, which for any continuous function f on the spectrum $\sigma(T)$ give a unique element $f(T)$ of $\mathcal{B}(H)$ that commutes with T and satisfies

$$\widehat{f(T)} = f \circ \widehat{T},$$

where the hat means the Gelfand transform with respect to the unital C^*-algebra generated by T. Recall that by Lemma 2.7.2 the spectrum of a normal operator T does not depend on the C^*-algebra. The map from $C(\sigma(T))$ to $\mathcal{B}(H)$ mapping f to $f(T)$ is the *continuous functional calculus*. In the next proposition, we summarize some important properties.

Proposition 5.1.1. *Let T be a normal bounded operator on the Hilbert space H and let $\mathcal{A} = C^*(T, 1)$ be the unital C^*-algebra generated by T.*

(a) *The map $f \mapsto f(T)$ is a unital isometric C^*-isomorphism from $C(\sigma(T))$ to $\mathcal{A}$, which sends the identity map $\mathrm{Id}_{\sigma(T)}$ to T.*

A. Deitmar, S. Echterhoff, *Principles of Harmonic Analysis*, Universitext,
DOI 10.1007/978-3-319-05792-7_5, © Springer International Publishing Switzerland 2014

(b) *Let $V \subset H$ be a closed subspace stable under T and T^*. Then V is stable under $\mathcal{A}$ and $f(T)|_V = f(T|_V)$.*

(c) *Let V be the kernel of $f(T)$. Then V is stable under T and T^*, and the spectrum of $f(T|_V)$ is contained in the zero-set of f.*

(d) *If $f(z) = \sum_{n=0}^{\infty} a_n z^n$ is a power series that converges for $z = \|T\|$, then $f(T) = \sum_{n=0}^{\infty} a_n T^n$.*

Proof The first assertion is a direct consequence of Theorem 2.7.3.

To show (b), note first that if V is stable under T and T^*, then V is $\mathcal{A}$-stable, since the linear combinations of operators of the form $T^k(T^*)^l$ are dense in $\mathcal{A}$. We therefore get a well defined $*$-homomorphism $\Psi : \mathcal{A} \to \mathcal{B}(V)$ mapping S to $S|_V$. The assertion in (b) is then a consequence of Corollary 2.7.5.

In (c), the space V is stable under T and T^* as these operators commute with $f(T)$. Further, using Corollary 2.7.5, one has

$$f(\sigma(T|_V)) = \sigma(f(T|_V)) = \sigma(f(T)|_V) = \{0\}.$$

Finally, part (d) is contained in Theorem 2.7.3, since convergence of the power series at $\|T\|$ implies uniform convergence on $\sigma(T) \subseteq B_{\|T\|}(0)$. $\square$

An important class of normal operators is formed by the self-adjoint operators, i.e., operators T with $T = T^*$. It is shown in Corollary 2.7.5 that $\sigma(T) \subseteq \mathbb{R}$ for every self-adjoint T. Another class of interesting normal operators consists of the *unitary operators*. These are operators $U \in \mathcal{B}(H)$ satisfying $UU^* = U^*U = 1$. Note that a normal operator $U \in \mathcal{B}(H)$ is unitary if and only if $\sigma(U) \subseteq \mathbb{T}$. This follows from functional calculus, because if U is normal, then $U^*U = 1$ if and only if $\bar{\text{Id}} \cdot \text{Id} = 1$ for $\text{Id} = \text{Id}_{\sigma(U)}$, which is equivalent to $\sigma(U) \subseteq \mathbb{T}$.

Recall the Schwartz space $\mathcal{S}(\mathbb{R})$ consisting of all functions $f : \mathbb{R} \to \mathbb{C}$ such that for any two integers $m, n \geq 0$ the function $x^n f^{(m)}(x)$ is bounded. So a *Schwartz function on $\mathbb{R}$* is a smooth function on $\mathbb{R}$, which, together with all its derivatives, is rapidly decreasing.

For $f \in \mathcal{S}(\mathbb{R})$ the *Fourier inversion formula* says that

$$f(x) = \int_{\mathbb{R}} \hat{f}(y)e^{2\pi i x y}\, dy,$$

where $\hat{f}(y) = \int_{\mathbb{R}} f(x)e^{-2\pi i x y}\, dx$ is the *Fourier transform* (See Exercise 3.14 or [Dei05] Sect. 3.4).

Proposition 5.1.2. *Let T be a self-adjoint bounded operator on the Hilbert space H. Then for every $f \in \mathcal{S}(\mathbb{R})$,*

$$f(T) = \int_{\mathbb{R}} \hat{f}(y)e^{2\pi i y T}\, dy,$$

where the unitary operator $e^{2\pi i y T}$ is defined by the continuous functional calculus and the integral is a vector-valued integral in the Banach space $\mathcal{B}(H)$ as in Sect. B.6.

Proof To see that the operator $e^{2\pi i y T}$ is unitary, we compute

$$\left(e^{2\pi i y T}\right)^* = e^{-2\pi i y T^*} = e^{-2\pi i y T} = \left(e^{2\pi i y T}\right)^{-1}.$$

The Bochner integral exists by Lemma B.6.2 and Proposition B.6.3. Next, let $\Phi :$ $C(\sigma(T)) \to C^*(T, 1), g \mapsto g(T)$ denote the isometric $*$-homomorphism underlying the Functional Calculus for T. The Fourier inversion formula implies that

$$f|_{\sigma(T)} = \int_{\mathbb{R}} \hat{f}(y) e^{2\pi i y \mathrm{Id}_{\sigma(T)}} \, dy.$$

By continuity of Φ we therefore get

$$f(T) = \Phi(f|_{\sigma(T)}) = \Phi\left(\int_{\mathbb{R}} \hat{f}(y) e^{2\pi i y \mathrm{Id}_{\sigma(T)}} \, dy\right)$$

$$= \int_{\mathbb{R}} \hat{f}(y) \Phi\left(e^{2\pi i y \mathrm{Id}_{\sigma(T)}}\right) dy = \int_{\mathbb{R}} \hat{f}(y) e^{2\pi i y T} \, dy,$$

where the last equation follows from Corollary 2.7.5, which implies $\Phi\left(e^{2\pi i y \mathrm{Id}_{\sigma(T)}}\right) = e^{2\pi i y \Phi(\mathrm{Id}_{\sigma(T)})} = e^{2\pi i T}$. □

Definition A self-adjoint operator $T \in \mathcal{B}(H)$ is called *positive* if

$$\langle Tv, v \rangle \geq 0 \quad \forall v \in H.$$

In what follows, we want to use the spectral theorem to compute a positive square root for any positive operator T. For this we need to know that positive operators have positive spectrum.

Theorem 5.1.3 *Let T be a self-adjoint bounded operator on the Hilbert space H. Then the following are equivalent:*

(a) *T is positive.*

(b) *The spectrum $\sigma(T)$ is contained in the interval $[0, \infty)$.*

(c) *There exists an operator $R \in \mathcal{B}(H)$ with $T = R^*R$.*

(d) *There exists a unique positive operator S with $T = S^2$. In this case we write $S = \sqrt{T}$.*

Proof The implications (d) $\Rightarrow$ (c) and (c) $\Rightarrow$ (a) are trivial. So it is enough to show that (a) $\Rightarrow$ (b) and (b) $\Rightarrow$ (d) hold.

For (a) $\Rightarrow$ (b) assume without loss of generality that $\|T\| = 1$. Then $\sigma(T) \subseteq [-1, 1]$ since T is self-adjoint. We show that $T_\mu \overset{\text{def}}{=} T + \mu 1$ is invertible for every $\mu > 0$, which will imply that there are no negative spectral values for T. By assumption we have

$$\|T_\mu v\| \|v\| \geq \langle T_\mu v, v \rangle = \langle Tv, v \rangle + \mu \langle v, v \rangle \geq \mu \|v\|^2,$$

which implies that $\|T_\mu v\| \geq \mu \|v\|$ for every $v \in H$. It follows that T_μ is injective. Since T_μ is self-adjoint we also get $(T_\mu(H))^\perp = \ker T_\mu = \{0\}$, since if $w \in (T_\mu(H))^\perp$, then $0 = \langle T_\mu v, w \rangle = \langle v, T_\mu w \rangle$ for every $v \in H$, which implies that $T_\mu w = 0$. Thus we get $\overline{T_\mu(H)} = H$ and for each $w \in H$ we find a sequence v_n in H with $Tv_n \to w$. Since $\|v_n - v_m\| \leq \frac{1}{\mu}\|T_\mu v_n - T_\mu v_m\|$ for all $n, m \in \mathbb{N}$, it follows that (v_n) is a Cauchy-sequence and hence converges to some $v \in H$. Then $T_\mu v = w$, which shows that T_μ is also surjective. The Open Mapping Theorem C.1.5 implies that T_μ^{-1} is continuous, so T_μ is invertible in $\mathcal{B}(H)$.

Assume finally that (b) holds. Then $t \to \sqrt{t}$ is a continuous function on $\sigma(T)$, and by functional calculus we can build the operator $S = \sqrt{T}$. Since $\sqrt{\cdot}$ is real and positive, it follows from Corollary 2.7.5 that S is self-adjoint, $\sigma(S) \subset [0, \infty)$, and $S^2 = T$. For uniqueness assume that $\tilde{S}$ is another such operator. Then T lies in the commutative C^*-algebra $C^*(\tilde{S}, 1)$. But then $S \in C^*(T, 1) \subseteq C^*(\tilde{S}, 1) \cong C(\sigma(\tilde{S}))$, and the result follows from the fact that a positive real function has a unique positive square root.   □

Definition Let T be a bounded operator on a Hilbert space H. Define the operator $|T|$ by $|T| \overset{\text{def}}{=} \sqrt{T^*T}$, which exists and is well defined by the above theorem.

Proposition 5.1.4 *Let T be a bounded operator on H. Then the norm of $|T|v$ coincides with $\|Tv\|$. There is an isometric operator U from the closure of $\mathrm{Im}(|T|)$ to the closure of $\mathrm{Im}(T)$ such that $T = U|T|$. This decomposition of T is called* polar de-composition. *It is unique in the following sense. If $T = U'P$, where P is self-adjoint and positive, and $U' : \overline{\mathrm{Im}(P)} \to H$ is isometric, then $U' = U$ and $P = |T|$.*

Proof For $v \in H$ the square of the norm $\|Tv\|^2$ equals

$$\langle Tv, Tv \rangle = \langle T^*Tv, v \rangle = \langle |T|^2 v, y \rangle = \langle |T|v, |T|v \rangle,$$

and the latter is $\||T|v\|^2$. For $v \in H$ we define $U(|T|v) = Tv$, then U is a well-defined isometry from $\mathrm{Im}(|T|)$ to $\mathrm{Im}(T)$, which extends to the closure, and satisfies the claim. For the uniqueness let $T = U|T| = U'P$. Extend U to a bounded operator on H by setting $U \equiv 0$ on $\mathrm{Im}(|T|)^\perp$ and do likewise for U'. Then U^*U is the orthogonal projection to $\overline{\mathrm{Im}(|T|)}$ and $(U')^*U'$ is the orthogonal projection to $\mathrm{Im}(P)$, so that $(U')^*U'P = P$. Note $|T| = \sqrt{T^*T} = \sqrt{(U'P)^*U'P} = \sqrt{P^*(U')^*U'P} = \sqrt{P^*P} = \sqrt{P^2} = P$. This also implies $U = U'$.   □

An important application of the functional calculus for operators on Hilbert space is Schur's Lemma, which we shall use quite frequently in the remaining part of this book. We first state

Lemma 5.1.5 *Let H be a Hilbert space, and let T be a bounded normal operator, the spectrum of which consists of a single point $\{\lambda\} \subset \mathbb{C}$. Then $T = \lambda\mathrm{Id}$.*

Proof If $\sigma(T) = \{\lambda\}$, then $\mathrm{Id}_{\sigma(T)} = \lambda 1_{\sigma(T)}$ and therefore $T = \mathrm{Id}_{\sigma(T)}(T) = \lambda \cdot \mathrm{Id}_H$. $\qquad\square$

Theorem 5.1.6 (Schur's Lemma) *Suppose that $A \subseteq \mathcal{B}(H)$ is a self-adjoint set of bounded operators on the Hilbert space H (i.e., $S \in A$ implies $S^* \in A$). Then the following are equivalent:*

(a) *A is topologically irreducible, i.e., if $\{0\} \neq L \subseteq H$ is any A-invariant closed subspace of H then $L = H$.*

(b) *If $T \in \mathcal{B}(H)$ commutes with every $S \in A$, then $T = \mu\mathrm{Id}$ for some $\mu \in \mathbb{C}$.*

Proof Assume first that the second assertion holds. Then, if $\{0\} \neq L \subseteq H$ is any A-invariant closed subspace of H, the orthogonal complement $L^\perp$ is A-invariant as well, for with $v \in L$, $u \in L^\perp$, and $S \in A$ we have

$$\langle v, Su \rangle = \langle \underbrace{S^*v}_{\in L}, u \rangle = 0.$$

So the orthogonal projection $P_L : H \to L$ commutes with A, so P_L must be a multiple of the identity. But this implies that $P_L = \mathrm{Id}$ and $L = H$.

For the converse, assume that (a) holds, and let $T \in \mathcal{B}(H)$ commute with A. Then also T^* commutes with A since A is self-adjoint. Thus, writing $T = \frac{1}{2}(T + T^*) - i\frac{1}{2}(iT - iT^*)$ we may assume without loss of generality that T is self-adjoint and $T \neq 0$. We want to show that the spectrum of T consists of a single point. Note that an operator S, which commutes with T, also commutes with $f(T)$ for every $f \in C(\sigma(T))$. Assume that there are $x, y \in \sigma(T)$ with $x \neq y$. Then there are two functions $f, g \in C(\sigma(T))$ with $f(x) \neq 0 \neq g(y)$ and $f \cdot g = 0$. Then $f(T) \neq 0 \neq g(T)$ and $f(T)g(T) = f \cdot g(T) = 0$. Since $g(T)$ commutes with A, the space $L = \overline{g(T)H}$ is a non-zero A-invariant subspace of H. By (a) we get $L = H$. But then $\{0\} \neq f(T)H = f(T)\overline{g(T)H} \subset \overline{f(T)g(T)H} = \{0\}$, a contradiction. $\qquad\square$

5.2 Compact Operators

An operator T on a Hilbert space H is called a *compact operator* if T maps bounded sets to relatively compact ones. It is clear from the definition that if T is compact and S a bounded operator on H, then ST and TS are compact. The definition can be rephrased as follows. An operator T is compact if and only if for a given bounded sequence $v_j \in H$ the sequence Tv_j has a convergent subsequence. If the v_j lie in a finite dimensional space, then this is true for every bounded operator. So one may restrict to sequences v_j that are linearly independent.

Definition A bounded linear map $F : H \to H$ on a Hilbert space H is said to be a *finite rank operator* if the image $F(H)$ is finite-dimensional.

Proposition 5.2.1 *For a bounded operator T on a Hilbert space H the following are equivalent.*

(a) *T is compact.*

(b) *For every orthonormal sequence e_j the sequence Te_j has a convergent subsequence.*

(c) *There exists a sequence F_n of finite rank operators such that $\|T - F_n\|_{\mathrm{op}}$ tends to zero, as $n \to \infty$.*

Proof The implication (a)$\Rightarrow$(b) is trivial. For (b)$\Rightarrow$(c) let $T : H \to H$ be compact and let $B \subset H$ denote the closed unit ball. Then $\overline{T(B)}$ is compact, hence has a vector v_1 of maximal norm. Next suppose the vectors $v_1, \ldots v_n$ are already constructed and let V_n be their span. Choose a vector v_{n+1} of maximal norm in $\overline{T(B)} \cap V_n^{\perp}$. The vectors $v_1, v_2, \ldots$ are pairwise orthogonal and for their norms we have $\|v_1\| \geq \|v_2\| \geq \cdots$. We claim that the sequence v_n tends to zero. Assume not, then there exists $\delta > 0$ such that $\|v_n\| \geq \delta$ for all n. For $i \neq j$ if follows $\|v_i - v_j\|^2 = \|v_i\|^2 + \|v_j\|^2 \geq 2\delta^2$, hence the sequence has no convergent subsequence, in contradiction to the compactness of $\overline{T(B)}$. So the sequence does tend to zero. Let P_n be the orthogonal projection onto V_n. Then

$$\|T - P_n T\| = \sup_{v \in T(B)} \|v - P_n v\| \leq \|v_n\| \to 0.$$

So with $F_n = P_n T$ the claim follows.

For (c)$\Rightarrow$(a) let v_j be a bounded sequence, and let T be the norm-limit of a sequence of finite rank operators F_n. We can assume $\|v_j\|, \|T\| \leq 1$. Then v_j has a subsequence v_j^1 such that $F_1(v_j^1)$ converges. Next, v_j^1 has a subsequence v_j^2 such that $F_2(v_j^2)$ converges, and so on. Let $w_j = v_j^j$. Then for every $n \in \mathbb{N}$, the sequence $(F_n(w_j))_{j \in \mathbb{N}}$ converges. As T is the uniform limit of the F_n, the sequence Tw_j converges as well. $\square$

Theorem 5.2.2 (Spectral Theorem). *Let T be a compact normal operator on the Hilbert space H. Then there exists a sequence λ_n of non-zero complex numbers, which is either finite or tends to zero, such that one has an orthogonal decomposition*

$$H = \ker(T) \oplus \overline{\bigoplus_n \mathrm{Eig}(T, \lambda_n)}.$$

Each eigenspace $\mathrm{Eig}(T, \lambda_n) = \{v \in H : Tv = \lambda_n v\}$ is finite dimensional, and the eigenspaces are pairwise orthogonal.

Proof We first show that a given compact normal operator $T \neq 0$ has an eigenvalue $\lambda \neq 0$. We show that it suffices to assume that T is self-adjoint. Note that $T =$

$\frac{1}{2}(T + T^*) - \frac{i}{2}(iT + (iT)^*) = T_1 + iT_2$ is a linear combination of two commuting compact self-adjoint operators. If $T_2 = 0$, then T is self-adjoint and we are done. Otherwise, T_2 has a non-zero eigenvalue $\nu \in \mathbb{R} \setminus \{0\}$. The corresponding eigenspace is left stable by T_1, which therefore induces a self-adjoint compact operator on that space, hence has an eigenvalue $\mu \in \mathbb{R}$. Then $\lambda = \mu + i\nu$ is a non-zero eigenvalue of T.

We have to show that a compact self-adjoint operator $T \neq 0$ has an eigenvalue $\lambda \neq 0$.

Lemma 5.2.3 *For a bounded self-adjoint operator T on a Hilbert space H we have*
$\|T\| = \sup\{|\langle Tv, v \rangle| : \|v\| = 1\}.$

Proof Let C be the right hand side. By the Cauchy-Schwarz inequality we have $C \leq \|T\|$. On the other hand, for $v, w \in H$ with $\|v\|, \|w\| \leq 1$ one has

$$C \geq \frac{1}{2} C \left(\|v\|^2 + \|w\|^2 \right) = \frac{1}{4} C \left(\|v + w\|^2 + \|v - w\|^2 \right)$$
$$\geq \frac{1}{4} |\langle T(v + w), v + w \rangle - \langle T(v - w), v - w \rangle|$$
$$= \frac{1}{2} |\langle Tv, w \rangle + \langle Tw, v \rangle| = \frac{1}{2} |\langle Tv, w \rangle + \langle w, Tv \rangle|$$
$$= |\operatorname{Re}\langle Tv, w \rangle|.$$

Replacing v with θv for some $\theta \in \mathbb{C}$ with $|\theta| = 1$ we get $C \geq |\langle Tv, w \rangle|$ for all $\|v\|, \|w\| \leq 1$ and so $\|T\| \leq C$. $\qquad\square$

We continue the proof that a compact self-adjoint operator $T \neq 0$ has an eigenvalue $\lambda \neq 0$. Indeed, we prove that either $\|T\|$ or $-\|T\|$ is an eigenvalue for T. By the lemma there is a sequence $v_j \in H$ with $\|v_j\| = 1$ and $\langle Tv_j, v_j \rangle \to \pm\|T\|$. Replacing T with $-T$ if necessary, we assume $\langle Tv_j, v_j \rangle \to \|T\|$. Since T is compact, there exists a norm-convergent subsequence, i.e., we can assume that $Tv_j \to u$ in norm. Then $\|u\| \leq \|T\|$ and we get

$$0 \leq \|Tv_j - \|T\| v_j\|^2 = \|Tv_j\|^2 - 2\|T\|\langle Tv_j, v_j \rangle + \|T\|^2 \|v_j\|^2$$
$$\to \|u\|^2 - \|T\|^2 \leq 0,$$

which implies that $\|Tv_j - \|T\| v_j\| \to 0$. Thus $v := \lim_j v_j = \frac{1}{\|T\|} u$ exists and $Tv = \lim_j Tv_j = u = \|T\| v$.

We have proven that every compact normal operator T has an eigenvalue $\lambda \neq 0$. Let $U \subset V$ be the closure of the sum of all eigenspaces of T corresponding to non-zero eigenvalues. By Lemma C.3.3 every eigenvector for T is also an eigenvector for T^*, so U is stable under T and T^* and hence the orthogonal complement $U^\perp$ is stable under T and T^* as well. The operator T induces a compact normal operator on $U^\perp$; as this operator cannot have a non-zero eigenvalue, it is zero and $U^\perp$ is the kernel of T. We have shown that H is a direct sum of eigenspaces of T.

It remains to show that every eigenspace for a non-zero eigenvalue is finite dimensional and that the eigenvalues do not accumulate away from zero. For this let f be a continuous function on $\mathbb{C}$ whose zero set is the closed ε-neighborhood $\bar{B}_\varepsilon(\lambda)$ of a given $\lambda \in \mathbb{C}$, where $0 < \varepsilon < |\lambda|$. Let V be the kernel of $f(T)$. By Proposition 5.1.1, the space V is stable under T and T^*, and $\sigma(T|_V) \subset \bar{B}_\varepsilon(\lambda)$. It follows from Functional Calculus that $\|T - \lambda\|_V = \|\mathrm{Id}_{\sigma(T|_V)} - \lambda 1_{\sigma(T|_V)}\|_{\sigma(T|_V)} \le \varepsilon$, which implies that for $v \in V$ one has $\|Tv\| \ge (|\lambda| - \varepsilon)\|v\|$. We want to deduce that V is finite dimensional. Assume the contrary, so there exists an orthonormal sequence $(f_j)_{j \in \mathbb{N}}$ in V. Then $\|f_i - f_j\| = \sqrt{2}$ for $i \ne j$ and so $\|Tf_j - Tf_i\| \ge (|\lambda| - \varepsilon)\sqrt{2}$, which means that no subsequence of (Tf_j) is a Cauchy sequence, hence (Tf_j) does not contain a convergent subsequence, a contradiction to the compactness of T. So V is finite dimensional, hence it is a direct orthogonal sum of T-eigenspaces. It now follows that no spectral values of T can accumulate away from zero, and all spectral values apart from zero are eigenvalues of finite multiplicity. Finally, the fact that the eigenspaces are pairwise orthogonal is in Lemma C.3.3. The theorem is proven. $\qquad \square$

Definition Let T be a compact operator on a Hilbert space H. Then T^*T is a self-adjoint compact operator with positive eigenvalues. The operator $|T| = \sqrt{T^*T}$ also is a compact operator. Let $s_j(T)$ be the family of non-zero eigenvalues of $|T|$ repeated with multiplicities and such that $s_{j+1}(T) \le s_j(T)$ for all j. These $s_j(T)$ are called the *singular values* of T.

Proposition 5.2.4 *Let T be a compact operator.*

(a) *We have $s_1(T) = \|T\|$ and*

$$s_{j+1}(T) = \inf_{v_1,\ldots,v_j \in H} \sup\{\|Tw\| : w \perp v_1,\ldots,v_j, \ \|w\| = 1\},$$

where the vectors $v_1,\ldots,v_j$ are unit eigenvectors for the eigenvalues $s_1(T),\ldots,s_j(T)$, respectively.

(b) *For any bounded operator S on H one has $s_j(ST) \le \|S\|s_j(T)$.*

Proof The formulas in (a) follow from the fact that the s_j are the eigenvalues of the self-adjoint operator $|T|$ and $\|T\| = \||T|\|$. Part (b) is a consequence of (a). We leave the details as an exercise (See Exercise 5.4). $\qquad \square$

5.3 Hilbert-Schmidt and Trace Class

Let $T \in \mathcal{B}(H)$, and let (e_j) be an orthonormal basis of H. The *Hilbert-Schmidt norm* $\|T\|_{\mathrm{HS}}$ of T is defined by

$$\|T\|_{\mathrm{HS}}^2 \stackrel{\mathrm{def}}{=} \sum_j \langle Te_j, Te_j \rangle.$$

This number is ≥ 0 but can be $+\infty$. It does not depend on the choice of the orthonormal basis, as we will prove now. Along the way we also show that $\|T\|_{HS} = \|T^*\|_{HS}$ holds for every bounded operator T. First recall that for any two vectors $v, w \in H$ and any orthonormal basis (e_j) one has

$$\langle v, w \rangle = \sum_j \langle v, e_j \rangle \langle e_j, w \rangle.$$

Let now (ϕ_α) be another orthonormal basis; then, not knowing the independence yet, we write $\|T\|_{HS}^2(e_i)$ and $\|T\|_{HS}^2(\phi_\alpha)$, respectively. We compute

$$\|T\|_{HS}^2(e_j) = \sum_j \sum_\alpha \langle Te_j, \phi_\alpha \rangle \langle \phi_\alpha, Te_j \rangle = \sum_j \sum_\alpha \langle e_j, T^*\phi_\alpha \rangle \langle T^*\phi_\alpha, e_j \rangle$$

$$= \sum_\alpha \sum_j \langle e_j, T^*\phi_\alpha \rangle \langle T^*\phi_\alpha, e_j \rangle = \|T^*\|_{HS}^2(\phi_\alpha).$$

The interchange of summation order is justified by the fact that all summands are positive. Applying this to $(e_j) = (\phi_\alpha)$ first and then to T^* instead of T we get $\|T\|_{HS}^2(e_j) = \|T^*\|_{HS}^2(e_j) = \|T\|_{HS}^2(\phi_\alpha)$, as claimed.

We say that the operator T is a *Hilbert-Schmidt operator* if the Hilbert-Schmidt norm $\|T\|_{HS}$ is finite.

Lemma 5.3.1 *For any two bounded operators S,T on H one has $\|ST\|_{HS} \leq \|S\|_{op}\|T\|_{HS}$, and $\|ST\|_{HS} \leq \|S\|_{HS}\|T\|_{op}$, as well as $\|T\|_{op} \leq \|T\|_{HS}$. For every unitary operator U we have $\|UT\|_{HS} = \|TU\|_{HS} = \|T\|_{HS}$.*

Proof Let (e_j) be an orthonormal basis. We have $\|ST\|_{HS}^2 = \sum_j \|STe_j\|^2 \leq \|S\|_{op}^2 \sum_j \|Te_j\|^2$, which implies the first estimate. The second follows by using $\|T\|_{HS} = \|T^*\|_{HS}$ and the same assertion for the operator norm.

Let $v \in H$ with $\|v\| = 1$. Then there is an orthonormal basis (e_j) with $e_1 = v$. We get $\|Tv\|^2 = \|Te_1\|^2 \leq \sum_j \|Te_j\|^2 = \|T\|_{HS}^2$. The invariance under multiplication by unitary operators is clear, since (Ue_j) is an orthonormal basis when (e_j) is. $\qquad\square$

Example 5.3.2 The main example we are interested in is the following. For a measure space $(X, \mathcal{A}, \mu)$ consider the Hilbert space $L^2(X)$. Assume that μ is either σ-finite or that X is locally compact and μ is an outer Radon measure, so that Fubini's Theorem holds with respect to the product measure $\mu \otimes \mu$ on $L^2(X \times X)$. Let k be a function in $L^2(X \times X)$. Then we call k an L^2-*kernel*.

Proposition 5.3.3 *Suppose $k(x, y)$ is an L^2-kernel on X. For $\phi \in L^2(X)$ define*

$$K\phi(x) \stackrel{\text{def}}{=} \int_X k(x, y)\phi(y) \, d\mu(y).$$

Then this integral exists almost everywhere in x. The function $K\phi$ lies in $L^2(X)$, and K extends to a Hilbert-Schmidt operator $K : L^2(X) \to L^2(X)$ with

$$\|K\|^2_{HS} = \int_X \int_X |k(x,y)|^2 \, d\mu(x) \, d\mu(y).$$

Proof To see that the integral exists for almost all $x \in X$ let ψ be any element in $L^2(X)$. Then $(x,y) \mapsto \psi(x)\phi(y)$ lies in $L^2(X \times X)$, and therefore the function $(x,y) \to k(x,y)\phi(y)\psi(x)$ is integrable over $X \times X$. By Fubini, it follows that

$$\int_X \psi(x)k(x,y)\phi(y) \, dy = \psi(x) \int_X k(x,y)\phi(y) \, dy$$

exists for almost all $x \in X$. Since $k(x,y)$ vanishes for every x outside some σ-finite subset A of X, we may let ψ run through the characteristic functions of an increasing sequence of finite measurable sets that exhaust A, to conclude that the integral $\int_X k(x,y)\phi(y) \, dy$ exists for almost all $x \in X$.

We use the Cauchy-Schwarz inequality to estimate

$$\|K\phi\|^2 = \int_X |K\phi(x)|^2 \, dx$$

$$= \int_X \left| \int_X k(x,y)\phi(y) \, dy \right|^2 dx$$

$$\leq \int_X \int_X |k(x,y)|^2 \, dx \, dy \int_X |\phi(y)|^2 \, dy$$

$$= \int_X \int_X |k(x,y)|^2 \, dx \, dy \, \|\phi\|^2.$$

So K extends to a bounded operator on $L^2(X)$. Let (e_j) be an orthonormal basis of $L^2(X)$. Then

$$\|K\|^2_{HS} = \sum_j \langle Ke_j, Ke_j \rangle = \sum_j \int_X Ke_j(x)\overline{Ke_j(x)} \, dx$$

$$= \sum_j \int_X \int_X k(x,y)e_j(y) \, dy \overline{\int_X k(x,y)e_j(y) \, dy} \, dx$$

$$= \sum_j \int_X \langle k(x,.), e_j \rangle \langle e_j, k(x,.) \rangle \, dx$$

$$= \int_X \sum_j \langle k(x,.), e_j \rangle \langle e_j, k(x,.) \rangle \, dx$$

$$= \int_X \langle k(x,.), k(x,.) \rangle \, dx = \int_X \int_X |k(x,y)|^2 \, dx \, dy. \qquad \square$$

Proposition 5.3.4 *The operator T is Hilbert-Schmidt if and only if it is compact and its singular values satisfy $\sum_j s_j(T)^2 < \infty$. Indeed, then one has $\sum_j s_j(T)^2 = \|T\|_{HS}^2$.*

Proof We show that a bounded operator T is Hilbert-Schmidt if and only if $|T| = \sqrt{T^*T}$ is. This follows from

$$\sum_j \langle Te_j, Te_j \rangle = \sum_j \langle T^*Te_j, e_j \rangle = \sum_j \langle |T|^2 e_j, e_j \rangle$$
$$= \sum_j \langle |T|e_j, |T|e_j \rangle.$$

Let T be Hilbert-Schmidt. To see that T is compact, it suffices to show that if e_j is an orthonormal sequence, then Te_j has a convergent subsequence. But indeed, extend e_j to an orthonormal basis, then the Hilbert-Schmidt criterion shows that Te_j tends to zero. So T is compact. The operator $|T|$ is Hilbert-Schmidt if and only if $\sum_j s_j(T)^2$ converges, as one sees by applying the Hilbert-Schmidt criterion to an orthonormal basis consisting of eigenvectors of $|T|$. Finally, it is clear that $\sum_j s_j(T)^2 = \||T|\|_{HS}^2$, but by the above computation the latter equals $\|T\|_{HS}^2$. □

A compact operator T is called a *trace class operator* if the *trace norm*,

$$\|T\|_{tr} \stackrel{\text{def}}{=} \sum_j s_j(T),$$

is finite. It follows that every trace class operator is also Hilbert-Schmidt.

Lemma 5.3.5 *Let T be a trace class operator and S a bounded operator.*

(a) *The norms $\|ST\|_{tr}, \|TS\|_{tr}$ are both $\leq \|S\| \|T\|_{tr}$.*

(b) *Let T be a compact operator on H. One has*

$$\|T\|_{tr} = \sup_{(e_i),(h_i)} \sum_i |\langle Te_i, h_i \rangle|,$$

where the supremum runs over all orthonormal bases (e_i) and (h_i).

Proof The inequality $\|ST\|_{tr} \leq \|S\| \|T\|_{tr}$ is a consequence of Proposition 5.2.4 (b). The other follows from $\|T\| = \|T^*\|$ and the same for the trace norm.

For the second part we use the Spectral Theorem for compact operators to find an orthonormal sequence (f_j) such that

$$|T|v = \sum_j s_j \langle v, f_j \rangle f_j.$$

We then write $T = U|T|$, where U is an isometric operator on the image of $|T|$ to get

$$Tv = U \sum_j s_j \langle v, f_j \rangle f_j = \sum_j s_j \langle v, f_j \rangle g_j,$$

where (g_j) is the orthonormal sequence $g_j = U f_j$. Therefore, we can use the Cauchy-Schwarz inequality to get for any two orthonormal bases e, h,

$$\sum_i |\langle T e_i, h_i \rangle| = \sum_i \left| \sum_j s_j \langle e_i, f_j \rangle \langle g_j, h_i \rangle \right|$$

$$\leq \sum_j s_j \sum_i |\langle e_i, f_j \rangle \langle g_j, h_i \rangle|$$

$$\leq \sum_j s_j \left(\sum_i |\langle e_i, f_j \rangle|^2 \right)^{\frac{1}{2}} \left(\sum_i |\langle g_j, h_i \rangle|^2 \right)^{\frac{1}{2}}$$

$$= \sum_j s_j \| f_j \| \| g_j \| = \sum_j s_j.$$

This implies the $\geq$ part of the claim. The other part is obtained by taking e to be any orthonormal basis that prolongs the orthonormal sequence f and h any orthonormal basis that prolongs the orthonormal sequence g, because then $\sum_i |\langle T e_i, h_i \rangle| = \sum_j s_j$.   □

Theorem 5.3.6 *For a trace class operator T the trace*

$$\mathrm{tr}\,(T) \overset{\text{def}}{=} \sum_j \langle T e_j, e_j \rangle$$

does not depend on the choice of an orthonormal base (e_j). If T is trace class and normal, we have $\mathrm{tr}\,(T) = \sum_n \lambda_n \dim \mathrm{Eig}(T, \lambda_n)$, *where the sum runs over the sequence of non-zero eigenvalues (λ_n) of T. The sum converges absolutely.*

Proof Let $T = U|T|$ be the polar decomposition of T. It follows from the Spectral Theorem that the image of the operator $S_2 = \sqrt{|T|}$ equals the image of $|T|$ and therefore we can define the operator $S_1 = U\sqrt{|T|}$. The operators S_1, S_2 and S_1^* are Hilbert-Schmidt operators, and $T = S_1 S_2$. Therefore $\sum_i \langle T e_i, e_i \rangle = \sum_i \langle S_2 e_i, S_1^* e_i \rangle$, and the latter does not depend on the choice of the orthonormal basis as can be seen in a similar way as in the beginning of this section. Choose a basis of eigenvectors to prove the second statement.   □

Theorem 5.3.7 *Let H be a Hilbert space, $\mathcal{F}$ the space of bounded operators T of finite rank (i.e., finite dimensional image), $\mathcal{T}$ the set of trace class operators, $\mathcal{HS}$ the set of Hilbert-Schmidt operators, and $\mathcal{K}$ the set of compact operators. Further,* *we write* $\mathcal{HS}^2$ *for the linear span of all operters of the form ST, where S and T are both in* $\mathcal{HS}$.

(a) *The spaces* $\mathcal{F}, \mathcal{T}, \mathcal{HS}$ *and* $\mathcal{K}$ *are ideals in the algebra* $\mathcal{B}(H)$, *which are stable under* $*$.

(b) *The space* $\mathcal{K}$ *is the norm closure of* $\mathcal{F}$.

(c) *One has*

$$\mathcal{F} \subset \mathcal{T} = \mathcal{HS}^2 \subset \mathcal{HS} \subset \mathcal{K},$$

where the inclusions are strict if $\dim(H) = \infty$.

Proof (a) The $*$-ideal property is clear for $\mathcal{F}$ and $\mathcal{K}$. The space $\mathcal{T}$ is an ideal by Lemma 5.3.5 and $\mathcal{HS}$ by Lemma 5.3.1. Part (b) of the theorem is contained in Proposition 5.2.1. The first inclusion of (c) is clear. Let T be in $\mathcal{T}$ and write $T = U|T|$ as in Proposition 5.1.4. With $S = \sqrt{|T|}$ one has $T = (US)S$ and by Proposition 5.3.4, the operators S and US are in $\mathcal{HS}$, so $T \in \mathcal{HS}^2$. If $S \in \mathcal{HS}$, then by definition $S^*S \in \mathcal{T}$ and by polarization we find $\mathcal{HS}^2 \subset \mathcal{T}$. The remaining inclusions are clear and we leave the strictness as an exercise. $\square$

5.4 Exercises

Exercise 5.1 Let A and B be bounded operators on a Hilbert space H. Show that $AB - BA \neq \mathrm{Id}$, where Id is the identity operator.

(Hint: Assume the contrary and show that $AB^n - B^nA = nB^{n-1}$ holds for every $n \in \mathbb{N}$. Then take norms.)

Exercise 5.2 Let H be a Hilbert space, and let $T \in \mathcal{B}(H)$ be a normal operator. Show that the map $\psi : t \mapsto \exp(tT)$ satisfies $\psi(t + s) = \psi(t)\psi(s)$, that it is differentiable as a map from $\mathbb{R}$ to the Banach space $\mathcal{B}(H)$, which satisfies $\psi(0) = \mathrm{Id}$ and $\psi'(t) = T\psi(t)$.

Exercise 5.3 Show that for a bounded operator T on a Hilbert space H the following are equivalent:

- T is compact,
- T^*T is compact,
- T^* is compact.

Exercise 5.4 Check the details of the proof of Proposition 5.2.4.

Exercise 5.5 Show that a continuous invertible operator T on a Hilbert space H can only be compact if H is finite dimensional.

Exercise 5.6 Show that $T \in \mathcal{B}(H)$ for a Hilbert space H is compact if and only if the image of the closed unit ball is compact (as opposed to relatively compact).

Exercise 5.7 Let H be a Hilbert space.

(a) Show that for a trace class operator T on H one has $\operatorname{tr}(T^*) = \overline{\operatorname{tr}(T)}$.

(b) Show that for two Hilbert-Schmidt operators S, T on H one has

$$\operatorname{tr}(ST) = \operatorname{tr}(TS).$$

Exercise 5.8 Let H be the real Hilbert space $\ell^2(\mathbb{N}, \mathbb{R})$ and let $(e_j)_{j \in \mathbb{N}}$ be the standard orthonormal basis. Define a linear operator T on H by

$$T(e_j) = \frac{(-1)^{j+1}}{j} e_{j+(-1)^{j+1}}.$$

Show that for every orthonormal basis (f_n) of H one has

$$\sum_n |\langle T f_n, f_n \rangle| \le \sum_{j=1}^{\infty} \frac{1}{2j(2j-1)}.$$

Exercise 5.9 Show that the set $\mathcal{HS}(V)$ of Hilbert-Schmidt operators on a given Hilbert space V becomes a Hilbert space with the inner product $\langle S, T \rangle = \operatorname{tr}(ST^*)$. Show that the map $\psi : V \hat{\otimes} V' \to \mathcal{HS}(V)$ given by $\psi(v \otimes \alpha)(w) = \alpha(w)v$ defines a Hilbert space isomorphism (Compare Appendix C.3 for the notation).

Exercise 5.10 Let H be a Hilbert space. For $p > 0$ let $S_p(H)$ be the set of all compact operators T on H such that

$$\|T\|_p \overset{\text{def}}{=} \left(\sum_j s_j(T)^p \right)^{\frac{1}{p}} < \infty.$$

Show that $S_p(H)$ is a vector space. It is called the p-th *Schatten class*.

Exercise 5.11 Let H be a Hilbert space. An operator $T \in \mathcal{B}(H)$ is called *nilpotent* if $T^k = 0$ for some $k \in \mathbb{N}$. Show that if T is nilpotent, then $\sigma(T) = \{0\}$. Show also that the converse is not generally true.

Exercise 5.12 Let H be a Hilbert space. Show that an operator T is invertible in $\mathcal{B}(H)$ if and only if $|T|$ is invertible.

Exercise 5.13 Let H be a Hilbert space, $T \in \mathcal{B}(H)$ invertible. Let $T = U|T|$ be the polar decomposition. Show that T is normal if and only if $U|T| = |T|U$.

Exercise 5.14 Let $G = \mathrm{SL}_n(\mathbb{R})$, and let H be the subgroup of upper triangular matrices in G. Let $K = \mathrm{SO}(n)$. Show that $G = HK$.

(Hint: For $g \in G$ apply the spectral theorem to the positive definite matrix $g^t g$.)

Chapter 6
Representations

In this chapter we introduce the basic concepts of representation theory of locally compact groups. Classically, a representation of a group G is an injective group homomorphism from G to some $GL_n(\mathbb{C})$, the idea being that the "abstract" group G is "represented" as a matrix group.

In order to understand a locally compact group, it is necessary to consider its actions on possibly infinite dimensional spaces like $L^2(G)$. For this reason, one considers infinite dimensional representations as well.

6.1 Schur's Lemma

For a Banach space V, let $GL_{cont}(V)$ be the set of bijective bounded linear operators T on V. It follows from the Open Mapping Theorem C.1.5 that the inverses of such operators are bounded as well, so that $GL_{cont}(V)$ is a group. Let G be a topological group. A *representation* of G on a Banach space V is a group homomorphism of G to the group $GL_{cont}(V)$, such that the resulting map $G \times V \to V$, given by $(g, v) \mapsto \pi(g)v$, is continuous.

Lemma 6.1.1 *Let π be a group homomorphism of the topological group G to $GL_{cont}(V)$ for a Banach space V. Then π is a representation if and only if*

(a) *the map $g \mapsto \pi(g)v$ is continuous at $g = 1$ for every $v \in V$, and*

(b) *the map $g \mapsto \|\pi(g)\|_{op}$ is bounded in a neighborhood of the unit in G.*

Proof Suppose π is a representation. Then (a) is obvious. For (b) note that for every neighborhood Z of zero in V there exists a neighborhood Y of zero in V and a neighborhood U of the unit in G such that $\pi(U)Y \subset Z$. This proves (b). For the converse direction write

A. Deitmar, S. Echterhoff, *Principles of Harmonic Analysis*, Universitext,
DOI 10.1007/978-3-319-05792-7_6, © Springer International Publishing Switzerland 2014

$$\|\pi(g)v - \pi(g_0)v_0\| \le \|\pi(g_0)\|_{op} \|\pi(g_0^{-1}g)v - v_0\|$$
$$\le \|\pi(g_0)\|_{op} \|\pi(g_0^{-1}g)(v - v_0)\|$$
$$+ \|\pi(g_0)\|_{op} \|\pi(g_0^{-1}g)v_0 - v_0\|$$
$$\le \|\pi(g_0)\|_{op} \|\pi(g_0^{-1}g)\|_{op} \|v - v_0\|$$
$$+ \|\pi(g_0)\|_{op} \|\pi(g_0^{-1}g)v_0 - v_0\|.$$

Under the assumptions given, both terms on the right are small if g is close to g_0, and v is close to v_0. $\square$

Examples 6.1.2

- For a continuous group homomorphism $\chi : G \to \mathbb{C}^\times$ define a representation π_χ on $V = \mathbb{C}$ by $\pi_\chi(g)v = \chi(g) \cdot v$.

- Let $G = SL_2(\mathbb{R})$ be the group of real 2×2 matrices of determinant one. This group has a natural representation on $\mathbb{C}^2$ given by matrix multiplication.

Definition Let V be a Hilbert space. A representation π on V is called a *unitary representation* if $\pi(g)$ is unitary for every $g \in G$. That means π is unitary if $\langle \pi(g)v, \pi(g)w \rangle = \langle v, w \rangle$ holds for every $g \in G$ and all $v, w \in V$.

Lemma 6.1.3 *A representation π of the group G on a Hilbert space V is unitary if and only if $\pi(g^{-1}) = \pi(g)^*$ holds for every $g \in G$.*

Proof An operator T is unitary if and only if it is invertible and $T^* = T^{-1}$. For a representation π and $g \in G$ the operator $\pi(g)$ is invertible and satisfies $\pi(g^{-1}) = \pi(g)^{-1}$. So π is unitary if and only if for every $g \in G$ one has $\pi(g^{-1}) = \pi(g)^{-1} = \pi(g)^*$. $\square$

Examples 6.1.4

- The representation π_χ defined by a continuous group homomorphism $\chi : G \to \mathbb{C}^\times$ is unitary if and only if χ maps into the compact torus $\mathbb{T}$.

- Let G be a locally compact group. On the Hilbert space $L^2(G)$ consider the *left regular representation* $x \mapsto L_x$ with

$$L_x\phi(y) = \phi(x^{-1}y), \qquad \phi \in L^2(G).$$

This representation is unitary, as by the left invariance of the Haar measure,

$$\langle L_x\phi, L_x\psi\rangle = \int_G L_x\phi(y)\overline{L_x\psi(y)}\,dy$$

$$= \int_G \phi(x^{-1}y)\overline{\psi(x^{-1}y)}\,dy$$

$$= \int_G \phi(y)\overline{\psi(y)}\,dy = \langle\phi,\psi\rangle.$$

Definition Let (π_1, V_1) and (π_2, V_2) be two unitary representations. On the direct sum $V = V_1 \oplus V_2$ one has the *direct sum representation* $\pi = \pi_1 \oplus \pi_2$. More generally, if $\{\pi_i : i \in I\}$ is a family of unitary representations acting on the Hilbert spaces V_i, we write $\bigoplus_{i \in I} \pi_i$ for the direct sum of the representations π_i, $i \in I$ on the Hilbert space $\widehat{\bigoplus}_{i \in I} V_i$. See also Exercise 6.3 and appendix C.3.

Example 6.1.5 Let $G = \mathbb{R}/\mathbb{Z}$, and let $V = L^2(\mathbb{R}/\mathbb{Z})$. Let π be the left regular representation. By the Plancherel Theorem, the elements of the dual group $\widehat{G} = \{e_k : k \in \mathbb{Z}\}$ with $e_k([x]) = e^{2\pi i kx}$ form an orthonormal basis of $L^2(\mathbb{R}/\mathbb{Z})$ so that π is a direct sum representation on $V = \widehat{\bigoplus}_{k \in \mathbb{Z}} \mathbb{C}e_k$, where $e_k(x) = e^{2\pi i kx}$ and G acts on $\mathbb{C}e_k$ through the character $\bar{e}_k$.

Definition A representation (π, V_π) is called a *subrepresentation* of a representation (η, V_η) if V_π is a closed subspace of V_η and π equals η restricted to V_π. So every closed subspace $U \subset V_\eta$ that is stable under η, i.e., $\eta(G)U \subset U$, gives rise to a subrepresentation.

A representation is called *irreducible* if it does not possess any proper subrepresentation, i.e., if for every closed subspace $U \subset V_\pi$ that is stable under π, one has $U = 0$ or $U = V_\pi$.

Example 6.1.6. Let $U(n)$ denote the group of unitary $n \times n$ matrices, so the group of all $u \in M_n(\mathbb{C})$ such that $uu^* = I$ (unit matrix), where $u^* = \bar{u}^t$. The natural representation of $U(n)$ on $\mathbb{C}^n$ is irreducible (See Exercise 6.5).

Definition Let (π, V_π) be a representation of G. A vector $v \in V_\pi$ is called a *cyclic vector* if the linear span of the set $\{\pi(x)v : x \in G\}$ is dense in V_π. In other words, v is cyclic if the only subrepresentation containing v is the whole of π. It follows that a representation is irreducible if and only if every nonzero vector is cyclic.

Lemma 6.1.7 *(Schur) Let (π, V_π) be a unitary representation of the topological group G. Then the following are equivalent*

(a) (π, V_π) *is irreducible.*

(b) *If T is a bounded operator on V_π such that $T\pi(g) = \pi(g)T$ for every $g \in G$, then $T \in \mathbb{C}\,\mathrm{Id}$.*

Proof Since $\pi(g^{-1}) = \pi(g)^*$, the set $\{\pi(g) : g \in G\}$ is a self-adjoint subset of $\mathcal{B}(V_\pi)$. Thus the result follows from Theorem 5.1.6.   □

Let $(\pi, V_\pi), (\eta, V_\eta)$ be representations of G. A continuous linear operator $T : V_\pi \to V_\eta$ is called a *G-homomorphism* or *intertwining operator* if

$$T\pi(g) = \eta(g)T$$

holds for every $g \in G$. We write $\text{Hom}_G(V_\pi, V_\eta)$ for the set of all G-homomorphisms from V_π to V_η. A nice way to rephrase the Lemma of Schur is to say that a unitary representation (π, V_π) is irreducible if and only if $\text{Hom}_G(V_\pi, V_\pi) = \mathbb{C}\,\text{Id}$.

Definition If π, η are unitary, they are called *unitarily equivalent* if there exists a unitary G-homomorphism $T : V_\pi \to V_\eta$.

Example 6.1.8. Let $G = \mathbb{R}$, and let $V_\pi = V_\eta = L^2(\mathbb{R})$. The representation π is given by $\pi(x)\phi(y) = \phi(x+y)$ and η is given by $\eta(x)\phi(y) = e^{2\pi i xy}\phi(y)$. By Theorem 3.3.1 in [Dei05] (see also Exercise 6.4), the Fourier transform $L^2(\mathbb{R}) \to L^2(\mathbb{R})$ is an intertwining operator from π to η.

Corollary 6.1.9 *Let (π, V_π) and (η, V_η) be two irreducible unitary representations. Then a G-homomorphism T from V_π to V_η is either zero or invertible with continuous inverse. In the latter case there exists a scalar $c > 0$ such that cT is unitary. The space $\text{Hom}_G(V_\pi, V_\eta)$ is zero unless π and η are unitarily equivalent, in which case the space is of dimension 1.*

Proof Let $T : V_\pi \to V_\eta$ be a G-homomorphism. Its adjoint $T^* : V_\eta \to V_\pi$ is also a G-homomorphism as is seen by the following calculation for $v \in V_\pi, w \in V_\eta$, and $g \in G$,

$$\begin{aligned}
\langle v, T^*\eta(g)w \rangle &= \langle Tv, \eta(g)w \rangle = \langle \eta(g^{-1})Tv, w \rangle \\
&= \langle T\pi(g^{-1})v, w \rangle = \langle \pi(g^{-1})v, T^*w \rangle \\
&= \langle v, \pi(g)T^*w \rangle.
\end{aligned}$$

This implies that T^*T is a G-homomorphism on V_π, and therefore it is a multiple of the identity $\lambda\,\text{Id}$ by the Lemma of Schur. If T is non-zero, T^*T is non-zero and positive semi-definite, so $\lambda > 0$. Let $c = \sqrt{\lambda^{-1}}$, then $(cT)^*(cT) = \text{Id}$. A similar argument shows that TT^* is bijective, which then implies that cT is bijective, hence unitary. The rest is clear.   □

Definition For a locally compact group G we denote by $\widehat{G}$ the set[1] of all equivalence classes of irreducible unitary representations of G. We call $\widehat{G}$ the *unitary dual* of

---

[1] There is a set-theoretic problem here, since it is not clear why the equivalence classes should form a set. It is, however, not difficult to show that there exists a cardinality α, depending on G, such that

G. It is quite common to make no notational difference between a given irreducible representation π and its unitary equivalence class $[\pi]$, and we will often do so in this book.

Example 6.1.10 If G is a locally compact abelian group, then every irreducible representation is one-dimensional, and therefore the unitary dual $\widehat{G}$ coincides with the Pontryagin dual of G. To see this, let (π, V_π) be any irreducible representation of G. Then $\pi(x)\pi(y) = \pi(xy) = \pi(yx) = \pi(y)\pi(x)$ for all $x, y \in G$, so it follows from Schur's Lemma that $\pi(x) = \lambda(x)\mathrm{Id}_{V_\pi}$ for some $\lambda(x) \in \mathbb{T}$. But this implies that every non-zero closed subspace of V_π is invariant, hence must be equal to V_π. This implies $\dim V_\pi = 1$.

6.2 Representations of $L^1(G)$

A unitary representation (π, V_π) of G induces an algebra homomorphism from the convolution algebra $L^1(G)$ to the algebra $\mathcal{B}(V_\pi)$, as the following proposition shows.

Proposition 6.2.1 *Let (π, V_π) be a unitary representation of the locally compact group G. For every $f \in L^1(G)$ there exists a unique bounded operator $\pi(f)$ on V_π such that*

$$\langle \pi(f)v, w \rangle = \int_G f(x)\langle \pi(x)v, w \rangle \, dx$$

holds for any two vectors $v, w \in V_\pi$. The induced map $\pi : L^1(G) \to \mathcal{B}(V_\pi)$ is a continuous homomorphism of Banach-*-algebras.

Proof Taking complex conjugates one sees that the claimed equation is equivalent to the equality $\langle w, \pi(f)v \rangle = \int_G \overline{f(x)}\langle w, \pi(x)v \rangle \, dx$. The map $w \mapsto \int_G \overline{f(x)}\langle w, \pi(x)v \rangle \, dx$ is linear. It is also bounded, since

$$\left| \int_G \overline{f(x)}\langle w, \pi(x)v \rangle \, dx \right| \leq \int_G |f(x)\langle w, \pi(x)v \rangle| \, dx$$

$$\leq \int_G |f(x)| \|w\| \|\pi(x)v\| \, dx$$

$$= \|f\|_1 \|w\| \|v\|.$$

---

every irreducible unitary representation (π, V_π) of G satisfies $\dim V_\pi \leq \alpha$. This means that one can fix a Hilbert space H of dimension α and each irreducible unitary representation π can be realized on a subspace of H. Setting the representation equal to 1 on the orthogonal complement one gets a representation on H, i.e., a group homomorphism $G \to \mathrm{GL}(H)$. Indeed, since every irreducible representation has a cyclic vector by Schur's lemma, one can choose α as the cardinality of G. Therefore, each equivalence class has a representative in the set of all maps from G to $\mathrm{GL}(H)$ and so $\widehat{G}$ forms a set.

Therefore, by Proposition C.3.1, there exists a unique vector $\pi(f)v \in V_\pi$ such that the equality holds. It is easy to see that the map $v \mapsto \pi(f)v$ is linear. To see that it is bounded, note that the above shows $\|\pi(f)v\|^2 = \langle \pi(f)v, \pi(f)v \rangle \leq \|f\|_1 \|v\| \|\pi(f)v\|$, and hence $\|\pi(f)v\| \leq \|f\|_1 \|v\|$. A straightforward computation finally shows $\pi(f * g) = \pi(f)\pi(g)$ and $\pi(f)^* = \pi(f^*)$ for $f, g \in L^1(G)$. $\square$

Alternatively, one can define $\pi(f)$ as the Bochner integral $\pi(f) = \int_G f(x)\pi(x)\,dx$ in the Banach space $\mathcal{B}(V_\pi)$. By the uniqueness in the above proposition, these two definitions agree.

The above proposition has a converse, as we shall see in Proposition 6.2.3 below.

Lemma 6.2.2 *Let (π, V_π) be a representation of G. Then for every $v \in V_\pi$ and every $\varepsilon > 0$ there exists a unit-neighborhood U such that for every Dirac function ϕ_U with support in U one has $\|\pi(\phi_U)v - v\| < \varepsilon$. In particular, for every Dirac net $(\phi_U)_U$ the net $(\pi(\phi_U)v)$ converges to v in the norm topology.*

Proof The norm $\|\pi(\phi_U)v - v\|$ equals $\|\int_G \phi_U(x)(\pi(x)v - v)\,dx\|$ and is therefore less than or equal to $\int_G \phi_U(x)\|\pi(x)v - v\|\,dx$. For given $\varepsilon > 0$ there exists a unit-neighborhood U_0 in G such that for $x \in U_0$ one has $\|\pi(x)v - v\| < \varepsilon$. For $U \subset U_0$ it follows $\|\pi(\phi_U)v - v\| < \varepsilon$. $\square$

Definition We say that a $*$-representation $\pi : L^1(G) \to \mathcal{B}(V)$ of $L^1(G)$ on a Hilbert space V is *non-degenerate* if the vector space

$$\pi(L^1(G))V \stackrel{\text{def}}{=} \text{span}\{\pi(f)v : f \in L^1(G), v \in V\}$$

is dense in V. It follows from the above lemma that every representation of $L^1(G)$ that comes from a representation (π, V_π) as in Proposition 6.2.1, is non-degenerate. The next proposition gives a converse to this.

Proposition 6.2.3 *Let $\pi : L^1(G) \to \mathcal{B}(V)$ be a non-degenerate $*$- representation on a Hilbert space V. Then there exists a unique unitary representation $(\tilde{\pi}, V)$ of G such that $\langle \pi(f)v, w \rangle = \int_G f(x)\langle \tilde{\pi}(x)v, w \rangle\,dx$ holds for all $f \in L^1(G)$ and all $v, w \in V$.*

Proof Note first that π is continuous by Lemma 2.7.1. We want to define an operator $\tilde{\pi}(x)$ on the dense subspace $\pi(L^1(G))V$ of V. This space is made up of sums of the form $\sum_{i=1}^n \pi(f_i)v_i$ for $f_i \in L^1(G)$ and $v_i \in V$. We propose to define $\tilde{\pi}(x)\sum_{i=1}^n \pi(f_i)v_i \stackrel{\text{def}}{=} \sum_{i=1}^n \pi(L_x f_i)v_i$. We have to show well-definedness, which amounts to show that if $\sum_{i=1}^n \pi(f_i)v_i = 0$, then $\sum_{i=1}^n \pi(L_x f_i)v_i = 0$ for every $x \in G$. For $x \in G$ and $f, g \in L^1(G)$ a short computation shows that $g^* * (L_x f) = (L_{x^{-1}}g)^* * f$. Based on this, we compute for $v, w \in V$ and $f_1, \ldots, f_n \in L^1(G)$,

$$\left\langle \sum_{i=1}^{n} \pi(L_x f_i)v, \pi(g)w \right\rangle = \sum_{i=1}^{n} \left\langle \pi(g^* * (L_x f_i))v, w \right\rangle$$

$$= \sum_{i=1}^{n} \left\langle \pi((L_{x^{-1}}g)^* * f_i)v, w \right\rangle = \left\langle \sum_{i=1}^{n} \pi(f_i)v, \pi(L_{x^{-1}}g)w \right\rangle.$$

Now for the well-definedness of $\tilde{\pi}(x)$ assume $\sum_{i=1}^{n} \pi(f_i)v_i = 0$, then the above computation shows that the vector $\sum_{i=1}^{n} \pi(L_x f_i)v_i$ is orthogonal to all vectors of the form $\pi(g)w$, which span the dense subspace $\pi(L^1(G))V$, hence $\sum_{i=1}^{n} \pi(L_x f_i)v_i = 0$ follows. The computation also shows that this, now well-defined operator $\tilde{\pi}(x)$ is unitary on the space $\pi(L^1(G))V$ and since the latter is dense in V, the operator $\tilde{\pi}(x)$ extends to a unique unitary operator on V with inverse $\tilde{\pi}(x^{-1})$, and we clearly have $\tilde{\pi}(xy) = \tilde{\pi}(x)\tilde{\pi}(y)$ for all $x, y \in G$. Since for each $f \in L^1(G)$ the map $G \to L^1(G)$ sending x to $L_x f$ is continuous by Lemma 1.4.2, it follows that $x \mapsto \tilde{\pi}(x)v$ is continuous for every $v \in V$. Thus $(\tilde{\pi}, V)$ is a unitary representation of G.

It remains to show that $\pi(f)$ equals $\tilde{\pi}(f)$ for every $f \in L^1(G)$. By continuity it is enough to show that $\langle \tilde{\pi}(f)\pi(g)v, w \rangle = \langle \pi(f)\pi(g)v, w \rangle$ for all $f, g \in C_c(G)$ and $v, w \in V$. Since $g \mapsto \langle \pi(g)v, w \rangle$ is a continuous linear functional on $L^1(G)$ we can use Lemma B.6.5 to get

$$\langle \tilde{\pi}(f)\pi(g)v, w \rangle = \int_G f(x)\langle \tilde{\pi}(x)(\pi(g)v), w \rangle \, dx$$

$$= \int_G \langle \pi(f(x)L_x g)v, w \rangle \, dx$$

$$= \left\langle \pi\left(\int_G f(x)L_x g \, dx\right)v, w \right\rangle$$

$$= \langle \pi(f * g)v, w \rangle = \langle \pi(f)\pi(g)v, w \rangle,$$

which completes the proof. □

Remark 6.2.4 If we define unitary equivalence and irreducibility for representations of $L^1(G)$ in the same way as we did for unitary representations of G, then it is easy to see that the one-to-one correspondence between unitary representations of G and non-degenerate $*$-representations of $L^1(G)$ preserves unitary equivalence and irreducibility in both directions. Note that an irreducible representation π of $L^1(G)$ is automatically non-degenerate, since the closure of $\pi(L^1(G))V_\pi$ is an invariant subspace of V_π. Thus, we obtain a bijection between the space $\widehat{G}$ of equivalence classes of irreducible representations of G and the set $L^1(G)\widehat{\,}$ of irreducible $*$-representations of $L^1(G)$.

Example 6.2.5 Consider the left regular representation on G. Then the corresponding representation $L : L^1(G) \to \mathcal{B}(L^2(G))$ is given by the convolution operators $L(f)\phi = f * \phi$ whenever the convolution $f * \phi$ makes sense.

6.3   Exercises

Exercise 6.1 Let G be a topological group and let V be a Banach space. We equip the group $\mathrm{GL}_{\mathrm{cont}}(V)$ with the topology induced by the operator norm. Show that any continuous group homomorphism $G \to \mathrm{GL}_{\mathrm{cont}}(V)$ is a representation but that not every representation is of this form.

Exercise 6.2 If π is a unitary representation of the locally compact group G, then $\|\pi(g)\| = 1$. Give an example of a representation π, for which the map $g \mapsto \|\pi(g)\|$ is not bounded on G.

Exercise 6.3 Let I be an index set, and for $i \in I$ let (π_i, V_i) be a unitary representation of the locally compact group G. Let $V = \bigoplus_{i \in I} V_i$ be the Hilbert direct sum (See Appendix C.3). Define the map $\pi : G \to \mathcal{B}(V)$ by

$$\pi(g) \sum_i v_i = \sum_i \pi_i(g) v_i.$$

Show that this is a unitary representation of the group G. It is called the *direct sum representation*.

Exercise 6.4 Show that the Fourier transform on $\mathbb{R}$ induces a unitary equivalence between the the the unitary representations π and η of $\mathbb{R}$ on $L^2(\mathbb{R})$ given by $\pi(x)\phi(y) = \phi(x + y)$ and $\eta(x)\phi(y) = e^{2\pi i x y}\phi(y)$.

Exercise 6.5 Show that the natural representation of $\mathrm{U}(n)$ on $\mathbb{C}^n$ is irreducible.

Exercise 6.6 (a) For $t \in \mathbb{R}$ let $A(t) = \left(\begin{smallmatrix} 1 & t \\ 0 & 1 \end{smallmatrix}\right)$. Show that $A(t)$ is not conjugate to a unitary matrix for $t \neq 0$.

(b) Let P be the group of upper triangular matrices in $\mathrm{SL}_2(\mathbb{R})$. The injection $\eta :$ $P \hookrightarrow \mathrm{GL}_2(\mathbb{C})$ can be viewed as a representation on $V = \mathbb{C}^2$. Show that η is not the sum of irreducible representations. Determine all irreducible subrepresentations.

Exercise 6.7 Let G be a locally compact group and H a closed subgroup. Let (π, V_π) be an irreducible unitary representation of G, and let

$$V_\pi^H = \{v \in V_\pi : \pi(h)v = v \,\forall h \in H\}$$

be the space of H-fixed vectors. Show: If H is normal in G, then V_π^H is either zero or the whole space V_π.

Exercise 6.8 Show that $G = \mathrm{SL}_2(\mathbb{R})$ has no finite dimensional unitary representations except the trivial one.

Instructions:

- For $m \in \mathbb{N}$ show

$$\begin{pmatrix} m & \\ & m^{-1} \end{pmatrix} A(t) \begin{pmatrix} m & \\ & m^{-1} \end{pmatrix}^{-1} = A(m^2 t) = A(t)^{m^2}.$$

Let $\phi : G \to U(n)$ be a representation. Show that the eigenvalues of $\phi(A(t))$ are a permutation of their m-th powers for every $m \in \mathbb{N}$. Conclude that they all must be equal to 1.

- Show that the normal subgroup of G generated by $\{A(t) : t \in \mathbb{R}\}$ is the whole group.

Exercise 6.9 Let (π, V_π) be a unitary representation of the locally compact group G. Let $f \in L^1(G)$. Show that the Bochner integral

$$\int_G f(x)\pi(x)\,dx \in \mathcal{B}(V_\pi)$$

exists and that the so defined operator coincides with $\pi(f)$ as defined in Proposition 6.2.1.

(Hint: Use Corollary 1.3.6 (d) and Lemma B.6.2 as well as Proposition B.6.3.)

Exercise 6.10 In Lemma 6.2.2 we have shown that for a representation π and Dirac functions ϕ_U the numbers $\|\pi(\phi_U)v - v\|$ become arbitrarily small for fixed $v \in V_\pi$. Give an example, in which $\|\pi(\phi_U) - \mathrm{Id}\|_{\mathrm{op}}$ does not become small as the support of the Dirac function ϕ_U shrinks.

Exercise 6.11 Give an example of a representation that possesses cyclic vectors without being irreducible.

Notes

As for abelian groups, one can associate to each locally compact group G the group C^*-algebra $C^*(G)$. It is defined as the completion of $L^1(G)$ with respect to the norm

$$\|f\|_{C^*} \overset{\mathrm{def}}{=} \sup\{\|\pi(f)\| : \pi \text{ a unitary representation of } G\},$$

which is finite since $\|\pi(f)\| \le \|f\|_1$ for every unitary representation π of G. By definition of the norm, every unitary representation π of G extends to a $*$-representation of $C^*(G)$, and, as for $L^1(G)$, this extension provides a one-to-one correspondence between the unitary representation of G to the non-degenerate $*$-representations of $C^*(G)$. Therefore, the rich representation theory of general C^*-algebras, as explained beautifully in Dixmier's classic book [Dix96] can be used for the study of unitary representations of G. For a more recent treatment of C^*-algebras related to locally compact groups we also refer to Dana William's book [Wil07].

Chapter 7
Compact Groups

In this chapter we will show that every unitary representation of a compact group is a direct sum of irreducibles, and that every irreducible unitary representation is finite dimensional. We further prove the Peter-Weyl theorem, which gives an explicit decomposition of the regular representation of the compact group K on $L^2(K)$.

The term *compact group* will always mean a compact topological group, which is a Hausdorff space.

7.1 Finite Dimensional Representations

Let K be a compact group, and let (τ, V_τ) be a finite dimensional representation, i.e., the complex vector space V_τ is finite dimensional.

Lemma 7.1.1 *On the space V_τ, there exists an inner product, such that τ becomes a unitary representation. If τ is irreducible, this inner product is uniquely determined up to multiplication by a positive constant.*

Proof Let $(\cdot, \cdot)$ be any inner product on V_τ. We define a new inner product $\langle v, w \rangle$ for $v, w \in V_\tau$ to be equal to $\int_K (\tau(k)v, \tau(k)w)\, dk$, where we have used the normalized Haar measure that gives K the measure 1. We have to show that this constitutes an inner product. Linearity in the first argument and anti-symmetry are clear. For the positive definiteness let $v \in V_\tau$ with $\langle v, v \rangle = 0$, i.e.,

$$0 = \langle v, v \rangle = \int_K (\tau(k)v, \tau(k)v)\, dk.$$

The function $k \mapsto (\tau(k)v, \tau(k)v)$ is continuous and positive, hence, by Corollary 1.3.6, the function vanishes identically, so in particular, $(v, v) = 0$, which implies $v = 0$ and $\langle \cdot, \cdot \rangle$ is an inner product. With respect to this inner product the representation τ is unitary, as for $x \in K$ one has

A. Deitmar, S. Echterhoff, *Principles of Harmonic Analysis*, Universitext,
DOI 10.1007/978-3-319-05792-7_7, © Springer International Publishing Switzerland 2014

$$\langle \tau(x)v, \tau(x)w \rangle = \int_K (\tau(k)\tau(x)v, \tau(k)\tau(x)w) \, dk$$

$$= \int_K (\tau(kx)v, \tau(kx)w) \, dk$$

$$= \int_K (\tau(k)v, \tau(k)w) \, dk = \langle v, w \rangle,$$

as K is unimodular.

Finally, assume that τ is irreducible, let $\langle \cdot, \cdot \rangle_1$ and $\langle \cdot, \cdot \rangle_2$ be two inner products that make τ unitary. Let (τ_1, V_1) and (τ_2, V_2) denote the representation (τ, V_τ) when equipped with the inner products $\langle \cdot, \cdot \rangle_1$ and $\langle \cdot, \cdot \rangle_2$, respectively. Since V_τ is finite dimensional, the identity Id $: V_1 \to V_2$ is a bounded non-zero intertwining operator for τ_1 and τ_2. By Corollary 6.1.9 there exists a number $c > 0$ such that $c \cdot$ Id is unitary. But this implies that $c^2 \langle v, w \rangle_2 = \langle v, w \rangle_1$ for all $v, w \in V_\tau$.   □

Proposition 7.1.2 *A finite dimensional representation of a compact group is a direct sum of irreducible representations.*

Proof Let (τ, V) be a finite dimensional representation of the compact group K. We want to show that τ is a direct sum of irreducibles. We proceed by induction on the dimension of V. If this dimension is zero or one, there is nothing to show. So assume the claim proven for all spaces of dimension smaller than dim V. By the last lemma, we can assume that τ is a unitary representation. If τ is irreducible itself, we are done. Otherwise, there is an invariant subspace $U \subset V$ with $0 \neq U \neq V$. Let $W = U^\perp$ be the orthogonal complement to U in V, so that $V = U \oplus W$. We claim that W is invariant as well. For this let $k \in K$ and $w \in W$. Then for every $u \in U$,

$$\langle \tau(k)w, u \rangle = \langle w, \underbrace{\tau(k^{-1})u}_{\in U} \rangle = 0.$$

This implies that $\tau(k)w \in U^\perp = W$, so W is indeed invariant. We conclude that τ is the direct sum of the subrepresentations on U and W. As both spaces have dimensions smaller than the one of V, the induction hypothesis shows that both are direct sums of irreducibles, and so is V.   □

Definition Let (τ, V_τ) be a finite dimensional representation of a compact group K. The dual space

$$V_\tau^* = \mathrm{Hom}(V_\tau, \mathbb{C})$$

of all linear functionals $\alpha : V_\tau \to \mathbb{C}$ carries a natural representation of K, the *dual representation* τ^* defined by

$$\tau^*(x)\alpha(v) = \alpha \left(\tau(x^{-1})v \right).$$

Suppose that V_τ is a Hilbert space. By the Riesz Representation Theorem for every $\alpha \in V_\tau^*$ there exists a unique vector v_α such that

$$\alpha(w) = \langle w, v_\alpha \rangle$$

holds for every $w \in V_\tau$. One instals a Hilbert space structure on the dual V_τ^* by setting

$$\langle \alpha, \beta \rangle = \langle v_\beta, v_\alpha \rangle.$$

Lemma 7.1.3 *If the representation τ is irreducible, then so is the dual representation τ^*. The same holds for the property of being unitary. For $x \in K$ and $\alpha \in V_\tau^*$ one gets the intertwining relation*

$$v_{\tau^*(x)\alpha} = \tau(x)v_\alpha,$$

so the map $\alpha \mapsto v_\alpha$ is an anti-linear intertwining operator between V_τ^* and V_τ.

Proof Suppose that $W^* \subset V_\tau^*$ is a subrepresentation. Then the space $(W^*)^\perp$ of all $v \in V_\tau$ with $\alpha(v) = 0$ for every $\alpha \in W^*$ is a subrepresentation of V_τ. If τ is irreducible the latter space is trivial and so then is W^*.

For the remaining assertions, we first show the claimed intertwining relation. For $w \in V_\tau$ we use unitarity of τ to get

$$\langle w, v_{\tau^*(x)\alpha} \rangle = \tau^*(x)\alpha(w) = \alpha(\tau(x^{-1})w)$$
$$= \langle \tau(x^{-1})w, v_\alpha \rangle = \langle w, \tau(x)v_\alpha \rangle.$$

Varying w, the relation follows. Now the unitarity of τ^* follows by transport of structure,

$$\langle \tau^*(x)\alpha, \tau^*(x)\beta \rangle = \langle v_{\tau^*(x)\beta}, v_{\tau^*(x)\alpha} \rangle = \langle \tau(x)v_\beta, \tau(x)v_\alpha \rangle$$
$$= \langle v_\beta, v_\alpha \rangle = \langle \alpha, \beta \rangle$$

The Lemma is proven. $\square$

7.2 The Peter-Weyl Theorem

Let K be a compact group, and let $\widehat{K}$ be the set of all equivalence classes of irreducible unitary representations of K. Let $\widehat{K}_{\text{fin}}$ be the subset of all finite dimensional irreducible representations. We want to show that $\widehat{K} = \widehat{K}_{\text{fin}}$.

A *matrix coefficient* for a unitary representation τ of K on V_τ is a function of the form $k \mapsto \langle \tau(k)v, w \rangle$ for some $v, w \in V_\tau$. The matrix coefficients are continuous functions, so they lie in the Hilbert space $L^2(K)$. We need to know that the set of matrix coefficients, where τ runs through all finite dimensional representations is closed under taking complex conjugates. To see this we use Lemma 7.1.3 for a finite dimensional unitary representation (τ, V_τ). So let $v, v' \in V_\tau$ and let $\alpha, \beta \in V_\tau^*$ be their Riesz duals, i.e., $v = v_\alpha$ and $v' = v_\beta$ in the notation of the last section. Then

$$\overline{\langle \tau(x)v, v' \rangle} = \langle v_\beta, \tau(x)v_\alpha \rangle = \langle v_\beta, v_{\tau^*(x)\alpha} \rangle = \langle \tau^*(x)\alpha, \beta \rangle$$

shows that the complex conjugate of a matrix coefficient is indeed a matrix coefficient.

Now, for every class in $\widehat{K}_{\text{fin}}$ choose a representative (τ, V_τ). Choose an orthonormal basis $e_1, \ldots, e_n$ of V_τ and write $\tau_{ij}(k) \overset{\text{def}}{=} \langle \tau(k)e_i, e_j \rangle$ for the corresponding matrix coefficient. It is easy to see that for every $v, w \in V_\tau$ the function $k \mapsto \langle \tau(k)v, w \rangle$ is a linear combination of the τ_{ij}, $1 \le i, j \le \dim V_\tau$. In what follows we shall write $\dim(\tau)$ for $\dim V_\tau$.

Theorem 7.2.1 (Peter-Weyl Theorem).

(a) *For $\tau \ne \gamma$ in $\widehat{K}_{\text{fin}}$ one has*

$$\langle \tau_{ij}, \gamma_{rs} \rangle = \int_K \tau_{ij}(k)\overline{\gamma_{rs}(k)}\, dk = 0.$$

So the matrix coefficients of non-equivalent representations are orthogonal.

(b) *For $\tau \in \widehat{K}_{\text{fin}}$ one has $\langle \tau_{ij}, \tau_{rs} \rangle = 0$, except for the case when $i = r$ and $j = s$. In the latter case the products are $\langle \tau_{ij}, \tau_{ij} \rangle = \frac{1}{\dim(\tau)}$. One can summarize this by saying that the family*

$$\left(\sqrt{\dim(\tau)}\, \tau_{ij} \right)_{\tau, i, j}$$

is an orthonormal system in $L^2(K)$.

(c) *It even is complete, i.e., an orthonormal basis.*

(d) *The translation-representations $(L, L^2(K))$ and $(R, L^2(K))$ decompose into direct sums of finite-dimensional irreducible representations.*

Proof For (a) let $\tau \ne \gamma$ in $\widehat{K}_{\text{fin}}$. Let $T : V_\tau \to V_\gamma$ be linear and set $S = S_T = \int_K \gamma(k^{-1})T\tau(k)\, dk$. Then one has $S\tau(k) = \gamma(k)S$, hence $S = 0$ by Corollary 6.1.9. Let (e_j) and (f_s) be orthonormal bases of V_τ and V_γ, respectively, and choose $T_{js} : V_\tau \to V_\gamma$ given by $T_{js}(v) = \langle v, e_j \rangle f_s$. Let $S_{js} = S_{T_{js}}$ as above. One gets

$$0 = \langle S_{js}e_i, f_r \rangle = \int_K \langle \gamma(k^{-1})T_{js}\tau(k)e_i, f_r \rangle\, dk$$

$$= \int_K \langle \gamma(k^{-1})\langle \tau(k)e_i, e_j \rangle f_s, f_r \rangle\, dk$$

$$= \int_K \langle \tau(k)e_i, e_j \rangle \underbrace{\langle \gamma(k^{-1})f_s, f_r \rangle}_{=\langle f_s, \gamma(k)f_r \rangle = \overline{\langle \gamma(k)f_r, f_s \rangle}}\, dk$$

$$= \int_K \tau_{ij}(k)\overline{\gamma_{rs}(k)}\, dk = \langle \tau_{ij}, \gamma_{rs} \rangle.$$

To prove (b), we perform the same computation for $\gamma = \tau$ to get

$$\langle S_{js}e_i, e_r \rangle = \langle \tau_{ij}, \tau_{rs} \rangle.$$

In this case the matrix S_{js} is a multiple of the identity $S_{js} = \lambda \mathrm{Id}$ for some $\lambda \in \mathbb{C}$, so if $i \neq r$ we infer $\langle S_{js}e_i, e_r \rangle = 0$, hence $\langle \tau_{ij}, \tau_{rs} \rangle = 0$. Assume $j \neq s$. We claim that $S_{js} = 0$, which implies the same conclusion, so in total we get the first assertion of (b). To show $S_{js} = 0$ recall that $S = S_{js} = \lambda \, \mathrm{Id}$, so the trace equals

$$\lambda \dim V_\tau = \mathrm{tr}\,(S) = \mathrm{tr}\, \left(\int_K \tau(k)^{-1} T \tau(k)\, dk \right)$$

$$= \int_K \mathrm{tr}\, \left(\tau(k)^{-1} T \tau(k) \right) dk = \int_K \mathrm{tr}\,(T)\, dk = \mathrm{tr}\,(T),$$

but as $j \neq s$, the trace of T is zero, hence S is zero and so is $\langle Se_i, e_i \rangle = \langle \tau_{ij}, \tau_{i,s} \rangle$. Finally, we consider the case $j = s$ and $i = r$. Then $S_{jj} = \lambda_j \mathrm{Id}$ for some $\lambda_j \in \mathbb{C}$. Our computation shows $\lambda_j = \langle \tau_{ij}, \tau_{ij} \rangle$, independent of i. But $\tau_{ij}(k) = \overline{\tau_{ji}(k^{-1})}$ and therefore, as K is unimodular we get

$$\langle \tau_{ij}, \tau_{ij} \rangle = \int_K \tau_{ij}(k)\overline{\tau_{ij}(k)}\, dk = \int_K \overline{\tau_{ji}(k^{-1})}\tau_{ji}(k^{-1})\, dk$$

$$= \int_K \overline{\tau_{ji}(k)}\tau_{ji}(k)\, dk = \langle \tau_{ji}, \tau_{ji} \rangle.$$

We conclude $\lambda_j = \langle \tau_{ij}, \tau_{ij} \rangle = \langle \tau_{ji}, \tau_{ji} \rangle = \lambda_i$. We call this common value λ and we have to show that $\lambda = \frac{1}{\dim(\tau)}$. Write $n = \dim V_\tau$ and note that $\mathrm{Id} = \sum_{j=1}^n T_{jj}$. Therefore $(n\lambda)\mathrm{Id} = \sum_{j=1}^n S_{jj} = \int_K \tau(k^{-1})\mathrm{Id}\tau(k)\, dk = \mathrm{Id}$ and the claim follows.

Finally, to show (c), let $\tau \in \widehat{K}_{\mathrm{fin}}$, and let M_τ be the subspace of $L^2(K)$ spanned by all matrix coeficients of the representation τ. If $h(k) = \langle \tau(k)v, w \rangle$, then one has

$$h^*(k) = \overline{h(k^{-1})} = \langle \tau(k)w, v \rangle \in M_\tau,$$

$$L_{k_0}h(k) = h(k_0^{-1}k) = \langle \tau(k)v, \tau(k_0)w \rangle \in M_\tau,$$

$$R_{k_0}h(k) = h(kk_0) = \langle \tau(k)\tau(k_0)v, w \rangle \in M_\tau.$$

This means that the finite-dimensional space M_τ is closed under adjoints, and left and right translations. Let M be the closure in $L^2(K)$ of the span of all M_τ, where $\tau \in \widehat{K}_{\mathrm{fin}}$. Then M decomposes into a direct sum of irreducible representations under the left or the right translation. By the discussion preceding the theorem, M is also closed under complex conjugation. We want to show that $L^2(K) = M$, or, equivalently, $M^\perp = 0$. So assume $M^\perp$ is not trivial. Our first claim is that $M^\perp$ contains a non-zero continuous function. Let $H \neq 0$ in $M^\perp$. Let $(\phi_U)_U$ be a Dirac net. Then the net $\phi_U * H$ converges to H in the L^2-norm. Since $M^\perp$ is closed under translation it follows that $\phi_U * H \in M^\perp$ for every U. As there must exist some U

with $\phi_U * H \neq 0$, the first claim follows. So let $F_1 \in M^\perp$ be continuous. After applying a translation and a multiplication by a scalar, we can assume $F_1(e) > 0$. Set $F_2(x) = \int_K F_1 \left(y^{-1}xy\right) dy$. Then $F_2 \in M^\perp$ is invariant under conjugation and $F_2(e) > 0$. Finally put $F(x) = F_2(x) + \overline{F_2 \left(x^{-1}\right)}$. Then the function F is continuous, $F \in M^\perp$, $F(e) > 0$, and $F = F^*$. Consider the operator $T(f) = f * F = R(F)f$ for $f \in L^2(K)$. Since $R : L^1(K) \to \mathcal{B}\left(L^2(K)\right)$ is a *-representation, T is self-adjoint. Further, as $Tf(x) = \int_K f(y)F \left(y^{-1}x\right) dy$, the operator T is an integral operator with continuous kernel $k(x, y) = F \left(y^{-1}x\right)$. By Proposition 5.3.3, $T = T^* \neq 0$ is a Hilbert-Schmidt operator, hence compact, and thus it follows that T has a real eigenvalue $\lambda \neq 0$ with finite dimensional eigenspace V_λ. We claim that V_λ is stable under left-translations. For this let $f \in V_\lambda$, so $f * F = \lambda f$. Then, for $k \in K$ one has $(L_k f) * F = L_k(f * F) = \lambda L_k f$. This implies that V_λ with the left translation gives a finite dimensional unitary representation of K, hence it contains an irreducible subrepresentation $W \subset V_\lambda \subset M^\perp$. Let $f, g \in W$, and let $h(k) = \langle L_k f, g \rangle$ be the corresponding matrix coefficient. One has $h(k) = \int_K f(k^{-1}x)\overline{g(x)} dx$, so $h = \overline{g} * f^* \in M^\perp$. On the other hand, $h \in M$, and so $\langle h, h \rangle = 0$, which is a contradiction. It follows that the assumption is wrong, so $M = L^2(K)$.

Above, we showed in particular that $L^2(K)$ decomposes as the closure of the direct sum $\bigoplus_{\tau \in \widehat{K}_{\text{fin}}} M_\tau$, where the the linear span M_τ of all matrix coefficients of τ has dimension $\dim(\tau)^2$. Since each M_τ is stable under left and right translations, this implies that $\left(L^2(K), L\right)$ and $\left(L^2(K), R\right)$ decompose as direct sums of finite dimensional representations. Hence (d) follows from Proposition 7.1.2 and the Peter-Weyl Theorem is proven.   □

Definition Let π be a finite dimensional representation of the compact group K. The function $\chi_\pi : K \to \mathbb{C}$ defined by $\chi_\pi(k) = \text{tr } \pi(k)$ is called the *character* of the representation π.

Corollary 7.2.2 *Let π, η be two finite-dimensional irreducible unitary representations of the compact group K. For their characters we have*

$$\langle \chi_\pi, \chi_\eta \rangle = \begin{cases} 1 & \text{if } \pi = \eta, \\ 0 & \text{otherwise.} \end{cases}$$

Here the inner product is the one of $L^2(K)$.

Proof The proof follows immediately from the Peter-Weyl Theorem. Note that it is shown in Exercise 7.10 that $\{\chi_\pi : \pi \in \widehat{K}\}$ even forms an orthonormal base of the space $L^2(K/\text{conj})$ of conjugacy invariant L^2-functions on K.   □

Let (π, V_π) be a representation of a locally compact group G. An *irreducible subspace* is a closed subspace $U \subset V_\pi$ which is stable under $\pi(G)$ such that the representation (π, U), obtained by restricting each $\pi(k)$ to U, is irreducible.

Theorem 7.2.3

(a) *Let K be a compact group. Then $\widehat{K} = \widehat{K}_{\mathrm{fin}}$, so every irreducible unitary representation of K is finite dimensional.*

(b) *Every unitary representation of the compact group K is an orthogonal sum of irreducible representations.*

Proof Let (π, V_π) be a unitary representation of K. We show that V_π can be written as a direct sum $V_\pi = \bigoplus_{i \in I} V_i$, where each V_i is a finite dimensional irreducible subspace of V_π. This proves (b) and if we apply this to a given irreducible representation V_π it also implies (a).

So let (π, V_π) be a given unitary representation of K. Consider the set S of all families $(V_i)_{i \in I}$, where each V_i is a finite dimensional irreducible subrepresentation of V_π and for $i \neq j$ in I we insist that V_i and V_j are orthogonal. We introduce a partial order on S given by $(V_i)_{i \in I} \leq (W_\alpha)_{\alpha \in A}$ if and only if $I \subset A$ and for each $i \in I$ we have $V_i = W_i$. The Lemma of Zorn yields the existence of a maximal element $(V_i)_{i \in I}$. We claim that the orthogonal sum $\bigoplus_{i \in I} V_i$ is dense in V_π. This is equivalent to the orthogonal space $W = \left(\bigoplus_{i \in I} V_i\right)^\perp$ being the zero space. Now assume that inside W we find a finite-dimensional irreducible subspace U, then we can extend I by one element i_0 and we set $V_{i_0} = U$ which contradicts the maximality of I. Therefore, it suffices to show that any given non-zero unitary representation (η, W_η) contains a finite-dimensional irreducible subspace. For this let $v, w \in W_\eta$, and let $\psi_{v,w}(x) = \langle \eta(x)v, w \rangle$ be the corresponding matrix coefficient. Then $\psi_{v,w} \in C(K) \subset L^2(K)$ and $\psi_{\eta(y)v,w}(x) = \langle \eta(xy)v, w \rangle = \psi_{v,w}(xy) = R_y \psi_{v,w}(x)$. In other words, for fixed w, the map $v \mapsto \psi_{v,w}$ is a K-homomorphism from V_η to $(R, L^2(K))$. We assume $\langle v, w \rangle \neq 0$. Then this map is non-zero. Since $(R, L^2(K))$ is a direct sum of finite dimensional irreducible representations, there exists an orthogonal projection $P : L^2(K) \to F$ to a finite dimensional irreducible subrepresentation, such that $P(\psi_{v,w}) \neq 0$. So there exists a non-zero K-homomorphism $T : V_\eta \to F$, which is surjective, hence induces an isomorphism from $U = (\ker(T))^\perp \subset V_\eta$ to F. Therefore U is the desired finite-dimensional irreducible subspace. □

We now give a reformulation of the Peter-Weyl Theorem. The group K acts on the space $L^2(K)$ by left and right translations, and these two actions commute, that is to say, we have a unitary representation η of the group $K \times K$ on the Hilbert space $L^2(K)$, given by

$$\eta(k_1, k_2) f(x) = f\left(k_1^{-1} x k_2\right).$$

On the other hand, for $(\tau, V_\tau) \in \widehat{K}$ the group $K \times K$ acts on the finite dimensional vector space $\mathrm{End}(V_\tau) = \mathrm{Hom}_K(V_\tau, V_\tau)$ by

$$\eta_\tau(k_1, k_2)(T) = \tau(k_1) T \tau\left(k_2^{-1}\right).$$

On End (V_τ) we have a natural inner product

$$\langle S, T \rangle = \dim(V_\tau) \operatorname{tr}(ST^*)$$

making the representation of $K \times K$ unitary (Compare with Exercise 5.8).

Theorem 7.2.4 (Peter-Weyl Theorem, second version). *There is a natural unitary isomorphism*

$$L^2(K) \cong \widehat{\bigoplus_{\tau \in \widehat{K}} \operatorname{End}(V_\tau)},$$

which intertwines the conjugation representation η of $K \times K$ on $L^2(K)$ with $\bigoplus_{\tau \in \widehat{K}} \eta_\tau$. This isomorphism maps a given $f \in L^2(K) \subset L^1(K)$ onto $\sum_{\tau \in \widehat{K}} \tau(f)$, where

$$\tau(f) = \int_K f(x) \tau(x) \, dx.$$

In particular, if for a given $f \in L^2(K)$ we define the map $\hat{f} : \widehat{K} \to \bigoplus_{\tau \in \widehat{K}} \operatorname{End}(V_\tau)$ by $\hat{f}(\tau) = \tau(f)$, then we get

$$\|f\| = \|\hat{f}\|$$

for every $f \in L^2(K)$. In this way the Peter-Weyl Theorem presents itself as a generalization of the Plancherel Formula.

Proof Since $\tau \mapsto \tau^*$ is a bijection from $\widehat{K}$ onto itself, the Peter-Weyl Theorem yields the orthonormal basis $\sqrt{\dim(\tau)}\tau_{kl}^*$, where the indices are taken with respect to the dual basis of a given orthonormal basis $\{e_1, \dots, e_{\dim(\tau)}\}$ of V_τ. For $f \in L^2(K)$ and indices i, j one has $\langle \tau(f)e_i, e_j \rangle = \int_K f(x)\tau_{ij}(x) \, dx = \langle f, \overline{\tau_{ij}} \rangle$. If we apply this formula to $f = \sigma_{kl}^* = \overline{\sigma_{kl}}$ for some $\sigma \in \widehat{K}$, we see that $\widehat{\sigma_{kl}^*}(\tau) = \tau(\sigma_{kl}) = 0$ for $\sigma \neq \tau$ and

$$\langle \widehat{\tau_{kl}^*}(\tau)e_i, e_j \rangle = \begin{cases} \dim(\tau) & \text{if } k = i \text{ and } l = j \\ 0 & \text{otherwise.} \end{cases}$$

Thus it follows that τ_{kl}^* is mapped to the operator $\frac{1}{\dim(\tau)} E_{kl}^\tau \in \operatorname{End}(V_\tau)$, where E_{kl}^τ denotes the endomorphism which sends e_k to e_l and all other basis elements to 0. Hence, the basis element $\sqrt{\dim(\tau)}\tau_{kl}^* \in M_{\tau^*}$ is mapped to $\sqrt{\dim(\tau)}E_{kl}^\tau$. It is trivial to check that these elements form an orthonormal basis of $\operatorname{End}(V_\tau)$ with respect to the given inner product. $\square$

Definition Let (τ, V_τ) and (γ, V_γ) be finite dimensional representations of the compact group K. There is a natural representation $\tau \otimes \gamma$ of the group $K \times K$ on the tensor product space $V_\tau \otimes V_\gamma$ given by

$$(\tau \otimes \gamma)(k_1, k_2) = \tau(k_1) \otimes \gamma(k_2).$$

Lemma 7.2.5 *For given $\tau \in \widehat{K}$, there is a natural unitary isomorphism*

$$\Psi : V_\tau \otimes V_{\tau^*} \to \text{End}(V_\tau),$$

which intertwines $\tau \otimes \tau^*$ with η_τ.

Show that the direct summand $\text{End}(V_\pi)$ of $L^2(K)$ equipped with the conjugation action η of $K \times K$ as in the second version of the Peter-Weyl Theorem is equivalent to the irreducible representation $\pi^* \otimes \pi$ of $K \times K$. (Compare with Exercise 5.8.)

Proof The map $\psi : V_\tau \otimes V_{\tau^*} \to \text{End}(V_\tau)$ given by

$$\psi(v \otimes \alpha) = [w \mapsto \alpha(w)v]$$

is linear and sends the simple tensors to the operators of rank one. Every operator of rank one is in the image, so the map is surjective as $\text{End}(V_\tau)$ is linearly generated by the operators of rank one. As the dimensions of the spaces agree, the map is bijective. It further is intertwining, as for $k, l \in K$ one has

$$
\begin{aligned}
\psi\left(\tau \otimes \tau^*(k, l)(v \otimes \alpha)\right)(w) &= \psi\left(\tau(k)v \otimes \tau^*(k)\alpha\right)(w) \\
&= \alpha(\tau(l^{-1})w)\tau(k)v \\
&= \tau(k)\psi(w \otimes \alpha)\tau(l^{-1})(w) \\
&= [\eta_\tau(k, l)\psi(w \otimes \alpha)](w).
\end{aligned}
$$

By Corollary 6.1.9 it follows that, modulo a scalar, ψ is unitary. Plugging in test vectors, one sees that $\Psi = \sqrt{\dim(V_\tau)}^{-1}\psi$ satisfies the lemma. $\qquad\square$

Corollary 7.2.6 *There is a natural unitary isomorphism*

$$L^2(K) \cong \widehat{\bigoplus_{\tau \in \widehat{K}}} V_\tau \otimes V_{\tau^*},$$

where each finite dimensional space $V_\tau \otimes V_{\tau^*}$ carries the tensor product Hilbert-space structure. This isomorphism intertwines the $K \times K$ representation η with the sum of the representations $\tau \otimes \tau^*$, where τ^* is the representation dual to τ. In particular, we get direct sum decompositions

$$L \cong \widehat{\bigoplus_{\tau \in \widehat{K}}} 1_{V_\tau} \otimes \tau^* \quad and \quad R \cong \widehat{\bigoplus_{\tau \in \widehat{K}}} \tau \otimes 1_{V_{\tau^*}}$$

for the left and right regular representations of K.

Proof The corollary is immediate from the theorem and the lemma. The assertion about the left and right translation operations follows from restricting to one factor of the group $K \times K$. $\qquad\square$

7.3 Isotypes

Let (π, V_π) be a unitary representation of the compact group K. For $(\tau, V_\tau) \in \widehat{K}$ we define the *isotype of* τ or the *isotypical component* of τ in π as the subspace

$$V_\pi(\tau) \stackrel{\text{def}}{=} \sum_{\substack{U \subset V_\pi \\ U \cong V_\tau}} U.$$

This is the sum of all invariant subspaces U, which are K-isomorphic to V_τ. Another description of the isotype is this: There is a canonical map

$$T_\tau : \text{Hom}_K(V_\tau, V_\pi) \otimes V_\tau \to V_\pi$$

$$\alpha \otimes v \mapsto \alpha(v).$$

This map intertwines the action $\text{Id} \otimes \tau$ on $\text{Hom}_K(V_\tau, V_\pi) \otimes V_\tau$ with π, from which it follows that the image of T_τ lies in $V_\pi(\tau)$. Indeed, the image is all of $V_\pi(\tau)$, since if $U \subset V_\pi$ is a closed subspace with $\pi|_U \cong \tau$ via $\alpha : V_\tau \to U$, then $U = T_\tau(\alpha \otimes V_\tau)$ by construction of T_τ. Note that if (τ, V_τ) and (σ, V_σ) are two non-equivalent irreducible representations, then $V_\pi(\tau) \perp V_\pi(\sigma)$, which follows from the fact that if $U, U' \subseteq V_\pi$ are subspaces with $U \cong V_\tau, U' \cong V_\sigma$, then the orthogonal projection $P : V_\pi \to U'$ restricts to a K-homomorphism $P|_U : U \to U'$, which therefore must be 0.

Lemma 7.3.1 *On the vector space* $\text{Hom}_K(V_\tau, V_\pi)$ *there is an inner product, making it a Hilbert space, such that* T_τ *is an isometry.*

Proof Let $v_0 \in V_\tau$ be of norm one. For $\alpha, \beta \in H = \text{Hom}_K(V_\tau, V_\pi)$ set $\langle \alpha, \beta \rangle \stackrel{\text{def}}{=} \langle \alpha(v_0), \beta(v_0) \rangle$. As by Corollary 6.1.9, any element of $\text{Hom}_K(V_\tau, V_\pi)$ is either zero or injective, it follows that $\langle \cdot, \cdot \rangle$ is indeed an inner product on H. We show that H is complete. For this let α_n be a Cauchy-sequence in H. Then $\alpha_n(v_0)$ is a Cauchy-sequence in V_π, so there exists $w_0 \in V_\pi$ such that $\alpha_n(v_0)$ converges to w_0. For $k \in K$ the sequence $\alpha_n(\tau(k)v_0) = \pi(k)\alpha_n(v_0)$ converges to $\pi(k)w_0$. Likewise, for $f \in L^1(K)$ the sequence $\alpha_n(\tau(f)v_0) = \pi(f)\alpha_n(v_0)$ converges to $\pi(f)w_0$. Let $I \subset L^1(K)$ be the annihilator of v_0, i.e., I is the set of all $f \in L^1(K)$ with $\tau(f)v_0 = 0$. It follows that every $f \in I$ also annihilates w_0. Therefore the map $\alpha : V_\tau \cong L^1(K)/I \to V_\pi$ mapping $\tau(f)v_0$ to $\pi(f)w_0$ is well-defined and a K-homomorphism. It follows that α is the limit of the sequence α_n, so H is complete. We now show that $T = T_\tau$ is an isometry. For fixed α the inner product on V_τ given by $(v, w) = \langle \alpha(v), \alpha(w) \rangle$ is K-invariant. Therefore, by Lemma 7.1.1, there is $c(\alpha) > 0$ such that $(v, w) = c(\alpha)\langle v, w \rangle$ for all $v, w \in V_\tau$. So we get $\langle T(\alpha \otimes v), T(\alpha \otimes v) \rangle = (v, v) = c(\alpha)\langle v, v \rangle$. Setting $v = v_0$, we conclude that $c(\alpha) = \langle \alpha, \alpha \rangle$, which proves that T_τ is indeed an isometry. $\square$

It follows from the above lemma that $V_\pi(\tau)$ is isometrically isomorphic to the Hilbert space tensor product $\text{Hom}_K(V_\tau, V_\pi) \hat{\otimes} V_\tau$ and that $\pi|_{V_\pi(\tau)}$ is unitarily equivalent to the representation $\mathrm{Id} \otimes \tau$ on this tensor product. If we choose an orthonormal base $\{\alpha_i : i \in I\}$ of $\mathrm{Hom}_K(V_\tau, V_\pi)$, then we get a canonical isomorphism

$$\mathrm{Hom}_K(V_\tau, V_\pi) \hat{\otimes} V_\tau \cong \widehat{\bigoplus_{i \in I}} V_\tau$$

given by sending an elementary tensor $\alpha \otimes v$ to $\sum_{i \in I} \langle \alpha, \alpha_i \rangle v$. Thus we see that $V_\pi(\tau)$ is unitarily equivalent to a direct sum of V_τ's with multiplicity $I = \dim \mathrm{Hom}_K(V_\tau, V_\pi)$.

Theorem 7.3.2

(a) $V_\pi(\tau)$ is a closed invariant subspace of V_π.

(b) $V_\pi(\tau)$ is K-isomorphic to a direct Hilbert sum of copies of V_τ.

(c) V_π is the direct Hilbert sum of the isotypes $V_\pi(\tau)$ where τ ranges over $\widehat{K}$.

Proof As $V_\pi(\tau)$ is an isometric image of a complete space, it is complete, hence closed. The space $V_\pi(\tau)$ is a sum of invariant spaces, hence invariant, so (a) follows. Now let $V_\pi = \bigoplus_i V_i$ be any decomposition into irreducibles. Set $\tilde{V}_\pi(\tau) = \widehat{\bigoplus}_{i: V_i \cong V_\tau} V_i$. Then it follows that $\tilde{V}_\pi(\tau) \subset V_\pi(\tau)$ as the latter contains the direct sum and is closed. Now clearly V_π is the direct Hilbert sum of the spaces $\tilde{V}_\pi(\tau)$, and hence it is also the direct Hilbert sum of the $V_\pi(\tau)$, as the latter are pairwise orthogonal. This implies (c) and a fortiori $\tilde{V}_\pi(\tau) = V_\pi(\tau)$ and thus (b). $\square$

Proposition 7.3.3 *Let (π, V_π) be a unitary representation of the compact group K. For $\tau \in \widehat{K}$ the orthogonal projection $P : V_\pi \to V_\pi(\tau)$ is given by*

$$P(v) = \dim(\tau) \int_K \overline{\chi_\tau(x)} \pi(x) v \, dx.$$

Proof We have to show that for any two vectors $v, w \in V_\pi$ one has $\langle Pv, w \rangle = \dim(\tau) \int_K \overline{\chi_\tau(x)} \langle \pi(x)v, w \rangle \, dx$. Let (v, w) denote the right hand side of this identity. Write $v = v_0 + v_1$, where $v_0 \in V_\pi(\tau)$ and $v_1 \in V_\pi(\tau)^\perp$. Likewise decompose w as $w_0 + w_1$. Then $\langle Pv, w \rangle = \langle v_0, w_0 \rangle$. The Peter-Weyl theorem implies that $(v_0, w_0) = \langle v_0, w_0 \rangle$. To see this, we decompose $V_\pi(\tau)$ into a direct sum of irreducibles, each equivalent to V_τ. It then suffices to assume that v_0, w_0 lie in the same summand, since otherwise we have $\langle Pv, w \rangle = 0 = \langle v_0, w_0 \rangle$. The result then follows from expressing v_0, w_0 in terms of an orthonormal basis of V_τ. The spaces $V_\pi(\tau)$ and its orthocomplement are invariant under π, therefore $(v_0, w_1) = 0 = (v_1, w_0)$. Finally, as $V_\pi(\tau)^\perp$ is a direct sum of isotypes different from τ, the Peter-Weyl theorem also implies that $(v_1, w_1) = 0$. As the map $(\cdot, \cdot)$ is additive in both components, we get

$$(v, w) = (v_0, w_0) + (v_0, w_1) + (v_1, w_0) + (v_1 + w_1)$$

$$= (v_0, w_0) = \langle v_0, w_0 \rangle = \langle Pv, w \rangle,$$

as claimed. $\square$

Example 7.3.4 It follows from the Peter-Weyl Theorem that the isotype $L^2(K)_R(\tau)$ of the right regular representation $(R, L^2(K))$ for the irreducible representation τ of the compact group K is the linear span of the functions $\tau_{ij}(x) = \langle \tau(x)e_i, e_j \rangle$. In particular, it follows that all functions in $L^2(K)_R(\tau)$ are continuous. Similarly, the isotype $L^2(K)_L(\tau)$ of the left regular representation $(L, L^2(K))$ is given by the linear span of the functions $\overline{\tau}_{ij}$, the complex conjugates of the τ_{ij}.

7.4   Induced Representations

Let K be a compact group, and let $M \subset K$ be a closed subgroup. Let (σ, V_σ) be a finite dimensional unitary representation of M. We now define the *induced representation* $\pi_\sigma = \mathrm{Ind}_M^K(\sigma)$ as follows. First define the Hilbert-space $L^2(K, V_\sigma)$ of all measurable functions $f : K \to V_\sigma$ satisfying $\int_K \|f(x)\|_\sigma^2 \, dk < \infty$ modulo nullfunctions, where $\|\cdot\|_\sigma$ is the norm in the space V_σ. This is a Hilbert-space with inner product $\langle f, g \rangle = \int_K \langle f(k), g(k) \rangle_\sigma \, dk$. Choosing an orthonormal basis of V_σ gives an isomorphism $L^2(K, V_\sigma) \cong L^2(K)^{\dim(V_\sigma)}$, which shows completeness of $L^2(K, V_\sigma)$.

The space of the representation π_σ is the space $\mathrm{Ind}_M^K(V_\sigma)$ of all $f \in L^2(K, V_\sigma)$ such that for every $m \in M$ the identity $f(mk) = \sigma(m)f(k)$ holds almost everywhere in $k \in K$. This is a closed subspace of $L^2(K, V_\sigma)$ as we have

$$\mathrm{Ind}_M^K(V_\sigma) = \bigcap_{m \in M} \ker T_m,$$

where for given $m \in M$ the continuous operator $f \mapsto L_{m^{-1}}f - \sigma(m)f$ is denoted by T_m. The representation π_σ is now defined by

$$\pi_\sigma(y)f(x) = f(xy).$$

The representation π_σ is clearly unitary.

It suffices to consider finite dimensional, indeed irreducible representations σ here, since an arbitrary representation σ of M decomposes as a direct sum $\sigma = \bigoplus_{i \in I} \sigma_i$ of irreducibles and there is a canonical isomorphism

$$\mathrm{Ind}_M^K \left(\bigoplus_{i \in I} \sigma_i \right) \cong \bigoplus_{i \in I} \mathrm{Ind}_M^K(\sigma_i).$$

So suppose that σ is irreducible. As K is compact, π_σ decomposes as a direct sum of irreducible representations $\tau \in \widehat{K}$, each occurring with some multiplicity $[\pi_\sigma : \tau] \overset{\text{def}}{=} \dim \mathrm{Hom}_K(V_\tau, V_{\pi_\sigma})$.

Theorem 7.4.1 (Frobenius reciprocity). *If σ is irreducible, the multiplicities $[\pi_\sigma : \tau]$ are all finite and can be given as*

$$[\pi_\sigma : \tau] = [\tau|_M : \sigma].$$

More precisely, for every irreducible representation (τ, U) there is a canonical isomorphism $\mathrm{Hom}_K\left(U, \mathrm{Ind}_M^K(V_\sigma)\right) \to \mathrm{Hom}_M\left(U|_M, V_\sigma\right).$

Proof Let V^c be the subspace of V_{π_σ} consisting of all continuous functions $f : K \to V_\sigma$ with $f(mk) = \sigma(m)f(k)$. The space V^c is stable under the K-action and dense in the Hilbert space V_{π_σ}, which can be seen by approximating any f in V_{π_σ} by $\pi_\sigma(\phi)f$ with Dirac functions ϕ in $C(K)$ of arbitrary small support. Let $\alpha \in \mathrm{Hom}_K\left(U, \mathrm{Ind}_M^K(V_\sigma)\right)$. We show that the image of α lies in V^c. For this recall that by the Peter-Weyl Theorem the space $L^2(K)$ decomposes into a direct sum of isotypes $L^2(K)(\gamma)$ for $\gamma \in \widehat{K}$. Here we consider the K-action by right translations only. Each isotype $L^2(K)(\gamma)$ is finite dimensional and consists of continuous functions. We have isometric K-homomorphisms,

$$\alpha : U \to \mathrm{Ind}_M^K(V_\sigma) \hookrightarrow L^2(K, V_\sigma) \xrightarrow{\cong} L^2(K) \otimes V_\sigma,$$

where K acts trivially on V_σ. This implies that $\alpha(U) \subset L^2(K)(\tau) \otimes V_\sigma$ consists of continuous functions. Let $\delta : V^c \to V_\sigma$ be given by $\delta(f) = f(1)$, and define $\psi : \mathrm{Hom}_K(U, \mathrm{Ind}_M^K(V_\sigma)) \to \mathrm{Hom}_M(U|_M, V_\sigma)$ by $\psi(\alpha)(u) = \delta(\alpha(u)) = \alpha(u)(1)$. We claim the ψ is a bijection. For injectivity assume that $\psi(\alpha) = 0$. Then for every $u \in U$ and $k \in K$ one has $\alpha(u)(k) = \pi_\sigma(k)\alpha(u)(1) = \alpha(\tau(k)u)(1) = \psi(\alpha)(\tau(k)u) = 0$, which means $\alpha = 0$.

For surjectivity let $\beta \in \mathrm{Hom}_M(U, V_\sigma)$ and define an element $\alpha \in \mathrm{Hom}_{\mathbb{C}}\left(U, \mathrm{Ind}_M^K(V_\sigma)\right)$ by $\alpha(u)(k) = \beta\left(\tau(k^{-1})u\right)$. By definition, α is a K-homomorphism and $\beta = \psi(\alpha)$. The theorem is proven. $\square$

Example 7.4.2 Let M be a closed subgroup of the compact group K. Then K/M carries a unique Radon measure μ that is invariant under the left translation action of the group K and is normalized by $\mu(K/M) = 1$. The group K acts on the Hilbert space $L^2(K/M, \mu)$ by left translations and this constitutes a unitary representation. This representation is isomorphic to the induced representation $\mathrm{Ind}_M^K(\mathbb{C})$ induced from the trivial representation. An isomorphism between these representations is given by the map $\Phi : L^2(K/M) \to \mathrm{Ind}_M^K(\mathbb{C})$, which maps $\psi \in L^2(K/M)$ to the function $\Phi(\psi) : K \to \mathbb{C}$ defined by

$$\Phi(\psi)(k) \stackrel{\text{def}}{=} \psi\left(k^{-1}M\right).$$

Now, for any $\tau \in \widehat{K}$ the multiplicity $[\tau|_M : 1]$ equals $\dim V_\tau^M$, where V_τ^M denotes the space of M-invariant vectors in V_τ. Thus by Frobenius we get

$$L^2(K/M) \cong \widehat{\bigoplus}_{\tau \in \widehat{K}} \dim\left(V_\tau^M\right) V_\tau$$

where $\dim\left(V_\tau^M\right) V_\tau$ denotes the $\dim\left(V_\tau^M\right)$-fold direct sum of the V_τ's.

7.5   Representations of SU(2)

In this section we consider the irreducible representations of the compact group SU(2). We use the description of these representations to construct decompositions of the Hilbert spaces $L^2(S^3)$ and $L^2(S^2)$, thus giving a glance into the harmonic analysis of the spheres. For this recall the *n-dimensional sphere*, $S^n = \{x \in \mathbb{R}^{n+1} : \|x\| = 1\}$, where $\|x\| = \sqrt{x_1^2 + \cdots + x_{n+1}^2}$ is the *euclidean norm* on $\mathbb{R}^{n+1}$. The set S^n inherits a topology from $\mathbb{R}^{n+1}$. For a subset $A \subset S^n$ let IA be the set of all ta, where $a \in A$ and $0 \le t \le 1$. The set $IA \subset \mathbb{R}^n$ is Borel measurable if and only if $A \subset S^n$ is, (Exercise 7.13). For a measurable set $A \subset S^n$, define the normalized Lebesgue measure as $\mu(A) \overset{\text{def}}{=} \frac{\lambda(IA)}{\lambda(IS^n)}$, where λ denotes the Lebesgue measure on $\mathbb{R}^{n+1}$. As a consequence of the transformation formula on $\mathbb{R}^{n+1}$, the Lebesgue measure λ is invariant under the action of the orthogonal group

$$O(n+1) \overset{\text{def}}{=} \left\{ g \in M_{n+1}(\mathbb{R}) : g^t g = 1 \right\}.$$

The group $O(n+1)$ can also be described as the group of all $g \in M_{n+1}(\mathbb{R})$ such that $\|gv\| = \|v\|$ holds for every $v \in \mathbb{R}^{n+1}$. It follows from this description, that $O(n+1)$ leaves stable the sphere S^n and that the measure μ is invariant under $O(n+1)$. We denote by $SO(n+1)$ the *special orthogonal group*, i.e., the group of all $g \in O(n+1)$ of determinant one. For the next lemma, we consider $O(n)$ as a subgroup of $O(n+1)$ via the embedding $g \mapsto \left(\begin{smallmatrix} 1 & 0 \\ 0 & g \end{smallmatrix} \right)$.

Lemma 7.5.1 *Let $n \in \mathbb{N}$, and let $e_1 = (1,0,\ldots,0)^t$ be the first standard basis vector of $\mathbb{R}^{n+1}$. The matrix multiplication $g \mapsto ge_1$ gives an identification*

$$S^n \cong O(n+1)/O(n) \cong SO(n+1)/SO(n).$$

This map is invariant under left translations and the normalized Lebesgue measure on S^n is the unique normalized invariant measure on this quotient space.

Proof The group $O(n)$ is the subgroup of $O(n+1)$ of all elements with first column equal to e_1, so it is the stabilizer of e_1 and one indeed gets a map $O(n+1)/O(n) \to S^n$. As the invariance of the measure is established by the transformation formula, we only need to show surjectivity. Now let $v \in S^n$. Then there always exists a rotation in $SO(n+1)$ that transforms e_1 into v. One simply chooses a rotation around an axis that is orthogonal to both e_1 and v. This proves surjectivity. The assertion on the measure is due to the uniqueness of invariant measures. $\qquad\square$

Recall that SU(2) is the group of all matrices $g \in M_2(\mathbb{C})$ that are unitary: $g^* g = gg^* = 1$ and satisfy $\det(g) = 1$. These conditions imply

$$SU(2) = \left\{ \begin{pmatrix} a & -\bar{b} \\ b & \bar{a} \end{pmatrix} : \begin{pmatrix} a \\ b \end{pmatrix} \in S^3 \right\},$$

where we realize the three sphere S^3 as the set of all $z \in \mathbb{C}^2$ with $|z_1|^2 + |z_2|^2 = 1$. From this description and the fact that the Lebesgue measure on S^3 is invariant under O(4) as well as the uniqueness of invariant measures we get the following lemma.

Lemma 7.5.2 *The map* SU(2) $\rightarrow$ S^3, *mapping the matrix* $g \in$ SU(2) *to its first column, is a homeomorphism. Via this homeomorphism, the normalized Lebesgue measure on* S^3 *coincides with the normalized Haar measure on* SU(2).

We want to obtain a convenient formula for computations with the Haar integral on SU(2) $\cong S^3$. Recall from Calculus that the gamma function $\Gamma : (0, \infty) \rightarrow \mathbb{R}$ is defined by the integral $\Gamma(x) = \int_0^\infty t^{x-1} e^{-t} \, dt$. Note that $\Gamma(1) = 1$ and that $\Gamma(x + 1) = x\Gamma(x)$ for every $x > 0$, which implies that $\Gamma(n) = (n - 1)!$ for every $n \in \mathbb{N}$. Moreover, via the substitution $t = r^2$ we get the alternative formula $\Gamma(x) = 2 \int_0^\infty r^{2x-1} e^{-r^2} \, dr$, which we shall use below.

Lemma 7.5.3 *Let* $f : S^3 \rightarrow \mathbb{C}$ *be any integrable function, and for each* $m \in \mathbb{N}_0$ *let* $F_m : \mathbb{C}^2 \rightarrow \mathbb{C}$ *be defined by* $F_m(rx) = r^m f(x)$ *for every* $x \in S^3$ *and* $r > 0$. *Further let* $c_m \overset{\text{def}}{=} \pi^{-2} \Gamma\left(\frac{m}{2} + 2\right)^{-1}$. *Then*

$$\int_{S^3} f(x) \, d\mu(x) = c_m \int_{\mathbb{C}^2} F_m(z) e^{-(\|z\|^2)} d\lambda(z),$$

where λ *stands for the Lebesgue measure on* $\mathbb{C}^2 \cong \mathbb{R}^4$, *and* $\|z\|^2 = |z_1|^2 + |z_2|^2$.

Proof Integration in polar coordinates on $\mathbb{C}^2$ implies

$$\int_{\mathbb{C}^2} F_m(z) e^{-(\|z\|^2)} \, d\lambda(z) = c \int_0^\infty r^3 \int_{S^3} F_m(rx) e^{-r^2} \, d\mu(x) \, dr$$

$$= c \left(\int_0^\infty r^{3+m} e^{-r^2} \, dr \right) \int_{S^3} f(x) \, d\mu(x)$$

$$= \frac{c}{2} \Gamma\left(\frac{m}{2} + 2\right) \int_{S^3} f(x) \, d\mu(x).$$

where c is some positive constant (the non-normalized volume of S^3). To compute the constant c let $f \equiv 1$ and $m = 0$. Since $\Gamma(2) = 1$ one gets

$$c = 2 \int_{\mathbb{C}^2} e^{-(\|z\|^2)} d\lambda(z) = 2 \left(\int_{\mathbb{C}} e^{-|z_1|^2} dz_1 \right) \left(\int_{\mathbb{C}} e^{-|z_2|^2} dz_2 \right) = 2\pi^2.$$

The lemma follows. $\square$

For $m \in \mathbb{N}_0$ let $\mathcal{P}_m$ denote the set of homogeneous polynomials on $\mathbb{C}^2$ of degree m. In other words, $\mathcal{P}_m$ is the space of all polynomial functions $p : \mathbb{C}^2 \rightarrow \mathbb{C}$ that satisfy $p(tz) = t^m p(z)$ for every $z \in \mathbb{C}^2$ and every $t \in \mathbb{C}$. Every $p \in \mathcal{P}_m$ can uniquely be written as $p(z) = \sum_{k=0}^m c_k z_1^k z_2^{m-k}$. For $p, \eta \in \mathcal{P}_m$ define

$$\langle p, \eta \rangle_m \overset{\text{def}}{=} \langle p|_{S^3}, \eta|_{S^3} \rangle_{L^2(S^3)} = \int_{S^3} p(x)\overline{\eta(x)}\,d\mu(x).$$

It then follows from Lemma 7.5.3 that

$$\langle p, \eta \rangle_m = c_{2m} \int_{\mathbb{C}^2} p(z)\overline{\eta(z)}e^{-\|z\|^2}\,d\lambda(z).$$

We define a representation π_m of SU(2) on $\mathcal{P}_m$ by

$$(\pi_m(g)p)(z) \overset{\text{def}}{=} p\left(g^{-1}(z)\right).$$

Theorem 7.5.4 *For every $m \geq 0$, the representation $(\pi_m, \mathcal{P}_m)$ is irreducible. Every irreducible unitary representation of SU(2) is unitarily equivalent to one of the representations $(\mathcal{P}_m, \pi_m)$. Thus*

$$\widehat{SU(2)} = \{[(\mathcal{P}_m, \pi_m)] : m \in \mathbb{N}_0\},$$

where $[(\mathcal{P}_m, \pi_m)]$ denotes the equivalence class of $(\mathcal{P}_m, \pi_m)$.

Proof This is Theorem 10.2.2 in [Dei05]. □

The next corollary follows from the Peter-Weyl Theorem.

Corollary 7.5.5 *The SU(2) representation on $L^2(S^3)$ is isomorphic to the orthogonal sum $\bigoplus_{m \geq 0}(m+1)\mathcal{P}_m$, where $\mathcal{P}_m$ is the space of homogeneous polynomials of degree m and each $\mathcal{P}_m$ occurs with multiplicity $m + 1$.*

We want to close this section with a study of the two-sphere S^2. For each $\lambda \in \mathbb{T}$ consider the matrix $g_\lambda \overset{\text{def}}{=} \text{diag}(\lambda, \bar{\lambda})$. Then we may regard $\mathbb{T} \cong \{g_\lambda : \lambda \in \mathbb{T}\}$ as a closed subgroup of SU(2). Recall that we can identify SU(2) $\cong S^3$. The map $p : S^3 \to S^2$ of the following lemma is known as the *Hopf fibration* of S^3.

Lemma 7.5.6 *Let us realize the two-sphere as*

$$S^2 = \left\{(v, x) : v \in \mathbb{C}, x \in \mathbb{R} \text{ and } |v|^2 + x^2 = 1\right\}.$$

Then the map $\eta : S^3 \to S^2$ defined by $\eta(a, b) = \left(2a\bar{b}, |a|^2 - |b|^2\right)$ factors through a homeomorphism SU(2)/$\mathbb{T} \cong S^2$, which maps the normalized SU(2)-invariant measure on SU(2)/$\mathbb{T}$ to the normalized Lebesgue measure on S^2.

Proof For $(a, b) \in S^3$ we compute $|2a\bar{b}|^2 + (|a|^2 - |b|^2)^2 = 4|ab|^2 + |a|^4 - 2|ab|^2 + |b|^4 = (|a|^2 + |b|^2)^2 = 1$, which shows that the image of η lies in S^2. To see that the map is surjective let $(v, x) \in S^2$ be given. Choose $a, b \in \mathbb{C}$ with $|a|^2 + |b|^2 = 1$ and $|a|^2 - |b|^2 = x$, which is possible since $|x| \leq 1$. Then $|v| = 2|ab|$, and there exists a complex number z of modulus one such that $v = 2za\bar{b}$. It then follows that $\eta(za, b) = (v, x)$.

We now claim that $\eta(a,b) = \eta(a',b')$ if and only if there exists $\lambda \in \mathbb{T}$ with $(a,b) = \lambda(a',b')$. The if direction is easy to check, so assume now that $(2a\bar{b}, |a|^2 - |b|^2) = (2a'\bar{b}', |a'|^2 - |b'|^2)$. Since $|a|^2 + |b|^2 = 1 = |a'|^2 + |b'|^2$ we get $1 - 2|b|^2 = |a|^2 - |b|^2 = |a'|^2 - |b'|^2 = 1 - 2|b'|^2$, from which it follows that $|b| = |b'|$. Hence $\lambda = \bar{b}'/\bar{b} \in \mathbb{T}$ with $b = \lambda b'$. Since $a\bar{b} = a'\bar{b}'$ it also follows that $a = \lambda a'$.

Finally, since

$$\begin{pmatrix} a & -\bar{b} \\ b & \bar{a} \end{pmatrix} \begin{pmatrix} \lambda & 0 \\ 0 & \bar{\lambda} \end{pmatrix} = \begin{pmatrix} \lambda a & -\overline{\lambda b} \\ \lambda b & \overline{\lambda a} \end{pmatrix}$$

it is then clear that η factorizes through a bijection $SU(2)/\mathbb{T} \cong S^2$. Since η is continuous and all spaces are compact, this bijection is also a homeomorphism.

We next need to show that the induced action of $g \in SU(2)$ on $S^2 \cong SU(2)/\mathbb{T}$ comes from some linear transformation. Since it maps S^2 to S^2 it is then automatically orthogonal. But if $g = \begin{pmatrix} a & -\bar{b} \\ b & \bar{a} \end{pmatrix}$ and $\begin{pmatrix} v \\ x \end{pmatrix} = p\begin{pmatrix} z \\ w \end{pmatrix} \in S^2$ then a short computation shows that

$$g \cdot \begin{pmatrix} v \\ x \end{pmatrix} = p\begin{pmatrix} az - \bar{b}w \\ bz + \bar{a}w \end{pmatrix} = \begin{pmatrix} 2(az - \bar{b}w)(\bar{b}\bar{z} + a\bar{w}) \\ |az - \bar{b}w|^2 - |bz + \bar{a}w|^2 \end{pmatrix}$$

$$= \begin{pmatrix} 2a\bar{b}x - a^2 v - \bar{b}^2\bar{v} \\ (|a|^2 - |b|^2)x - abv - \overline{ab}\bar{v} \end{pmatrix}.$$

This expression is obviously $\mathbb{R}$-linear in v and x. This implies that $SU(2)$ acts on S^2 through orthogonal transformations. As the normalized Lebesgue measure on S^2 is invariant under such transformations, it is invariant under the action of $SU(2)$, so it coincides with the unique invariant measure on $SU(2)/\mathbb{T}$.   □

The left translation on $SU(2)/\mathbb{T}$ induces a unitary representation π of $SU(2)$ on $L^2(S^2) \cong L^2(SU(2)/\mathbb{T})$. We will now make use of the Frobenius reciprocity to give an explicit decomposition of this representation.

Proposition 7.5.7 *The representation π of $SU(2)$ on $L^2(S^2)$ is isomorphic to the direct sum $\bigoplus_{m \geq 0} \pi_{2m}$.*

Proof As $SU(2)$ is a compact group, Theorem 7.2.3 implies that π is a direct sum of irreducibles. By Theorem 7.5.4, it is a direct sum of copies of the π_m. It remains to show that for $m \geq 0$ the irreducible representation π_m has multiplicity 1 in π if m is even and 0 otherwise. Example 7.4.2 shows that π is equivalent to the induced representation $\mathrm{Ind}_{\mathbb{T}}^{SU(2)}(1)$, hence the Frobenius reciprocity applies. So by Theorem 7.4.1 we can compute the multiplicity as $[\pi : \pi_m] = [\pi_m|_{\mathbb{T}} : 1]$. Using the basis $z_1^m, z_1^{m-1}z_2, \ldots, z_2^m$ one sees that

$$[\pi_m|_{\mathbb{T}} : 1] = \begin{cases} 1 & m \text{ even,} \\ 0 & \text{otherwise.} \end{cases}$$

The proposition is proven.   □

7.6   Exercises

Exercise 7.1   Let $\mathcal{A}$ be an abelian subgroup of $U(n)$. Show that there is $S \in GL_2(\mathbb{C})$ such that $S\mathcal{A}S^{-1}$ consists of diagonal matrices only.

Exercise 7.2   Let G be a finite group.

(a) Show that the number of elements of G equals $\sum_{\tau \in \widehat{G}} \dim(V_\tau)^2$.

(b) Show that the space of conjugation-invariant functions has a basis $(\chi_\tau)_{\tau \in \widehat{G}}$. Conclude that the number of irreducible representations of G equals the number of conjugacy classes.

Exercise 7.3   Let (π, V_π) and (η, V_η) be two unitary representations of the compact group K. Suppose there exists a bijective bounded linear operator $T : V_\pi \to V_\eta$ that intertwines π and η, i.e., $T\pi(k) = \eta(k)T$ holds for every $k \in K$. Show that there already exists a unitary intertwining operator $S : V_\pi \to V_\eta$.

Exercise 7.4   For a compact group K write $L^2(\widehat{K}) := \bigoplus_{\tau \in \widehat{K}} \mathrm{End}(V_\tau)$ equipped with the inner product as in the second version of the Peter-Weyl Theorem and let

$$\mathcal{F} : L^2(K) \to L^2(\widehat{K}), \quad \mathcal{F}(f) = \hat{f}$$

be the Plancherell-isomorphism as given in that theorem. Show that the inverse of this isomorphism is given by the inverse Fourier transform

$$\widehat{\mathcal{F}} : L^2(\widehat{K}) \to L^2(K); \widehat{\mathcal{F}}(g)(x) = \sum_{\tau \in \widehat{K}} \dim(\tau)\,\mathrm{tr}\left(g(\tau)\tau(x^{-1})\right),$$

for $(\tau \mapsto g(\tau)) \in L^2(\widehat{K})$.

Exercise 7.5   Let K and L be compact groups, and let $\tau \in \widehat{K}$ and $\eta \in \widehat{L}$. Show that $\tau \otimes \eta(k, l) = \tau(k) \otimes \tau(l)$ defines an element of $\widehat{K \times L}$ and that the map $(\tau, \eta) \mapsto \tau \otimes \eta$ is a bijection from $\widehat{K} \times \widehat{L}$ to $\widehat{K \times L}$.

Exercise 7.6   Let K be a compact group, let $(\pi, V_\pi) \in \widehat{K}$, and let χ_π be its character. Show that for $x \in K$, $\int_K \pi(kxk^{-1})\,dk = \frac{\chi_\pi(x)}{\dim V_\pi}\,\mathrm{Id}$. Conclude that $\chi_\pi(x)\chi_\pi(y) = \dim(V_\pi)\int_K \chi_\pi\left(kxk^{-1}y\right)\,dk$ holds for all $x, y \in K$.

Exercise 7.7   Let (π, V_π) be a finite dimensional representation of the compact group K with character χ_π. Let π^* be the *dual representation* on $V_{\pi^*} = V_\pi^*$. Show that $\chi_{\pi^*} = \overline{\chi_\pi}$.

Exercise 7.8   Keep the notation of the last exercise. The representation π is called a *self-dual representation* if $\pi \cong \pi^*$. Show that the following are equivalent.

(a) π is self-dual,

(b) there exists an antilinear bijective K-homomorphism $C : V_\pi \to V_\pi$,

(c) there exists a real sub vector space $V_\mathbb{R}$ of V_π such that V_π is the orthogonal sum of $V_\mathbb{R}$ and $i V_\mathbb{R}$, and $\pi(K)V_\mathbb{R} = V_\mathbb{R}$.

(d) χ_π takes only real values,

Exercise 7.9 Show that every unitary representation of SU(2) is self dual. (Hint: Use Theorem 7.2.3.)

Exercise 7.10 Let the compact group K act on $L^2(K)$ by conjugation, i.e., $k.f(x) = f(k^{-1}xk)$. Show that the space of K-invariants $L^2(K/\text{conj})$ is closed and that $(\chi_\pi)_{\pi \in \widehat{K}}$ is an orthonormal basis of the Hilbert space $L^2(K/\text{conj})$.

Exercise 7.11 Let (π, V_π) be a representation of a compact group K on a Banach space V_π. Show that $V_\pi = \bigoplus_{i \in I} V_i$, where each V_i is a finite-dimensional irreducible subspace.

(Hint: Use matrix coefficients as in the proof of Theorem 7.2.3 to get a map $T : V_\pi \to W$, where W is a finite-dimensional irreducible representation. Then fix a complementary space of $\ker(T)$ inside V_π and apply a projection operator as in Proposition 7.3.3.)

Exercise 7.12 Let K be a compact group. Show that the following are equivalent.

- Every character χ_π for $\pi \in \widehat{K}$ is real valued.

- For every $k \in K$ there exists $l \in K$ such that $lkl^{-1} = k^{-1}$.

Exercise 7.13 For a subset $A \subset S^n$ let IA be defined as in the beginning of Sect. 7.5. Show that A is Borel measurable as a subset of S^n if and only if IA is measurable as a subset of $\mathbb{R}^n$.

Exercise 7.14 Consider the map $\phi : \text{SU}(2) \times \mathbb{T} \to \text{U}(2)$ that sends a pair (g, z) to the matrix zg. Show that ϕ is a surjective homomorphism. Compute $\ker \phi$ and $\widehat{\text{U}(2)}$.

Exercise 7.15 Via Lemma 7.5.6, the group SU(2) acts on $S^2 \cong \text{SU}(2)/\mathbb{T}$. Show that this action determines a surjective homomorphism $\psi : \text{SU}(2) \to \text{SO}(3)$ such that $\ker \psi = \{\pm I\}$. In particular, this gives an isomorphism $\text{SO}(3) \cong \text{SU}(2)/\{\pm I\}$. Use this to compute $\widehat{\text{SO}(3)}$.

Chapter 8
Direct Integrals

Direct integrals are a generalization of direct sums. For a compact group every representation is a direct sum of irreducibles. This property fails in general for non-compact groups. The best one can get for general groups is a direct integral decomposition into factor representations. The latter is a notion more general than irreducibility. For nice groups these notions coincide, and then every unitary representation is a direct integral of irreducible representations.

8.1 Von Neumann Algebras

Let H be a Hilbert space. For a subset M of the space of bounded operators $\mathcal{B}(H)$ on H, define the *commutant* to be

$$M^\circ \stackrel{\text{def}}{=} \{T \in \mathcal{B}(H) : Tm = mT \ \forall m \in M\}.$$

So the commutant is the centralizer of M in $\mathcal{B}(H)$. If $M \subset N \subset \mathcal{B}(H)$, then $N^\circ \subset M^\circ$. We write $M^{\circ\circ}$ for the *bi-commutant*, i.e., the commutant of M°. For a subset M of $\mathcal{B}(H)$ we define its *adjoint set* to be the set M^* of all adjoints m^* where m is in M. The set M is called a *self-adjoint set* if $M = M^*$.

We define a *von Neumann algebra* to be a sub-*-algebra $\mathcal{A}$ of $\mathcal{B}(H)$ that satisfies $\mathcal{A}^{\circ\circ} = \mathcal{A}$. A von Neumann algebra is closed in the operator norm, and so every von Neumann algebra is a C^*-algebra. The converse does not hold (See Exercise 8.6).

For a subset $M \subset \mathcal{B}(H)$, one has $M \subset M^{\circ\circ}$ and hence $M^{\circ\circ\circ} \subset M^\circ$. Since, on the other hand, also $M^\circ \subset (M^\circ)^{\circ\circ} = M^{\circ\circ\circ}$, it follows $M^\circ = M^{\circ\circ\circ}$, so M° is a von Neumann algebra if M is a self-adjoint set. In particular, for a self-adjoint set M the algebra $M^{\circ\circ}$ is the smallest von Neumann algebra containing M, called the *von Neumann algebra generated by M*.

Let $\mathcal{A} \subset \mathcal{B}(H)$ be a von Neumann algebra. Then $Z(\mathcal{A}) = \mathcal{A} \cap \mathcal{A}^\circ$ is the *center* of $\mathcal{A}$, i.e., the set of elements a of $\mathcal{A}$ that commute with every other element of $\mathcal{A}$. A von Neumann algebra $\mathcal{A}$ is called a *factor* if the center is trivial, i.e., if $Z(\mathcal{A}) = \mathbb{C} \, \mathrm{Id}$.

A. Deitmar, S. Echterhoff, *Principles of Harmonic Analysis*, Universitext,
DOI 10.1007/978-3-319-05792-7_8, © Springer International Publishing Switzerland 2014

Examples 8.1.1

- $\mathcal{A} = \mathcal{B}(H)$ is a factor, this is called a *type-I factor*.

- $\mathcal{A} = \mathbb{C}\,\mathrm{Id}$ is a factor.

- The algebra of diagonal matrices in $M_2(\mathbb{C}) \cong \mathcal{B}(\mathbb{C}^2)$ is a von Neumann algebra, which is not a factor.

- Let V, W be two Hilbert spaces. The algebra $\mathcal{B}(V) \otimes \mathcal{B}(W)$ acts on the Hilbert tensor product $V \hat{\otimes} W$ via $A \otimes B(v \otimes w) = A(v) \otimes B(w)$. Then the von Neumann algebra generated by the image of $\mathcal{B}(V) \otimes \mathcal{B}(W)$ is the entire $\mathcal{B}(V \hat{\otimes} W)$ (See Exercise 8.2).

8.2   Weak and Strong Topologies

Let H be a Hilbert space. On $\mathcal{B}(H)$ one has the topology induced by the operator norm, called the *norm topology*. There are other topologies as well. For instance, every $v \in H$ induces a seminorm on $\mathcal{B}(H)$ through $T \mapsto \|Tv\|$. The topology given by this family of seminorms is called the *strong topology* on $\mathcal{B}(H)$. Likewise, any two $v, w \in H$ induce a seminorm by $T \mapsto |\langle Tv, w \rangle|$. The topology thus induced is called the *weak topology*. It is clear that norm convergence implies strong convergence and that strong convergence implies weak convergence. Therefore, for a set $\mathcal{A} \subset \mathcal{B}(H)$ one has

$$\mathcal{A} \subset \overline{\mathcal{A}}^n \subset \overline{\mathcal{A}}^s \subset \overline{\mathcal{A}}^w,$$

where $\overline{\mathcal{A}}^n$ denotes the closure of $\mathcal{A}$ in the norm topology, or norm closure, $\overline{\mathcal{A}}^s$ the strong closure, and $\overline{\mathcal{A}}^w$ the weak closure. In general, these closures will differ from each other. It is easy to see that $\overline{\mathcal{A}}^s, \overline{\mathcal{A}}^w \subset \mathcal{A}^{\circ\circ}$ since multiplication in $\mathcal{B}(H)$ is easily seen to be separately continuous with respect to the weak topology. Hence every von Neumann algebra is strongly and weakly closed.

Theorem 8.2.1 (von Neumann's Bicommutant Theorem). *Let H be a Hilbert space, and let $\mathcal{A}$ be a unital *-subalgebra of $\mathcal{B}(H)$. Then $\overline{\mathcal{A}}^s = \overline{\mathcal{A}}^w = \mathcal{A}^{\circ\circ}$.*

Proof It suffices to show that $\mathcal{A}^{\circ\circ} \subset \overline{\mathcal{A}}^s$. Let $T \in \mathcal{A}^{\circ\circ}$. We want to show that T lies in the strong closure of $\mathcal{A}$. A neighborhood base of zero in the strong topology is given by the system of all sets of the form $\{S \in \mathcal{B}(H) : \|Sv_j\| < \varepsilon, \ j = 1, \ldots, n\}$ where $v_1, \ldots, v_n$ are arbitrary vectors in H and $\varepsilon > 0$. So it suffices to show that for given $v_1, \ldots, v_n \in H$ and $\varepsilon > 0$ there is $a \in \mathcal{A}$ with $\|Tv_j - av_j\| < \varepsilon$ for $j = 1, \ldots n$. For this let $\mathcal{B}(H)$ act diagonally on H^n. The commutant of $\mathcal{A}$ in $\mathcal{B}(H^n)$ is the algebra of all $n \times n$ matrices with entries in $\mathcal{A}^\circ$, and the bicommutant of $\mathcal{A}$ in $\mathcal{B}(H^n)$ is the algebra $\mathcal{A}^{\circ\circ}I$, where $I = I_n$ denotes the $n \times n$ unit matrix. Consider the vector $v = (v_1, \ldots, v_n)^t$ in H^n. The closure of $\mathcal{A}v$ in H^n is a closed, $\mathcal{A}$-stable subspace of H^n. As $\mathcal{A}$ is a *-algebra, the orthogonal complement $(\mathcal{A}v)^\perp$ is $\mathcal{A}$-stable as well;

therefore the orthogonal projection P onto the closure of $\mathcal{A}v$ is in the commutant of $\mathcal{A}$ in $\mathcal{B}(H^n)$. It follows that $T \in \mathcal{A}^{\circ\circ}I$ commutes with P and leaves $\overline{\mathcal{A}v}$ stable. One concludes $Tv \in \overline{\mathcal{A}v}$, and so there is, to given $\varepsilon > 0$, an element a of $\mathcal{A}$ such that $\|Tv - av\| < \varepsilon$, which implies the desired $\|Tv_j - av_j\| < \varepsilon$ for $j = 1, \ldots, n$. $\quad\square$

The Bicommutant Theorem says that for a *-subalgebra $\mathcal{A}$ of $\mathcal{B}(H)$ the von Neumann algebra generated by $\mathcal{A}$ equals the weak or strong closure of $\mathcal{A}$.

Lemma 8.2.2 *A von Neumann algebra $\mathcal{A}$ is generated by its unitary elements.*

Proof Let $\mathcal{A}$ be a von Neumann algebra in $\mathcal{B}(H)$. Let $\mathcal{A}_{\mathbb{R}}$ be the real vector space of self-adjoint elements, then $\mathcal{A} = \mathcal{A}_{\mathbb{R}} + i\mathcal{A}_{\mathbb{R}}$. Let $T \in \mathcal{A}_{\mathbb{R}}$, and let $f \in \mathcal{S}(\mathbb{R})$ be such that $f(x) = x$ for x in the (bounded) spectrum of T (see Exercise 8.1). By Proposition 5.1.2,

$$T = f(T) = \int_{\mathbb{R}} \hat{f}(y) e^{2\pi i y T} \, dy.$$

The unitary elements $e^{2\pi i y T} \in \mathcal{B}(H)$ are power series in T, so belong to the von Neumann algebra $\mathcal{A}$, and every operator that commutes with the $e^{2\pi i y T}$ will commute with T, so T belongs to the von Neumann algebra generated by the unitaries $e^{2\pi i y T}$. $\quad\square$

Let B_1 be the unit ball in $\mathcal{B}(H)$, i.e., the set of all $T \in \mathcal{B}(H)$ with $\|T\|_{\mathrm{op}} \le 1$.

Lemma 8.2.3 B_1 *is weakly compact.*

Proof For $r \ge 0$ and $z \in \mathbb{C}$ let $\bar{B}_r(z)$ be the closed ball around z of radius r. For $T \in B_1$ and $v, w \in H$, one has $|\langle Tv, w\rangle| \le \|v\|\|w\|$, so the map

$$\psi : B_1 \rightarrow \prod_{v,w \in H} \bar{B}_{\|v\|\|w\|}(0)$$

with $\psi(T)_{v,w} = \langle Tv, w\rangle$ embeds B_1 into the Hausdorff space on the right, which is compact by Tychonov's Theorem A.7.1. The weak topology is induced by ψ, so B_1 is weakly compact if we can show that the image of ψ is closed. We claim that this image equals the set A of all elements x of the product such that $(v, w) \mapsto x_{v,w}$ is linear in v and conjugate linear in w. Since convergence in the product space is component-wise, this set is closed. Given $x \in A$ and $w \in H$, the map $\alpha_v : w \mapsto \overline{x_{v,w}}$ is a linear functional on H with $\|\alpha_v\| \le \|v\|$ and hence there exists an element $Tv \in H$ such that $x_{v,w} = \langle Tv, w\rangle$ for all $w \in H$. One then checks that $v \mapsto Tv$ defines an element in B_1 such that $\psi(T) = x$. $\quad\square$

8.3 Representations

A unitary representation (π, V_π) of a locally compact group G is called a *factor representation* if the von Neumann algebra $\mathrm{VN}(\pi)$ generated by $\pi(G) \subset \mathcal{B}(V_\pi)$ is a factor. So π is a factor representation if and only if $\pi(G)^\circ \cap \pi(G)^{\circ\circ} = \mathbb{C}\,\mathrm{Id}$.

Lemma 8.3.1 *Every irreducible representation is a factor representation.*

Proof It follows from the Lemma of Schur 6.1.7 that $VN(\pi) = \mathcal{B}(V_\pi)$ for every irreducible representation π. □

Definition Two unitary representations π_1, π_2 of G are called *quasi-equivalent* if there is an isomorphism of *-algebras

$$\phi : VN(\pi_1) \ \rightarrow \ VN(\pi_2)$$

satisfying $\phi(\pi_1(x)) = \pi_2(x)$ for every $x \in G$.

Example 8.3.2 A given unitary representation π is quasi-equivalent to the direct sum representation $\pi \oplus \pi$. This follows from the general fact that any von-Neumann algebra $\mathcal{A} \subset \mathcal{B}(H)$ is isomorphic to $\mathcal{A} \begin{pmatrix} 1 \\ & 1 \end{pmatrix} \subseteq \mathcal{B}(H^2)$. (Compare with the proof of von Neumann's Bicommutant Theorem.)

Lemma 8.3.3 *Two irreducible unitary representations of a locally compact group are quasi-equivalent if and only if they are unitarily equivalent.*

Proof Let the unitary representations (π, V_π) and (η, V_η) be unitarily equivalent, i.e., there is a unitary intertwining operator $T : V_\pi \rightarrow V_\eta$. Then T induces an isomorphism $VN(\pi) \rightarrow VN(\eta)$ by mapping S to TST^{-1}. This shows that π and η are also quasi-equivalent. Conversely, let (π, V_π) and (η, V_η) be two irreducible unitary representations of G, and let $\phi : VN(\pi) \rightarrow VN(\eta)$ be an isomorphism of C^*-algebras such that $\phi(\pi(x)) = \eta(x)$ for all $x \in G$. For $u, v \in V_\pi$ let $T_{u,v} : V_\pi \rightarrow V_\pi$ be given by $T_{u,v}(x) \overset{\text{def}}{=} \langle x, u \rangle v$. Then $T_{u,v} T_{w,z} = \langle z, u \rangle T_{w,v}$, and $T_{u,v}^* = T_{v,u}$. Let $(e_j)_{j \in I}$ be an orthonormal basis of V_π. For each $j \in I$ the map $P_j = T_{e_j, e_j}$ is the orthogonal projection onto the one dimensional space $\mathbb{C}e_j$ and T_{e_j, e_k} is an isometry from $\mathbb{C}e_j$ to $\mathbb{C}e_k$ and is zero on $\mathbb{C}e_i$ for $i \neq j$. The P_j are pairwise orthogonal projections that add up to the identity in the strong topology. The same holds for the images $\phi(P_j)$. Let $V_{\eta,j} = \phi(P_j)V_\eta$. Then V_η is the direct orthogonal sum of the $V_{\eta,j}$. We claim that $\phi(T_{e_j, e_k})$ is an isometry from $V_{\eta,j}$ to $V_{\eta,k}$ and zero on $V_{\eta,i}$ for $i \neq j$. For this let $x, y \in V_{\eta,j}$, then

$$\langle \phi(T_{e_j,e_k})x, \phi(T_{e_j,e_k})y \rangle = \langle \phi(T_{e_k,e_j} T_{e_j,e_k})x, y \rangle$$
$$= \langle \phi(T_{e_j,e_j})x, y \rangle = \langle x, y \rangle.$$

Now fix some $j_0 \in I$ and choose $f_{j_0} \in V_{\eta,j_0}$ of norm one. For $j \neq j_0$ set $f_j = \phi(T_{e_{j_0},e_j})f_{j_0}$. Consider the isometry $S : V_\pi \rightarrow V_\eta$ given by $S(e_j) = f_j$. It then follows that $ST_{e_j,e_k} = \phi(T_{e_j,e_k})S$. The C^*-algebra $VN(\pi) = \mathcal{B}(V_\pi)$ is generated by the T_{e_j,e_k}, so S is an intertwining operator onto a closed subspace of V_η. As η is irreducible, S must be surjective, i.e., unitary. □

Definition A factor representation π is called a *type-I representation* if π is quasi-equivalent to a representation π_1 whose von Neumann algebra $\mathrm{VN}(\pi_1)$ is a type-I factor. Then π is of type I if and only if π is quasi-equivalent to an irreducible representation.

Example 8.3.4 We here give an example of a factor representation, which is not of type I. Let Γ be a non-trivial group with the property that every conjugacy class in Γ is infinite or trivial. So the only finite conjugacy class in Γ is $\{1\}$. An example of this instance is the free group F_2 generated by two elements. Another example is the group $\mathrm{SL}_2(\mathbb{Z})/\pm 1$.

Consider the regular right representation R of Γ on the Hilbert space $H = \ell^2(\Gamma)$. Let $\mathrm{VN}(R)$ be the von Neumann algebra generated by $R(\Gamma) \subset \mathcal{B}(\ell^2(\Gamma))$.

Proposition 8.3.5 $\mathrm{VN}(R)$ *is a factor, which is not of type I.*

Proof We show that the commutant $\mathrm{VN}(R)^\circ$ is the von Neumann algebra generated by the regular left representation L of Γ. For this consider the natural orthonormal basis $(\delta_\gamma)_{\gamma \in \Gamma}$, which is defined by $\delta_\gamma(\tau) = 1$ if $\gamma = \tau$ and zero otherwise. One has $R_\gamma \delta_{\gamma_0} = \delta_{\gamma_0 \gamma^{-1}}$ and $L_\gamma \delta_{\gamma_0} = \delta_{\gamma \gamma_0}$. Let $T \in \mathrm{VN}(R)^\circ$, so $T R_\gamma = R_\gamma T$ for every $\gamma \in \Gamma$. Then $T(\delta_1) = \sum_\gamma c_\gamma \delta_\gamma$ for some coefficients $c_\gamma \in \mathbb{C}$ satisfying $\sum_\gamma |c_\gamma|^2 < \infty$. For $\gamma_0 \in \Gamma$ arbitrary one gets

$$
T(\delta_{\gamma_0}) = T\left(R_{\gamma_0^{-1}}\delta_1\right) = R_{\gamma_0^{-1}} T(\delta_1)
$$
$$
= R_{\gamma_0^{-1}} \sum_\gamma c_\gamma \delta_\gamma = \sum_\gamma c_\gamma \delta_{\gamma \gamma_0}
$$
$$
= \sum_\gamma c_\gamma L_\gamma(\delta_{\gamma_0}),
$$

so $T = \sum_\gamma c_\gamma L_\gamma$, where the sum converges in the strong topology. Hence $T \in \mathrm{VN}(L)$. As trivially $\mathrm{VN}(L) \subset \mathrm{VN}(R)^\circ$ we get $\mathrm{VN}(R)^\circ = \mathrm{VN}(L)$. This means that $\mathrm{VN}(L)$ and $\mathrm{VN}(R)$ are each other's commutants. In particular, it follows that each element of $\mathrm{VN}(L)$ can be written as a point-wise convergent sum of the form $\sum_\gamma c_\gamma L_\gamma$, and likewise each element of $\mathrm{VN}(R)$ can be written as a sum of the form $\sum_\gamma d_\gamma R_\gamma$. We show that $\mathrm{VN}(R)$ is a factor. For this we have to show that the intersection of $\mathrm{VN}(R)$ and $\mathrm{VN}(L)$ is trivial. So let $T \in \mathrm{VN}(L) \cap \mathrm{VN}(R)$. Then we have two representations

$$
\sum_\gamma c_\gamma L_\gamma = T = \sum_\gamma d_\gamma R_\gamma.
$$

In particular, $\sum_\gamma c_\gamma \delta_\gamma = T(\delta_1) = \sum_\gamma d_\gamma \delta_{\gamma^{-1}}$, which implies $d_\gamma = c_{\gamma^{-1}}$, so for $\alpha \in \Gamma$, on the one hand,

$$
T(\delta_\alpha) = \sum_\gamma c_\gamma L_\gamma \delta_\alpha = \sum_\gamma c_\gamma \delta_{\gamma \alpha} = \sum_\gamma c_{\gamma \alpha^{-1}} \delta_\gamma
$$

and on the other,

$$T(\delta_\alpha) = \sum_\gamma c_\gamma R_{\gamma^{-1}} \delta_\alpha = \sum_\gamma c_\gamma \delta_{\alpha\gamma} = \sum_\gamma c_{\alpha^{-1}\gamma} \delta_\gamma.$$

This means that the function $\gamma \mapsto c_\gamma$ is constant on conjugacy classes. Since the sums must converge, this function can only be supported on finite conjugacy classes. As there is only one of them, it follows that $c_\gamma = 0$ except for $\gamma = 1$, so $T \in \mathbb{C}\,\mathrm{Id}$.

Finally we show that $\mathrm{VN}(R)$ is not of type I. For this consider the map $\sigma : \mathrm{VN}(R) \to \mathbb{C}$; $T \mapsto \langle T\delta_1, \delta_1\rangle$. This map is evidently continuous with respect to the strong and weak topologies. We show $\sigma(ST) = \sigma(TS)$ for all $S, T \in \mathrm{VN}(R)$. By continuity it suffices to show this for $S = R_\gamma$ and $T = R_\tau$, where $\gamma, \tau \in \Gamma$. Then we have

$$\sigma(ST) = \sigma(R_\gamma R_\tau) = \sigma(R_{\gamma\tau}) = \langle \delta_{\gamma\tau}, \delta_1\rangle = \begin{cases} 1 & \text{if } \gamma\tau = 1, \\ 0 & \text{otherwise.} \end{cases}$$

The last condition is symmetric in γ and τ, since in the group Γ we have $\gamma\tau = 1 \Leftrightarrow \tau\gamma = 1$, so the same calculation gives $\sigma(ST) = \sigma(TS)$ as claimed.

We now show that for every selfadjoint projection $P \neq 0$ in $\mathrm{VN}(R)$ one has $0 < \sigma(P) \leq 1$. We first observe that for $T = \sum_{\gamma \in \Gamma} c_\gamma R_\gamma \in \mathrm{VN}(R)$ one has $\sigma(T) = c_1$. Next let P be a selfadjoint projection, which is the same as an orthogonal projection. So it satisfies $P^* = P = P^2$. We write $P = \sum_{\gamma \in \Gamma} c_\gamma R_\gamma$ and we get

$$\sum_\gamma c_\gamma R_\gamma = P = P^2 = \sum_\gamma \left(\sum_\delta c_\delta c_{\delta^{-1}\gamma} \right) R_\gamma.$$

So in particular $c_1 = \sum_\delta c_\delta c_{\delta^{-1}}$. The condition $P = P^* = \sum_\gamma \overline{c_{\gamma^{-1}}} R_\gamma$ implies $c_{\gamma^{-1}} = \overline{c_\gamma}$ and therefore $\sigma(P) = c_1 = \sum_\gamma |c_\gamma|^2$. This implies $c_1 > 0$ and $c_1 \geq c_1^2$, so $1 \geq c_1$.

Now assume there is a *-isomorphism $\phi : \mathcal{B}(H) \to \mathrm{VN}(R)$ for some Hilbert space H. Since $\mathrm{VN}(R)$ is infinite-dimensional, the space H is infinite-dimensional. So let $(e_j)_{j \in \mathbb{N}}$ be an orthogonal sequence in H. Let Q_j be the orthogonal projection with image $\mathbb{C}e_j$ and let $P_j = \phi(Q_j)$. Then P_j is a selfadjoint projection. Further Q_j is conjugate to Q_k in $\mathcal{B}(H)$, since there are unitary operators interchanging e_j and e_k. Then P_j and P_k are conjugate in $\mathrm{VN}(R)$ and therefore $\sigma(P_j) = \sigma(P_k)$ is a fixed number $c > 0$. Now $Q_1 + \cdots + Q_n$ again is a selfadjoint projection, so the same holds for $P_1 + \cdots + P_n$. So we have

$$1 \geq \sigma(P_1 + \cdots + P_n) = \sigma(P_1) + \cdots + \sigma(P_n) = nc,$$

Since this holds foe every n, we get $c = 0$, a contradiction! Hence ϕ does not exist and $\mathrm{VN}(R)$ is not of type I. $\square$

8.4   Hilbert Integrals

A family of vectors $(\xi_i)_{i \in I}$ in a Hilbert space H is called a *quasi-orthonormal basis* if the non-zero members of the family form an orthonormal basis of H.

Let X be a set and $\mathcal{D}$ a σ-algebra of subsets of X. A *Hilbert bundle* over X is a family of Hilbert spaces $(H_x)_{x \in X}$ and a family of maps $\xi_i : X \to \bigcup_{x \in X} H_x$ (disjoint union) with $\xi_i(x) \in H_x$, such that for each $x \in X$ the family $(\xi_i(x))$ is a quasi-orthonormal basis of H_x, and for each $i \in I$ the set of all $x \in X$ with $\xi_i(x) = 0$ is measurable.

A *section* is a map $s : X \to \bigcup_{x \in X} H_x$ with $s(x) \in H_x$ for every $x \in X$. A section is called *measurable section* if for every $j \in I$ the function $x \mapsto \langle s(x), \xi_j(x) \rangle$ is measurable on X, and there exists a countable set $I_s \subset I$, such that the function $x \mapsto \langle s(x), \xi_i(x) \rangle$ vanishes identically for every $i \notin I_s$.

Let μ be a measure on $\mathcal{D}$. A measurable section s is called a *nullsection* if it vanishes outside a set of measure zero. The *direct Hilbert integral* is the vector space of all measurable sections s, which satisfy

$$\|s\|^2 \overset{\text{def}}{=} \int_X \|s(x)\|^2 \, d\mu(x) < \infty$$

modulo the space of nullsections.

This space, written as $H = \int_X H_x \, d\mu(x)$, is a Hilbert space with the inner product $\langle s, t \rangle = \int_X \langle s(x), t(x) \rangle \, d\mu(x)$. To show the completeness, for $i \in I$ let X_i be the set of all $x \in X$ with $\xi_i(x) \neq 0$. We get a map $P_i : H \to L^2(X_i)$ given by $P_i(s)(x) = \langle s(x), \xi_i(x) \rangle$. These maps combine to give a unitary isomorphism,

$$H = \int_X H_x \, d\mu(x) \overset{\cong}{\longrightarrow} \widehat{\bigoplus_{i \in I}} L^2(X_i).$$

Example 8.4.1   Direct sums are special cases of direct integrals. Let $H = \bigoplus_{j \in I} H_j$ be a direct sum of separable Hilbert spaces. This space equals the direct integral $\int_X H_x \, d\mu(x)$ with $X = I$ and μ the counting measure on X.

Let (H_x, ξ_j) be a Hilbert bundle and μ a measure on X. Let G be a locally compact group, and for every $x \in X$ let η_x be a unitary representation of G on H_x, such that for every $g \in G$ and all $i, j \in I$ the map $x \mapsto \langle \eta_x(g)\xi_i(x), \xi_j(x) \rangle$ is measurable. Then $(\eta(g)s)(x) \overset{\text{def}}{=} \eta_x(g)s(x)$ defines a unitary representation of G on $H = \int_X H_x \, d\mu(x)$.

Example 8.4.2   Let A be a locally compact abelian group with dual group $\widehat{A}$ equipped with the Plancherel measure. Each character $\chi : A \to \mathbb{T} = U(\mathbb{C})$ determines a one-dimensional representation of A on $H_\chi = \mathbb{C}$. Consider the constant section

$\xi_1(\chi) = 1 \in \mathbb{C} = H_\chi$. Let $\eta_\chi(y) = \chi(y)$. Then the direct integral satisfies

$$\int_{\widehat{A}} H_\chi \, d\chi \cong L^2(\widehat{A})$$

with $(\eta(y)\xi)(\chi) = \chi(y)\xi(\chi)$. It follows then from the Plancherel Theorem 3.4.8 that $(\eta, L^2(\widehat{A}))$ is unitarily equivalent to the left regular representation $(L, L^2(A))$ of A via the Fourier transform.

8.5   The Plancherel Theorem

A locally compact group G is called a *type-I group* if every factor representation of G is of type I, i.e., is quasi-equivalent to an irreducible one.

Examples 8.5.1

- Abelian groups are of type I. For an abelian group A and a unitary representation π of A, the von Neumann algebra $\mathrm{VN}(\pi)$ is commutative. So, if $\mathrm{VN}(\pi)$ is a factor, it must be isomorphic to $\mathbb{C}$, which means that π is quasi-equivalent to a one-dimensional representation.

- Compact groups are of type I. For a compact group any unitary representation is a direct sum of irreducible representations.

- Nilpotent Lie groups are of type I. See [BCD+72] Chapter VI.

- Semisimple Lie groups are of type I. See [HC76].

- A discrete group is of type I if and only if it contains a normal abelian subgroup of finite index. See [Tho68].

Let G and H be locally compact groups, and let (π, V_π), (σ, V_σ) be unitary representations of G and H, respectively. On the Hilbert tensor product $V_\pi \widehat{\otimes} V_\sigma$ (see Appendix C.3) we define a representation $\pi \otimes \sigma$ of the product group $G \times H$ by linear extension of

$$v \otimes w \mapsto \pi(x)v \otimes \sigma(y)w$$

for $(x, y) \in G \times H$, $v \in V_\pi$, and $w \in V_\sigma$.

Recall that the unitary dual $\widehat{G}$ consists of all equivalence classes of irreducible unitary representations of G. On $\widehat{G}$ we will install a natural σ-algebra in the case that G has a countable dense subset.

Lemma 8.5.2 *Assume that G has a countable dense subset. Then every irreducible unitary representation (π, V_π) has countable dimension, i.e., the Hilbert space V_π has a countable orthonormal system.*

Proof Let (π, V_π) be an irreducible unitary representation of G. A subset $\mathcal{T} \subset V_\pi$ is called *total*, if the linear span of $\mathcal{T}$ is dense in V_π. By the orthonormalization scheme it suffices to show that there is a countable total set in V_π. Let $0 \neq v \in V_\pi$. Then the set $\pi(G)v$ is total in V_π, as V_π is irreducible. Let $D \subset G$ be a countable dense subset. Then the set $\pi(D)v$ is dense in $\pi(G)v$, hence also total in V_π.

Assume that G has a dense countable subset. For a countable cardinal $n = 1, 2, \ldots \aleph_0$, let H_n denote a fixed Hilbert space of dimension n. For each class C in $\widehat{G}$ we fix a representative $\pi \in C$ with representation space H_n, which exists by Lemma 8.5.2. The cardinal n is uniquely determined by $C = [\pi]$ and is called the *dimension* of the representation. Let $\widehat{G}_n$ be the subset of $\widehat{G}$ consisting of all classes $[\pi]$ of dimension n. On $\widehat{G}_n$ we install the smallest σ-algebra making all maps $[\pi] \mapsto \langle \pi(g)v, w \rangle$ measurable, where g ranges in G and v, w range over H_n. On $\widehat{G} = \bigcup_n \widehat{G}_n$ we install the union σ-algebra.

The prescription $\eta(x, y) = L_x R_y$ defines a unitary representation of $G \times G$ on the Hilbert space $L^2(G)$. Note that if G is second countable, then it contains a dense countable subset, i.e., is separable.

Theorem 8.5.3 *Let G be a second countable, unimodular, locally compact group of type I. There is a unique measure μ on $\widehat{G}$ such that for $f \in L^1(G) \cap L^2(G)$ one has*

$$\|f\|_2^2 = \int_{\widehat{G}} \|\pi(f)\|_{\mathrm{HS}}^2 \, d\mu(\pi).$$

The map $f \mapsto (\pi(f))_\pi$ extends to a unitary $G \times G$ equivariant map

$$L^2(G) \cong \int_{\widehat{G}} \mathrm{HS}(V_\pi) \, d\mu(\pi),$$

where the representation of η_π of $G \times G$ on the space of Hilbert-Schmidt operators $HS(V_\pi)$ is given by $\eta_\pi(x, y)(T) = \pi(x)T\pi(y^{-1})$ for each $\pi \in \widehat{G}$ and $x, y \in G$.

The proof is in [Dix96], 18.8.1.

This Plancherel Theorem generalizes the Plancherel Theorem in the abelian case, Theorem 3.4.8, as well as the Peter-Weyl Theorem in the compact case, Theorems 7.2.1 and 7.2.4. Concrete examples for groups, which are neither abelian nor compact will be given in Theorem 10.3.1 and Theorem 11.3.1.

8.6 Exercises

Exercise 8.1 For $S > 0$ show that there exists a function $f : \mathbb{R} \to \mathbb{R}$ which is infinitely differentiable, of compact support and satisfies $f(x) = x$ for $|x| \leq S$.

(Hint: Let $g(x) = 1$ for $|x| \leq S + 1$ and $g(x) = 0$ otherwise. Let $h = \phi * g$ for some smooth Dirac function with support in $[-1, 1]$. Set $f(x) = xh(x)$.)

Exercise 8.2 Let V, W be Hilbert spaces.

(a) For $A \in \mathcal{B}(V)$ set $\alpha(A)(v \otimes w) = A(v) \otimes w$. Show that α is a norm-preserving
    *-homomorphism from $\mathcal{B}(V)$ to $\mathcal{B}(V \hat{\otimes} W)$.

(b) Let β be the analogous map on the second factor. Show that $\alpha \otimes \beta$ defines a
    *-homomorphism from $\mathcal{B}(V) \otimes \mathcal{B}(W)$ to $\mathcal{B}(V \hat{\otimes} W)$.

(c) Show that the von Neumann algebra generated by the image of $\alpha \otimes \beta$ equals the
    entire $\mathcal{B}(V \hat{\otimes} W)$.

Exercise 8.3 Give an example of a set of bounded operators on a Hilbert space, for
which the norm-closure differs from the strong closure. Also give an example, for
which the strong and weak closures differ.

Exercise 8.4 Let H be a Hilbert space, and let $\mathcal{P}$ be the set of all orthogonal projections on H. Let $\mathcal{T}_s$ and $\mathcal{T}_w$ be the restrictions of the strong and weak topologies to
the set $\mathcal{P}$. Show that $\mathcal{T}_s = \mathcal{T}_w$.

Exercise 8.5 Let H be a Hilbert space, and let $\mathcal{U}$ be the set of all unitary operators
on H. Let $\mathcal{T}_s'$ and $\mathcal{T}_w'$ be the restrictions of the strong and weak topologies to the set
$\mathcal{U}$. Show that $\mathcal{T}_s' = \mathcal{T}_w'$.

Exercise 8.6 Show that not every unital C^*-algebra is isomorphic to a von Neumann
algebra.

(Hint: Consider an infinite dimensional Hilbert space H and the space $\mathcal{K}$ of compact
operators. The algebra $\mathcal{A} = \mathcal{K} + \mathbb{C}\mathrm{Id}$ is a C^*-algebra.)

Exercise 8.7 Let H be a Hilbert space, and let $M \subset \mathcal{B}(H)$ be self-adjoint and
commutative, i.e., for $S, T \in M$ one has $S^*, T^* \in M$ and $ST = TS$. Show that the
bicommutant $M^{\circ\circ}$ is commutative.

Exercise 8.8 For a von Neumann algebra $\mathcal{A} \subset \mathcal{B}(H)$ let $\mathcal{A}^+$ be the set of all finite
sums of elements of the form aa^* for some $a \in \mathcal{A}$. Show:

(a) $\mathcal{A}^+$ is a proper cone, i.e.:

$$\mathcal{A}^+ + \mathcal{A}^+ \subset \mathcal{A}^+, \quad \mathbb{R}^+\mathcal{A}^+ \subset \mathcal{A}^+, \quad \mathcal{A}^+ \cap (-\mathcal{A}^+) = 0.$$

(b) For $a \in \mathcal{A}$ one has

$$a \in \mathcal{A}^+ \quad \Leftrightarrow \quad \exists b \in \mathcal{A} : a = bb^*, \quad \Leftrightarrow \quad a \geq 0.$$

Exercise 8.9 Let $\mathcal{A} \subset \mathcal{B}(H)$ be a von Neumann algebra. A *finite trace* is a linear
map $\tau : \mathcal{A} \to \mathbb{C}$ with $\tau(\mathcal{A}^+) \subset \mathbb{R}^+$ and $\tau(ab) = \tau(ba)$ for all $a, b \in \mathcal{A}$. Show:

(a) Let τ be a finite trace on $\mathcal{A} = M_n(\mathbb{C})$. Then $\tau(a) = c\,\mathrm{tr}\,(a)$ for some $c \geq 0$.

(b) Let $\mathcal{A} = \mathcal{B}(H)$, where H is an infinite-dimensional Hilbert space. Then there is no finite trace on $\mathcal{A}$.

(c) Let Γ be a discrete group, and let $\mathcal{A} = \mathrm{VN}(R)$. Then $\tau\left(\sum_{\gamma \in \Gamma} c_\gamma R_\gamma\right) = c_1$ is a finite trace on $\mathcal{A}$.

Exercise 8.10 Show that a von Neumann algebra $\mathcal{A}$ is generated by all orthogonal projections it contains.

Exercise 8.11 Let G be a locally compact group. For a unitary representation (π, V_π) let its *matrix coefficients* be all continuous functions on G of the form

$$g \mapsto \psi_{v,w}(g) \overset{\mathrm{def}}{=} \langle \pi(g)v, w\rangle, \qquad v, w \in V_\pi.$$

Let G be of type I. Let π be a unitary representation such that all its matrix coefficients are in $L^2(G)$. Show that π is a direct sum of irreducible representations.

Chapter 9
The Selberg Trace Formula

In this chapter we introduce the Selberg trace formula, which is a natural generalization of the Poisson summation formula to non-abelian groups. Applications of the trace formula will be given in the next two chapters.

9.1 Cocompact Groups and Lattices

We will study the following situation. G will be a locally compact group, and H will be a closed subgroup, which is unimodular. We say that a subgroup H of a topological group G is a *cocompact subgroup* if the quotient G/H is a compact space.

Examples 9.1.1

- A classical example is $G = \mathbb{R}$ and $H = \mathbb{Z}$.

- For a unimodular group H and a compact group K together with a group homomorphism $\eta : K \to \mathrm{Aut}(H)$ such that the ensuing map $K \times H \to H$ given by $(k, h) \mapsto \eta(k)(h)$ is continuous, one can form the *semi-direct product* $G = H \rtimes K$. As a topological space, one has $G = H \times K$. The multiplication is $(h, k)(h_1, k_1) = (h \, \eta(k)(h_1), kk_1)$. Then H is a cocompact subgroup of G.

- For $G = \mathrm{SL}_2(\mathbb{R})$, one uses hyperbolic geometry [Bea95] to construct discrete subgroups $\Gamma = H \subset G$, which provide examples of the situation considered here.

Proposition 9.1.2 *Let G be a locally compact group. If G admits a unimodular closed cocompact subgroup, then G is unimodular itself.*

Proof Let $H \subset G$ be a unimodular closed cocompact subgroup. The inversion on G induces a homeomorphism $G/H \to H\backslash G$ by $gH \mapsto Hg^{-1}$. Therefore $H\backslash G$ is compact as well. We install a Radon measure μ on the compact Hausdorff space $H\backslash G$ using Riesz's representation theorem as follows. For $f \in C(H\backslash G)$ choose $g \in C_c(G)$ with $^Hg = f$, where $^Hg(x) \stackrel{\text{def}}{=} \int_H g(hx)\,dh$. In this case define $\int_{H\backslash G} f(x)\,d\mu(x)$ by $\int_G g(x)\,dx$. For this to be well-defined, we have to show

A. Deitmar, S. Echterhoff, *Principles of Harmonic Analysis*, Universitext, DOI 10.1007/978-3-319-05792-7_9, © Springer International Publishing Switzerland 2014

that $^Hg = 0$ implies $\int_G g(x)\,dx = 0$. Note that, as H is unimodular, we have $\int_H g(h^{-1}x)\,dh = \int_H g(hx)\,dh = 0$. Let $\phi \in C_c(G)$ with $^H\phi \equiv 1$. One gets

$$\int_G g(x)\,dx = \int_G \int_H \phi(hx)g(x)\,dh\,dx$$

$$= \int_H \int_G \phi(hx)g(x)\,dx\,dh$$

$$= \int_H \int_G \phi(x)g(h^{-1}x)\,dx\,dh$$

$$= \int_G \phi(x)\int_H g(h^{-1}x)\,dh\,dx = 0.$$

This measure μ on $H\backslash G$ satisfies

$$\int_{H\backslash G} f(xy)\,d\mu(x) = \int_G g(xy)\,dx = \Delta(y^{-1})\int_G g(x)\,dx$$

$$= \Delta(y^{-1})\int_{H\backslash G} f(x)\,d\mu(x).$$

So in particular, for $f \equiv 1 \in C(H\backslash G)$ one gets

$$0 < \int_{H\backslash G} f(x)\,d\mu(x) = \int_{H\backslash G} f(xy)\,d\mu(x) = \Delta(y^{-1})\int_{H\backslash G} f(x)\,d\mu(x),$$

which implies $\Delta(y^{-1}) = 1$. $\square$

Definition A subgroup $\Gamma \subset G$ of a topological group is called a *discrete subgroup* if the subspace-topology on Γ is the discrete topology.

Lemma 9.1.3

(a) *A subgroup $\Gamma \subset G$ is discrete if and only if there is a unit-neighborhood $U \subset G$ with $\Gamma \cap U = \{1\}$.*

(b) *A discrete subgroup is closed in G.*

Proof As Γ is a topological group, it is discrete if and only if $\{1\}$ is open, which is equivalent to the existence of an open set U with $\Gamma \cap U = \{1\}$. This implies part (a). For part (b) let Γ be a discrete subgroup, and let $\gamma_j \to x$ be a net in Γ, convergent in G. We have to show that $x \in \Gamma$. Let U be a unit-neighborhood in G such that $\Gamma \cap U = \{1\}$ and let V be a symmetric unit-neighborhood such that $V^2 \subset U$. As xV is a neighborhood of x, there exists an index j_0 such that for $i, j \geq j_0$ one has $\gamma_i, \gamma_j \in xV$, therefore $\gamma_i^{-1}\gamma_j \in V^2 \subset U$, hence $\gamma_i = \gamma_j$, so the net is eventually constant, hence $x = \gamma_{j_0} \in \Gamma$. $\square$

Example 9.1.4

- The subgroup $\mathbb{Z}$ of $\mathbb{R}$ is a discrete subgroup.

- Let G be Hausdorff and compact. Then a subgroup $\Gamma \subset G$ is discrete if and only if it is finite.

- Let $G = SL_2(\mathbb{R})$ be the group of real 2×2 matrices of determinant one. Then $\Gamma = SL_2(\mathbb{Z})$ is a discrete subgroup of G.

Definition Let G be a locally compact group. A discrete subgroup Γ such that G/Γ carries an invariant Radon measure μ with $\mu(G/\Gamma) < \infty$ is called a *lattice* in G. A cocompact lattice is also called a *uniform lattice*.

Proposition 9.1.5 *Let G be a locally compact group. A discrete, cocompact subgroup Γ is a uniform lattice.*

Proof Let $\Gamma \subset G$ as in the proposition. Then G is unimodular by Proposition 9.1.2. As Γ is unimodular, by Theorem 1.5.3 there exists an invariant Radon measure on G/Γ, and, as the latter space is compact, its volume is finite, so Γ is a lattice. $\square$

Remark There are lattices, which are not uniform, like $\Gamma = SL_2(\mathbb{Z})$ in $G = SL_2(\mathbb{R})$ (See [Ser73]).

Theorem 9.1.6 *Let G be a locally compact group. If G admits a lattice, then G is unimodular.*

Proof Let Δ be the modular function of G, and let $H = \ker(\Delta)$. Then H is unimodular by Corollary 1.5.5. Let $\Gamma \subset G$ be a lattice. As there is an invariant measure on G/Γ, Theorem 1.5.3 implies that Δ is trivial on Γ, so $\Gamma \subset H$. On G/H and on H/Γ there are invariant measures by Theorem 1.5.3, so by the uniqueness of these respective measures we conclude that

$$\infty > \mathrm{vol}(G/\Gamma) = \int_{G/\Gamma} 1\, dx = \int_{G/H} \int_{H/\Gamma} 1\, dx\, dy.$$

Therefore G/H has finite volume. Being a locally compact group, it follows that G/H is compact. By Proposition 9.1.2. the group G is unimodular. $\square$

9.2 Discreteness of the Spectrum

Let G be a locally compact group, and let $\Gamma \subset G$ be a cocompact lattice. It is a convention that one uses the right coset space $\Gamma \backslash G$ instead of the left coset space. This quotient carries a Radon measure μ that is invariant under right translations by elements of G and has finite volume. On the ensuing Hilbert space $L^2(\Gamma\backslash G)$ the right translations give a unitary representation R of G by

$$R(x)\phi(y) \overset{\text{def}}{=} \phi(yx).$$

Example 9.2.1 Let $G = \mathbb{R}$, and $\Gamma = \mathbb{Z}$ then Γ is a closed cocompact subgroup and the theory of Fourier series, [Dei05] Chap. 1, implies that the representation R decomposes into the direct sum $\bigoplus_{n\in\mathbb{Z}}\chi_n$, where $\chi_n : \mathbb{R} \to \mathbb{T}$ is the character given by $\chi_n(x) = e^{2\pi i n x}$. Thus R decomposes into a discrete sum of irreducible representations. This is a general phenomenon, as the next theorem shows.

For a representation π and a natural number N we write $N\pi$ for the N-fold direct sum $\pi \oplus \pi \cdots \oplus \pi$.

Theorem 9.2.2 *Let G be a locally compact group, and let $\Gamma \subset G$ be a unimodular closed cocompact subgroup. The representation R on $L^2(\Gamma\backslash G)$ decomposes as a direct sum of irreducible representations with finite multiplicities, i.e.,*

$$L^2(\Gamma\backslash G) \cong \bigoplus_{\pi\in\widehat{G}} N_\Gamma(\pi)\pi,$$

where the sum runs over the unitary dual $\widehat{G}$ of G, and $N_\Gamma(\pi) \in \mathbb{N}_0$ is a finite multiplicity for $\pi \in \widehat{G}$.

Proof We need a lemma that tells us that for $f \in C_c(G)$ the operator $R(f)$ is given by a continuous integral kernel. For later use we will extend this to a greater class of functions f. Let U be a compact unit-neighborhood in G. For a continuous function f on G let $f_U : G \to [0, \infty)$ be defined by

$$f_U(y) \overset{\text{def}}{=} \sup_{x,z\in U} |f(xyz)|.$$

Lemma 9.2.3 *The function f_U is continuous.*

Proof It suffices to show that for $a \geq 0$ the sets $f_U^{-1}((a, \infty))$ and $f_U^{-1}([0, a))$ are open. For the former assume $f_U(x) > a$, then there exist $u_1, u_2 \in U$ with $|f(u_1 x u_2)| > a$. As the function $y \mapsto f(u_1 y u_2)$ is continuous, there exists an open neighborhood V of x such that $|f(u_1 v u_2)| > a$ for every $v \in V$. This implies $f_U(v) > a$ for every $v \in V$.

The second assertion is equivalent to saying that $f_U^{-1}([a, \infty))$ is closed. So let (x_j) be a net in this set, convergent to some $x \in G$. This means that $f_U(x_j) \geq a$ and we have to show that $f_U(x) \geq a$. For each j fix some $u_j, v_j \in U$ such that $f_U(x_j) = |f(u_j x_j v_j)|$. By switching to a subnet, we may assume that the nets $(u_j), (v_j)$ are both convergent in the compact set U with limits u and v respectively. It follows that

$$f_U(x) \geq |f(uxv)| = \lim_j |f(u_j x_j v_j)| \geq a. \qquad \square$$

Definition We say that a continuous function f is a *uniformly integrable function* if there exists a compact unit-neighborhood U such that f_U is in $L^1(G)$. If f is uniformly integrable, then $f \in L^1(G)$ as $|f| \le f_U$. Let $C_{\text{unif}}(G)$ be the set of all continuous functions f on G that are uniformly integrable.

Example 9.2.4

- If $f \in C_c(G)$, then f is uniformly integrable, as $f_U \in C_c(G)$ again.
- Every Schwartz-function on $\mathbb{R}$ is uniformly integrable (Exercise 9.1).

Lemma 9.2.5 *Let G be unimodular. Every uniformly integrable function vanishes at infinity, and the space $C_{\text{unif}}(G)$ is an algebra under convolution.*

Proof Assume that f is uniformly integrable and let U be a compact symmetric unit-neighborhood in G such that f_U is integrable. If f does not vanish at ∞, there exists an $\varepsilon > 0$ such that for every compact set $K \subset G$ there is $x \in G \setminus K$ with $|f(x)| \ge \varepsilon$. Let $x_1 \in G$ be any element with $|f(x_1)| \ge \varepsilon$. Then there exists $x_2 \notin x_1 U^2$ such that $|f(x_2)| \ge \varepsilon$. As U is symmetric, it follows that $x_1 U \cap x_2 U = \emptyset$. Next we pick an element x_3 outside of $x_1 U^2 \cup x_2 U^2$ with $|f(x_3)| \ge \varepsilon$. Repeating this argument, we find a sequence $\{x_n : n \in \mathbb{N}\}$ in G such that $x_n U \cap x_m U = \emptyset$ for all $n \ne m$ and $|f(x_n)| \ge \varepsilon$ for every n. But then $f_U \ge \varepsilon$ on $x_n U$ for every n, which contradicts the integrability of f_U.

Since integrable functions in $C_0(G)$ are square integrable, we get

$$C_{\text{unif}}(G) \subset L^2(G).$$

Let $f, g \in C_{\text{unif}}(G)$. We can write $f * g(x) = \langle f, L_x g^* \rangle$. The map $x \mapsto L_x g^*$ is continuous as a map from G to $L^2(G)$. The inner product is continuous, hence so is $f * g$. Finally, choose a unit-neighborhood U such that f_U and g_U are both integrable. Then

$$(f * g)_U(y) = \sup_{x,z \in U} \left| \int_G f(\xi) g(\xi^{-1} xyz) \, d\xi \right|$$

$$= \sup_{x,z \in U} \left| \int_G f(x\xi) g(\xi^{-1} yz) \, d\xi \right|$$

$$\le \sup_{x,z \in U} \int_G |f(x\xi) g(\xi^{-1} yz)| \, d\xi$$

$$\le \int_G f_U(\xi) g_U(\xi^{-1} y) \, d\xi = f_U * g_U(y).$$

This implies that $(f * g)_U$ is integrable over G. $\qquad\square$

Lemma 9.2.6 *For $f \in C_{\text{unif}}(G)$ and $\phi \in L^2(\Gamma \backslash G)$ one has*

$$R(f)\phi(x) = \int_{\Gamma \backslash G} k(x, y)\phi(y) \, dy,$$

where $k(x, y) = \sum_{\gamma \in \Gamma} f(x^{-1}\gamma y) \, dh$. The kernel k is continuous on $\Gamma \backslash G \times \Gamma \backslash G$.

Proof Let $f \in L^1(G)$. For $\phi \in L^2(\Gamma \backslash G)$, one computes with the quotient integral formula

$$R(f)\phi(x) = \int_G f(y)R(y)\phi(x) \, dy = \int_G f(y)\phi(xy) \, dy$$

$$= \int_G f(x^{-1}y)\phi(y) \, dy$$

$$= \int_{\Gamma \backslash G} \sum_{\gamma \in \Gamma} f(x^{-1}\gamma y)\phi(\gamma y) \, dy = \int_{\Gamma \backslash G} \sum_{\gamma \in \Gamma} f(x^{-1}\gamma y)\phi(y) \, dy,$$

as claimed, and the integral converges almost everywhere in x. In particular, for $f \in C_{\text{unif}}(G)$ this argument works with f replaced by f_U for a suitable symmetric unit-neighborhood U to get a kernel k_U. We choose $g \in C_c(G)$ with $g \geq 0$ and $\sum_{\gamma \in \Gamma} g(\gamma x) = 1$ for all $x \in G$ and use the quotient integral formula to compute

$$\int_{\Gamma \backslash G \times \Gamma \backslash G} k_U(x, y) \, dx \, dy = \int_{\Gamma \backslash G} \int_{\Gamma \backslash G} \sum_{\gamma \in \Gamma} g(\gamma x) \sum_{\tau \in \Gamma} f_U(x^{-1}\tau y) \, dx \, dy$$

$$= \int_{\Gamma \backslash G} \int_G g(x) \sum_{\tau \in \Gamma} f_U(x^{-1}\tau y) \, dx \, dy$$

$$= \int_G \int_G g(x) f_U(x^{-1}y) \, dx \, dy = \|g * f_U\|_1 < \infty.$$

By the quotient integral theorem, the sum $\sum_{\gamma \in \Gamma} f_U(x^{-1}\gamma y)$ converges almost everywhere in (x, y), so it converges on a dense set of (x, y). Let (x_0, y_0) be such a point of convergence. We show that $k(x, y)$ is continuous in the set $x_0 U \times y_0 U$. For given $\varepsilon > 0$ there exists a finite set $S \subset \Gamma$ such that $\sum_{\gamma \notin S} f_U(x_0^{-1}\gamma y_0) < \varepsilon/2$. This means that for $(x, y) \in x_0 U \times y_0 U$ one has $\sum_{\gamma \notin S} |f(x^{-1}\gamma y)| < \varepsilon/2$. This implies for $(x', y') \in x_0 U \times y_0 U$,

$$|k(x, y) - k(x', y')| \leq \sum_{\gamma \in S} |f(x^{-1}\gamma y) - f(x'^{-1}\gamma y')| + \varepsilon,$$

from which the continuity of k follows   □

As $k \in C(\Gamma \backslash G \times \Gamma \backslash G) \subset L^2(\Gamma \backslash G \times \Gamma \backslash G)$, the operator $R(f)$ is a Hilbert-Schmidt operator, hence compact. The theorem follows from the next lemma.

Lemma 9.2.7 *Let A be a *-closed subspace of $C_c(G)$. Let (η, V_η) be a unitary representation of G such that for every $f \in A$, the operator $\eta(f)$ is compact and such that for every non-zero $v \in V_\eta$ the space $\eta(A)v$ is non-zero. Then η is a direct sum of irreducible representations with finite multiplicities.*

Note that the condition $\eta(A)v \neq 0$ is always satisfied if A contains a Dirac-net $(\phi_U)_U$.

Proof Zorn's Lemma provides us with a subspace E, maximal with the property that it decomposes as a sum of irreducibles. The assumption of the Lemma also holds for the orthocomplement $E^\perp$ of E in $V = V_\eta$. This orthocomplement cannot contain any irreducible subspace. We have to show that it is zero. In other words, we have to show that a representation η as in the assumption, always contains an irreducible subspace.

The space A is generated by its self-adjoint elements. Let $f \in A$ be self-adjoint. Then $\eta(f)$ is self-adjoint and compact. By the spectral theorem for compact operators 5.2.2 the space $V = V_\eta$ decomposes,

$$V = V_{f,0} \oplus \bigoplus_{j=1}^{\infty} V_{f,j},$$

where $V_{f,0}$ is the kernel of $\eta(f)$ and $V_{f,j}$ is the eigenspace of $\eta(f)$ for an eigenvalue $\lambda_j \neq 0$. The sequence λ_j tends to zero and each $V_{f,j}$ is finite dimensional for $j > 0$. For every closed invariant subspace $E \subset V$ one has a similar decomposition

$$E = E'_{f,0} \oplus \bigoplus_{j=1}^{\infty} E'_{f,j},$$

and $E'_{f,j} \subset V_{f,j}$ for every $j \geq 0$. It follows that every non-zero closed invariant subspace E has a non-zero intersection with one of the $V_{f,i}$ for some $i > 0$ and some $f \in A$, since otherwise $E \subset \ker \eta(f)$ for every $f \in A$. Fix f and j and consider the set of all non-zero intersections $V_{f,j} \cap E$, where E runs over all closed invariant subspaces. Among these intersections choose one $W = V_{f,j} \cap E$ of minimal dimension $\neq 0$, which is possible as $V_{f,j}$ is finite-dimensional. Let $E^1 = \bigcap_{E:E \cap V_{f,j}=W} E$, where the intersection runs over all closed invariant subspaces E with $E \cap V_{f,j} = W$. Then E^1 is a closed invariant subspace. We claim that it is irreducible. For this assume that $E^1 = F \oplus F'$ with closed invariant subspaces F, F'. Then by minimality, $W \subset F$ or $W \subset F'$, which implies that one of the spaces F or F' is zero, so E^1 is indeed irreducible. We have shown that V indeed has an irreducible subspace E^1. As shown above this implies that V decomposes as a sum of irreducibles.

It remains to show that the multiplicities are finite. For this note first that if τ and σ are unitarily equivalent representations, then λ is an eigenvalue of $\tau(f)$ if and only if it is one for $\sigma(f)$. Thus finiteness of the multiplicities follows from the fact that any collection of orthogonal subspaces, which give rise to unitarily equivalent representations, must all have non-trivial intersection with the same eigenspaces $V_{f,j}$. Since the $V_{f,j}$ are finite dimensional, there can only exist finitely many of them. $\quad\square$

9.3   The Trace Formula

Let X be a locally compact Hausdorff space, and let μ be a Radon measure on X. A continuous L^2-kernel k on X is called *admissible kernel* if there exists a function $g \in C(X) \cap L^2(X)$ such that $|k(x,y)| \leq g(x)g(y)$. Note that if X is compact, then every continuous kernel is admissible.

An operator $S : L^2(X) \to L^2(X)$ is called an *admissible operator* if there exists an admissible kernel k such that

$$S\phi(x) = \int_X k(x,y)\phi(y)\,dy,$$

where we have written dy for $d\mu(y)$.

Proposition 9.3.1 *Let X be a locally compact Hausdorff space equipped with a Radon measure. Assume that X is first countable or compact. Let T be an integral operator with continuous L^2-kernel on X. Assume that there exists admissible operators S_1, S_2 with $T = S_1 S_2$.*

Then T is of trace class and

$$\operatorname{tr}(T) = \int_X k(x,x)\,dx.$$

Proof Replacing S_2 with S_2^* we can assume $T = S_1 S_2^*$ in the proposition. The map $\sigma : (S_1, S_2) \mapsto S_1 S_2^*$ is sesquilinear, so it obeys the polarization rule,

$$\sigma(S,R) = \frac{1}{4}\left(\sigma(S+R) - \sigma(S-R) + i\left(\sigma(S+iR) - \sigma(S-iR)\right)\right),$$

where we have written $\sigma(S+R)$ for $\sigma(S+R, S+R)$. This implies that, in order to prove the proposition, it suffices to assume that $T = SS^*$ for some admissible operator S. As S is a Hilbert-Schmidt operator, $T = SS^*$ is trace class and also $\operatorname{tr}(T) = \|S\|_{\mathrm{HS}}^2$. Let $l(x,y)$ be the admissible kernel of S. For $\phi \in C_c(X)$ we compute

$$SS^*\phi(x) = \int_X l(x,u)S^*\phi(u)\,du$$

$$= \int_X \int_X l(x,u)\overline{l(y,u)}\phi(y)\,dy\,du$$

$$= \int_X \left(\int_X l(x,u)\overline{l(y,u)}\,du\right)\phi(y)\,dy.$$

So the operator SS^* has the kernel $l * l^*(x,y) = \int_X l(x,z)\overline{l(y,z)}\,dz$. If l is admissible and X is first countable, it suffices to check convergence with sequences, and therefore the Theorem of Dominated Convergence implies that the kernel $l * l^*$ is continuous. If X is compact, the Radon measure is finite and the function l is uniformly continuous in the following sense: For every $x_0 \in X$ and every $\varepsilon > 0$ there exists a neighborhood U of x_0 such that for all $x \in U$ and all $z \in X$ we have $|l(x,z) - l(x_0,z)| < \varepsilon$. This implies the continuity of $l * l^*$. For $\phi, \psi \in C_c(X)$ we have

$$\int_X \int_X l * l^*(x,y)\phi(y)\psi(x)\,dy\,dx = \langle SS^*\phi, \overline{\psi}\rangle$$

$$= \int_X \int_X k(x,y)\phi(y)\psi(x)\,dy\,dx.$$

Varying ϕ and ψ we conclude that the continuous kernels k and $l * l^*$ coincide. Therefore,

$$\int_X k(x,x)\,dx = \int_X \int_X |l(x,u)|^2\,du\,dx = \|S\|_{\mathrm{HS}} = \mathrm{tr}\,(T).$$

The proposition follows. □

Recall that for $\pi \in \widehat{G}$ the number $N_\Gamma(\pi) \geq 0$ is the multiplicity of π as a subrepresentation of $(R, L^2(\Gamma \backslash G))$. Let $\widehat{G}_\Gamma$ denote the set of all $\pi \in \widehat{G}$ with $N_\Gamma(\pi) > 0$.

Definition We write $C_{\mathrm{unif}}(G)^2$ for the space of all linear combinations of functions of the form $g * h$ with $g, h \in C_{\mathrm{unif}}(G)$. Moreover, if Γ is a lattice in G and $\gamma \in \Gamma$, then we denote by $[\gamma]$ the conjugacy class of γ in Γ, $G_\gamma \overset{\mathrm{def}}{=} \{x \in G : x\gamma x^{-1} = \gamma\}$ denotes the centralizer of γ in G, and $\Gamma_\gamma \overset{\mathrm{def}}{=} G_\gamma \cap \Gamma$ denotes the centralizer of γ in Γ. Note that the map from Γ to $[\gamma]$, which sends ν to $\nu^{-1}\gamma\nu$, factors through a bijection $\Gamma_\gamma \backslash \Gamma \cong [\gamma]$.

Theorem 9.3.2 (Trace Formula) *Let G be a locally compact group and $\Gamma \subset G$ a uniform lattice. Let $\widehat{G}_\Gamma$ denote the set of all $\pi \in \widehat{G}$ which appear as subrepresentations of the representation R on $L^2(\Gamma \backslash G)$ and let $f \in C_{\mathrm{unif}}(G)^2$. For every $\pi \in \widehat{G}_\Gamma$ the operator $\pi(f)$ is of trace class and*

$$\sum_{\pi \in \widehat{G}_\Gamma} N_\Gamma(\pi)\,\mathrm{tr}\,\pi(f) = \sum_{[\gamma]} \mathrm{vol}(\Gamma_\gamma \backslash G_\gamma)\,\mathcal{O}_\gamma(f),$$

where the summation on the right runs over all conjugacy classes $[\gamma]$ in the group Γ, and $\mathcal{O}_\gamma(f)$ denotes the orbital integral,

$$\mathcal{O}_\gamma(f) = \int_{G_\gamma \backslash G} f(x^{-1}\gamma x)\,dx.$$

We shall see in Lemma 9.3.3 below that the centralizer G_γ is unimodular and that $\Gamma_\gamma \backslash G_\gamma$ has finite measure for every $\gamma \in \Gamma$. The expression $\mathrm{vol}(\Gamma_\gamma \backslash G_\gamma)\mathcal{O}_\gamma(f)$ is therefore well-defined. It does not depend on the choice of Haar measure on G_γ.

Proof of the trace formula The algebra $C_{\mathrm{unif}}(G)^2$ consists of all finite linear combinations of convolution products of the form $g * h^*$ with $g, h \in C_{\mathrm{unif}}(G)$. So it suffices to show the trace formula for $f = g * h^*$. According to Lemma 9.2.6, the operators $R(g)$ and $R(h)$ are integral operators with continuous kernels

$$k_g(x, y) = \sum_{\gamma \in \Gamma} g(x^{-1} \gamma y) \quad \text{and} \quad k_h(x, y) = \sum_{\gamma \in \Gamma} h(x^{-1} \gamma y).$$

By Proposition 9.3.1 the operator $R(f)$ is of trace class and $\mathrm{tr}(R(f)) = \int_{\Gamma \backslash G} k_f(x, x) \, dx$. With $R(f)$, all its restrictions to subrepresentations are of trace class. It follows that

$$\sum_{\pi \in \widehat{G}_\Gamma} N_\Gamma(\pi) \mathrm{tr}\, \pi(f) = \mathrm{tr}(R(f)) = \int_{\Gamma \backslash G} k_f(x, x) \, dx$$

$$= \int_{\Gamma \backslash G} \sum_{\gamma \in \Gamma} f(x^{-1} \gamma x) \, dx.$$

We order the sum according to conjugacy classes $[\gamma]$ in Γ, interchange integration and summation and use the quotient integral formula to get

$$\mathrm{tr}(R(f)) = \int_{\Gamma \backslash G} \sum_{[\gamma]} \sum_{\sigma \in \Gamma_\gamma \backslash \Gamma} f(x^{-1} \sigma^{-1} \gamma \sigma x) \, dx$$

$$= \sum_{[\gamma]} \int_{\Gamma \backslash G} \sum_{\sigma \in \Gamma_\gamma \backslash \Gamma} f((\sigma x)^{-1} \gamma \sigma x) \, dx$$

$$= \sum_{[\gamma]} \int_{\Gamma_\gamma \backslash G} f(x^{-1} \gamma x) \, dx.$$

Lemma 9.3.3 *For every $\gamma \in \Gamma$, the centralizer G_γ is unimodular and $\Gamma_\gamma \backslash G_\gamma$ has finite invariant measure* $\mathrm{vol}(\Gamma_\gamma \backslash G_\gamma)$.

Proof The above calculation shows that for every f as in the theorem, which is positive, i.e., for $f \geq 0$, one has $\int_{\Gamma_\gamma \backslash G} f(x^{-1} \gamma x) \, dx < \infty$ for every $\gamma \in \Gamma$. Consider the projection $p : \Gamma_\gamma \backslash G \to G_\gamma \backslash G$. Since G is unimodular by Theorem 9.1.6, the space $\Gamma_\gamma \backslash G$ carries an invariant Radon measure ν. Let μ be the image under p of ν, i.e., we define $\int_{G_\gamma \backslash G} f(x) \, d\mu(x) = \int_{\Gamma_\gamma \backslash G} f(p(y)) \, d\nu(y)$ for every $f \in C_c(G_\gamma \backslash G)$. We need to show that this is finite. For this let $0 \leq f \in C_c(G_\gamma \backslash G)$, and let $\Phi : G_\gamma \backslash G \to G$ be given by $\Phi(G_\gamma x) = x^{-1} \gamma x$. Then $K = \Phi(\mathrm{supp} f) \subseteq G$ is compact in G and by Tietze's extension theorem we can find a function $0 \leq \tilde{f} \in C_c(G)$ such that $\tilde{f}(y^{-1} \gamma y) = f(G_\gamma y)$ for every $y \in G$ with $f(G_\gamma y) > 0$. Choose $0 \leq F \in C_c(G)^2$ such that $\tilde{f} \leq F$. To show the existence of such a function, let $g \geq 0$ in $C_c(G)$ such that $g > 0$ in a neighborhood of $\mathrm{supp}(f)$. There exists a unit-neighborhood U, such that $\phi_U * g$ is > 0 on the set $\mathrm{supp}(f)$, where ϕ_U is a Dirac function of support in U. Set $F = c\phi_U * g$ for some sufficiently large $c > 0$, then $f \in C_c(G)^2$ and $F \geq f$. Further F satisfies the conditions of the theorem, so we get

$$\int_{G_\gamma \backslash G} f(x)\,dx = \int_{\Gamma_\gamma \backslash G} f(p(y))\,dv(y) = \int_{\Gamma_\gamma \backslash G} \tilde{f}(y^{-1}\gamma y)\,dv(y)$$

$$\leq \int_{\Gamma_\gamma \backslash G} F(y^{-1}\gamma y)\,dv(y),$$

which is finite by computations preceding the lemma. Thus μ is a well defined Radon measure. One easily checks that it is invariant. By Theorem 1.5.3 it follows that the modular function of G_γ agrees with the one of G. As G is unimodular, so is G_γ. Finally, since for every function in $C_c(G_\gamma \backslash G)$ the function $f \circ p$ is invariant under G_γ, it follows from the finiteness of $\int_{\Gamma_\gamma \backslash G} f(p(y))\,dv(y)$ that $\Gamma_\gamma \backslash G_\gamma$ has finite invariant volume. $\qquad \square$

By the lemma, we can continue the calculation above to arrive at

$$\mathrm{tr}\, R(f) = \sum_{[\gamma]} \int_{G_\gamma \backslash G} \int_{\Gamma_\gamma \backslash G_\gamma} f((\sigma x)^{-1}\gamma \sigma x)\,d\sigma\,dx$$

$$= \sum_{[\gamma]} \mathrm{vol}(\Gamma_\gamma \backslash G_\gamma)\, \mathcal{O}_\gamma(f).$$

This proves the theorem. $\qquad \square$

Example 9.3.4 Consider the case $G = \mathbb{R}$ and $\Gamma = \mathbb{Z}$. Every $t \in \mathbb{R}$ gives a character $x \mapsto e^{2\pi i t x}$, and in this way we identify $\widehat{G}$ with $\mathbb{R}$. The subset $\widehat{G}_\Gamma$ is mapped to $\mathbb{Z}$ and the multiplicities $N_\Gamma(\pi)$ are each equal to one. Therefore the spectral side of the trace formula equals $\sum_{k \in \mathbb{Z}} \hat{f}(k)$, and the geometric side equals $\sum_{k \in \mathbb{Z}} f(k)$. In other words, the trace formula is the same as the Poisson summation formula.

For applications of the trace formula, the following lemma will be important.

Lemma 9.3.5 *Let G be a locally compact group, and let A be a $*$-subalgebra of $C_c(G)$ which contains a Dirac net $(\phi_U)_U$. Assume further that A is stable under left translations L_y, $y \in G$. Let (π, V) and (σ, W) be unitary representations of G such that for every $f \in A$ the operators $\pi(f)$ and $\sigma(f)$ are trace class and that $\mathrm{tr}\, \pi(f) = \mathrm{tr}\, \sigma(f)$. Then π and σ are both direct sums of irreducible representations with finite multiplicities, and they are equivalent.*

Proof The fact that π and σ are both direct sums of irreducible representations with finite multiplicities is Lemma 9.2.7. There is a subrepresentation of π, maximal with the property of being isomorphic to a subrepresentation of σ. Restricting to the orthogonal space we can assume that π and σ do not have isomorphic subrepresentations. We have to show that they both are zero.

Let $V = \bigoplus_{\alpha \in I} V_\alpha$ be a decomposition in pairwise orthogonal subrepresentations and let $v_{\alpha,\mu} \in V_\alpha$ be vectors, such that

$$\sum_\alpha \sum_\mu \|\pi(f)v_{\alpha,\mu}\|^2 < \infty$$

for every $f \in A$. Choose a vector $0 \neq w \in W$. We claim that for every $\varepsilon > 0$ there exists some $f \in A$, such that

$$\sum_\alpha \sum_\mu \|\pi(f)v_{\alpha,\mu}\|^2 < \varepsilon \|\sigma(f)w\|^2.$$

If this were not the case, the map

$$\sum_\alpha \sum_\mu \pi(f)v_{\alpha,\mu} \mapsto \sigma(f)w, \qquad f \in A,$$

would be well-defined and could be extended to a non-trivial intertwining map from the closure L of the space

$$\left\{ \sum_\alpha \sum_\mu \pi(f)v_{\alpha,\mu} : f \in A \right\} \subset \bigoplus_\alpha \bigoplus_\mu V_{\alpha,\mu}$$

to W. Here every $V_{\alpha,\mu}$ is a new copy of the space V_α. The space L decomposes as a direct sum of irreducibles and it only contains isotypes, which don't occur in W, so such an intertwiner cannot exist.

Now assume $\sigma \neq 0$. As A contains Dirac functions of arbitrary small supports, there is a function h in A, such that $\sigma(h) \neq 0$. Let $f = h * h^*$. Then $\sigma(f)$ is of trace class and positive. Therefore $\sigma(f)$ possesses a largest eigenvalue and we can scale h in such a way that this eigenvalue is equal to 1. Let $w \in W$ of norm one with $\sigma(f)w = w$. Let $\lambda > 0$ be the largest eigenvalue of $\pi(f)$. For every α, let $(v_{\alpha,\mu})$ be an orthonormal basis of V_α consisting of eigenvectors of $\pi(f)$, say $\pi(f)v_{\alpha,\mu} = \lambda_{\alpha,\mu}v_{\alpha,\mu}$. For every $g \in A$, the sum $\sum_\alpha \sum_\mu \|\pi(g)v_{\alpha,\mu}\|^2$ equals the square of the Hilbert-Schmidt norm of $\pi(g)$ and therefore is finite. By the above remark there is a $g \in A$ with

$$\sum_\alpha \sum_\mu \|\pi(g)v_{\alpha,\mu}\|^2 < \frac{1}{\lambda^2} \|\sigma(g)w\|^2.$$

The trace of $\pi(g * f)^* \pi(g * f)$ equals

$$\sum_\alpha \sum_\mu \|\pi(g * f)v_{\alpha,\mu}\|^2 = \sum_\alpha \sum_\mu \lambda_{\alpha,\mu}^2 \|\pi(g)v_{\alpha\mu}\|^2$$

$$\leq \lambda^2 \sum_\alpha \sum_\mu \|\pi(g)v_{\alpha,\mu}\|^2.$$

The right hand side is strictly smaller than $\|\sigma(g)w\|^2 = \|\sigma(g * f)w\|^2$, which again is smaller than

$$\mathrm{tr}\left(\sigma(g * f)^*\sigma(g * f)\right) = \mathrm{tr}\left(\pi(g * f)^*\pi(g * f)\right),$$

a contradiction!

This implies $\sigma = 0$ and, by symmetry, also $\pi = 0$. $\qquad\square$

9.4 Locally Constant Functions

Let G be a locally compact group with a cocompact lattice $\Gamma \subset G$. A function f on G is called *locally constant*, if for every $x \in G$ there exists a neighborhood U_x of x such that f is constant on U_x. There are not many locally constant functions if G is connected, but there are many if G is totally disconnected.

The function f is called *uniformly locally constant*, if there exists a unit-neighborhood U such that f is constant on every set of the form UxU, $x \in G$. For example, if f is locally constant and of compact support, then f is uniformly locally constant.

Proposition 9.4.1 *Let G be totally disconnected and f a uniformly locally constant and integrable function on G. Then $f \in C_{\mathrm{unif}}(G)^2$. Hence the trace formula is valid for f.*

Proof Let U denote a compact open subgroup such that f is constant on every set of the form UxU. Then $f_U = |f|$ and therefore $f \in C_{\mathrm{unif}}(G)$. The same holds for the function $g = \frac{1}{\mathrm{vol}(U)}1_U$ and one sees that $f = g * f$, so $f \in C_{\mathrm{unif}}(G)^2$ as claimed. $\qquad\square$

9.5 Lie Groups

In this section, we shall present a simple class of functions which satisfy the trace formula in the case of a Lie-group, i.e., a group which is a smooth manifold such that the group operations are smooth maps. We shall freely make use of the notion of a smooth manifold as in [War83].

Theorem 9.5.1 *Let G be a Lie group of dimension n and let $\Gamma \subset G$ be a cocompact lattice. Let f be a continuous integrable function on G, such that the sum*

$$k(x, y) = \sum_{\gamma \in \Gamma} f(x^{-1}\gamma y)$$

converges uniformly and the kernel k is 2r-times continuously differentiable, where r is the smallest integer with $r > n/2$. Then the trace formula is valid for f.

In particular, the trace formula holds for every $f \in C_c^r(G)$.

Proof   To show the theorem, we need a partition of unity with a smooth square-root. The easiest way to achieve that is to adapt the classical construction of a partition of unity as is done in the following proposition.

Lemma 9.5.2 *Let M be a smooth manifold and let $(U_i)_{i \in I}$ be an open covering of M. Then there are smooth functions $u_i : M \to [0, 1]$, such that the support of u_i in contained in U_i and that*

$$\sum_{i \in I} u_i \equiv 1,$$

where the sum is locally finite, i.e., for every $p \in M$ there is a neighborhood U, such that the set

$$\{i \in I : u_i|_U \neq 0\}$$

is finite. The family (u_i) is called a smooth partition of unity subordinate to the covering (U_i). One can choose the u_i in a way that for each $i \in I$ the function $\sqrt{u_i}$ is smooth as well.

Proof   Except for the smoothness of the square-root, this is Theorem 1.11 of [War83], the proof of which is adapted to give the lemma. In loc.cit., one constructs a sequence $(\psi_j)_{j \geq 1}$ of smooth functions, which are ≥ 0 and such that the sets $\{p : \psi_j(p) > 0\}$ form a locally-finite covering of M, subordinate to the covering (U_i). So the function

$$\psi = \sum_{j \geq 1} \psi_j^2$$

is a smooth function with $\psi(p) > 0$ for every $p \in M$. The functions

$$u_j = \frac{\psi_j^2}{\psi}$$

are smooth with smooth square root and satisfy $\sum_j u_j = 1$.   □

A Borel measure v on $\mathbb{R}^n$ is called a *smooth measure*, if it has a smooth, strictly positive density with respect to the Lebesgue measure λ, i.e., if there is a smooth function

$$d : \mathbb{R}^n \to (0, \infty)$$

such that

$$v(A) = \int_A d(x) \, d\lambda(x)$$

holds for every Borel set $A \subset \mathbb{R}^n$.

A measure μ on a smooth manifold M is called a *smooth measure*, if for every smooth chart $\phi : U \to \mathbb{R}^n$ the induced measure $\nu = \phi_*\mu$ on $\mathbb{R}^n$, given by $\mu(A) = \nu(\phi^{-1}(A))$ is smooth.

Proposition 9.5.3 *Let M be a compact smooth manifold of dimension n with a smooth measure μ. Let $k : M \times M \to \mathbb{C}$ be continuous and $2r$-times continuously differentiable in the first argument, where r is the smallest integer $r > n/2$. Then the induced integral operator $T_k : L^2(M) \to L^2(M)$,*

$$T_k(\phi)(x) = \int_M k(x, y)\phi(y)\, dy,$$

is of trace class and

$$\mathrm{tr}\,(T_k) = \int_M k(x, x)\, dx.$$

Proof We first prove the assertion in the case $M = \mathbb{R}^n/\mathbb{Z}^n$ and μ being the Lebesgue measure λ. In this case we define

$$l(x, y) = \sum_{k \in \mathbb{Z}^n} \left(\frac{1}{1 + 4\pi^2\|k\|^2} \right)^r e_k(x)\overline{e_k(y)},$$

where $e_k(x) = e^{2\pi i(x_1 k_1 + x_2 k_2 + \cdots + x_n k_n)}$ and $\|k\|^2 = k_1^2 + \cdots + k_n^2$. As the corresponding integral converges, this sum converges absolutely uniformly and the kernel $l(x, y)$ therefore is continuous. Let Δ be the Laplace operator,

$$\Delta = -\frac{\partial^2}{\partial x_1^2} - \cdots - \frac{\partial^2}{\partial x_n^2}.$$

Lemma 9.5.4 *For $\phi \in C^{2r}(\mathbb{R}^n/\mathbb{Z}^n)$ one has*

$$T_l(1 + \Delta)^r \phi = \phi.$$

Proof Both sides are continuous functions, so it suffices to show that they agree as L^2-functions. But for all e_k, $k \in \mathbb{Z}^n$, we have $(1 + \Delta)^r e_k = (1 + 4\pi^2\|k\|^2)^r e_k$ and $T_l e_k = (1 + 4\pi^2\|k\|^2)^{-r} e_k$, hence $T_l(1 + \Delta)^r e_k = e_k$. Since $\{e_k : k \in \mathbb{Z}^n\}$ is an orthonormal basis of $L^2(\mathbb{R}^n/\mathbb{Z}^n)$ we see that $T_l(1 + \Delta)^r = \mathrm{Id}_{L^2(\mathbb{R}^n/\mathbb{Z}^n)}$, and the result follows. $\qquad\square$

We now show the proposition in the case $M = \mathbb{R}^n/\mathbb{Z}^n$ equipped with the Lebesgue measure. Let k be as in the theorem. Then for $\phi \in L^2(M)$ the function $T_k\phi$ is in $C^{2r}(M)$, so by the lemma,

$$T_l(1+\Delta)^r T_k\phi = T_k\phi.$$

One has $(1+\Delta)^r T_k = T_{k'}$, where k' is the continuous kernel

$$k'(x, y) = (1+\Delta_x)^r k(x, y).$$

So $T_k = T_l T_{k'}$ is a product of two Hilbert-Schmidt operators, hence of trace class. As both, T_l and $T_{k'}$ are admissible, the claim follows from Lemma 9.3.1.

Next let M be an arbitrary smooth compact manifold of dimension n. Let $(U_i)_{i=1}^s$ be an open covering by chart sets, where we choose the chart maps ψ_i not as maps to $\mathbb{R}^n$ but instead to $\mathbb{R}^n/\mathbb{Z}^n$, which is possible as well. So for every i the map ψ_i is a homeomorphism of U_i to some open set $V_i \subset \mathbb{R}^n/\mathbb{Z}^n$. The sets $V_1, \ldots, V_s \subset \mathbb{R}^n/\mathbb{Z}^n$ can be chosen pairwise disjoint. Let (u_i) be a smooth partition of unity with smooth square root subordinate to the covering (U_i). For $1 \le i, j \le s$ let

$$k_{i,j}(x, y) = \sqrt{u_i(x)}k(x, y)\sqrt{u_j(y)}.$$

Then $k_{i,j}$ is a continuous kernel on $M \times M$. For given $1 \le i, j \le s$, define a continuous kernel $\tilde{k}_{i,j}$ on $\mathbb{R}^n/\mathbb{Z}^n \times \mathbb{R}^n/\mathbb{Z}^n$ by

$$\tilde{k}_{i,j}(x, y) = \begin{cases} \sqrt{d_i(x)}k_{i,j}(\psi_i^{-1}(x), \psi_j^{-1}(y))\sqrt{d_j(y)} & (x, y) \in V_i \times V_j, \\ 0 & \text{otherwise.} \end{cases}$$

Here d_j denotes the density of the measure $(\psi_j)_*\mu$ with respect to the Lebesgue measure, i.e. $\int_{V_i} f(x)d_i(x)\,d\lambda(x) = \int_{U_i} f(\psi_i(y))d\mu(y)$ for all $f \in C_c(V_i)$. We define $\tilde{k} = \sum_{i,j} \tilde{k}_{i,j}$ and for $\phi \in L^2(M)$ we set

$$\phi_j(x) = \phi(x)\sqrt{u_j(x)}.$$

We define $\tilde{\phi}_j \in L^2(\mathbb{R}^n/\mathbb{Z}^n)$ by

$$\tilde{\phi}_j(x) = \begin{cases} \phi_j(\psi_j^{-1}(x))\sqrt{d_j(x)} & x \in V_j, \\ 0 & \text{otherwise.} \end{cases}$$

Finally set $\tilde{\phi} = \sum_j \tilde{\phi}_j$.

Lemma 9.5.5 *The map* $\Psi : \phi \mapsto \tilde{\phi}$ *is a linear isometry* $L^2(M) \hookrightarrow L^2(\mathbb{R}^n/\mathbb{Z}^n)$ *and one has*

$$\Psi(T_k\phi) = T_{\tilde{k}}\Psi(\phi)$$

for every $\phi \in L^2(M)$. *The operator* $T_{\tilde{k}}$ *equals* $PT_{\tilde{k}}P$, *where P is the orthogonal projection* $L^2(\mathbb{R}^n/\mathbb{Z}^n) \to \text{Im}(\Psi)$. *Finally we have*

$$\int_M k(x, x)\,d\mu(x) = \int_{\mathbb{R}^n/\mathbb{Z}^n} \tilde{k}(x, x)\,dx.$$

Proof The map Ψ is linear. For $\phi \in L^2(M)$ we compute

$$\|\Psi(\phi)\|^2 = \int_{\mathbb{R}^n/\mathbb{Z}^n} |\tilde{\phi}(x)|^2 \, d\lambda(x) = \sum_j \int_{V_j} |\tilde{\phi}_j(x)|^2 \, d\lambda(x)$$

$$= \sum_j \int_{V_j} |\phi_j(\overset{=z}{\overbrace{\psi_j^{-1}(x)}})|^2 \underbrace{d_j(x) \, d\lambda(x)}_{d(\psi_j)_* \mu(x)}$$

$$= \sum_j \int_{U_j} |\phi_j(z)|^2 \, d\mu(z) = \sum_j \int_M |\phi(z)|^2 u_j(x) \, d\mu(z)$$

$$= \int_M |\phi(z)|^2 \, d\mu(x) = \|\phi\|^2,$$

so Ψ is an isometry. In order to show $\Psi T_k = T_{\tilde{k}} \Psi$, we compute

$$T_{\tilde{k}} \Psi(\phi)(x) = T_{\tilde{k}} \tilde{\phi}(x) = \sum_j T_{\tilde{k}} \tilde{\phi}_j(x) = \sum_j \int_{V_j} \tilde{k}(x, y) \tilde{\phi}_j(y) \, d\lambda(y)$$

$$= \sum_{i,j} \sqrt{d_i(x)} \int_{V_j} k_{i,j}(\psi_i^{-1}(x), \psi_j^{-1}(y)) \phi_j(\psi_j^{-1}(y)) \underbrace{d_j(y) \, d\lambda(y)}_{=d(\psi_j)_* \mu(y)}$$

$$= \sum_{i,j} \sqrt{d_i(x)} \int_{U_j} k_{i,j}(\psi_i^{-1}(x), z) \phi_j(z) \, d\mu(z)$$

$$= \sum_{i,j} \sqrt{d_i(x)} \int_{U_j} \sqrt{u_i(\psi_i^{-1}(x))} k(\psi_i^{-1}(x), z) u_j(z) \phi(z) \, d\mu(z)$$

$$= \sum_i \sqrt{d_i(x)} \sqrt{u_i(\psi_i^{-1}(x))} \int_M k(\psi_i^{-1}(x), z) \phi(z) \, d\mu(z)$$

$$= \sum_i \sqrt{d_i(x)} \sqrt{u_i(\psi_i^{-1}(x))} T_k \phi(\psi_i^{-1}(x)) = \Psi T_k \phi(x).$$

To show $T_{\tilde{k}} = P T_{\tilde{k}} P$ one has to show that for $g \in \mathrm{Im}(\Psi)^\perp$,

$$T_{\tilde{k}}(g) = 0 \quad \text{and} \quad \langle T_{\tilde{k}} h, g \rangle = 0$$

holds for every $h \in L^2(\mathbb{R}^n/\mathbb{Z}^n)$. The first of these assertions follows from the fact that for fixed x, the map $y \mapsto \tilde{k}(x, y)$ lies in the image of Ψ. The second assertion follows similarly from the observation, that for fixed y the map $x \mapsto \tilde{k}(x, y)$ lies in the image of Ψ.

We finish the proof of the lemma with

$$\int_{\mathbb{R}^n/\mathbb{Z}^n} \tilde{k}(x,x)\,d\lambda(x) = \sum_{i,j} \underbrace{\int_{\mathbb{R}^n/\mathbb{Z}^n} \tilde{k}_{i,j}(x,x)\,d\lambda(x)}_{=0\,\text{if}\,i\neq i}$$

$$= \sum_j \int_{V_j} k_{j,j}(\psi_j^{-1}(x),\psi_j^{-1}(x))d_j(x)\,d\lambda(x)$$

$$= \sum_j \int_{U_j} k_{j,j}(x,x)\,d\mu(x)$$

$$= \sum_j \int_{U_j} k(x,x)u_j(x)\,d\mu(x)$$

$$= \int_M k(x,x)\,d\mu(x). \qquad\square$$

The lemma implies

$$\operatorname{tr} T_k = \operatorname{tr} T_{\tilde{k}} = \int_{\mathbb{R}^n/\mathbb{Z}^n} \tilde{k}(x,x)\,dx = \int_M k(x,x)\,dx$$

and the proposition follows. $\qquad\square$

Now Theorem 9.5.1 follows from Proposition 9.5.3 in the same way as Theorem 9.3.2, i.e., the trace formula, follows from Proposition 9.3.1. $\qquad\square$

9.6 Exercises

Exercise 9.1 Show that every Schwartz function on $\mathbb{R}$ is uniformly integrable.

Exercise 9.2 Let B be the group of upper triangular matrices in $SL_2(\mathbb{R})$. Show that B does not contain a lattice.

Exercise 9.3 Let Γ be a subgroup of the locally compact group G. Show that Γ is discrete if and only if for every compact subset $C \subset G$ the intersection $\Gamma \cap C$ is finite.

Exercise 9.4 A group Γ acts *discontinuously* on a topological space X if for every $x \in X$ there is a neighborhood U of x, such that the set of all $\gamma \in \Gamma$ with $U \cap \gamma U \neq \emptyset$ is finite.

Let $K \subset G$ be a compact subgroup of the locally compact group G. Show that a subgroup $\Gamma \subset G$ is discrete if and only if it acts discontinuously on the space G/K.

Exercise 9.5 Let G be a finite group, and let H be a subgroup. For each choose the counting measure as Haar measure. For $x \in H$ let H_x, G_x be the centralizers of x in

H and G, respectively. For $\pi \in \widehat{G}$ let $\chi_\pi : G \to \mathbb{C}$ be defined by $\chi_\pi(x) = \operatorname{tr} \pi(x)$. For $x \in H$ let $[x]_H$ be the H-conjugacy class of x, i.e., $[x]_H = \{hxh^{-1} : h \in H\}$. Show:

$$\sum_{\pi \in \widehat{G}} (\dim V_\pi^H) \chi_\pi = \sum_{[h]_H} \frac{|G_h|}{|H_h|} \mathbf{1}_{[h]_G},$$

where V_π^H denotes the space of H-invariant vectors in V_π.

(Hint: Let g be one of the sides of the equation. For $f \in L^2(G)$ consider the inner product $\langle f, g \rangle$ and use the trace formula.)

Exercise 9.6 Let $f \in \mathcal{S}(\mathbb{R})$ be a Schwartz function. Apply the trace formula to f with $G = \mathbb{R}$ and $\Gamma = \mathbb{Z}$ to give a proof of the classical Poisson summation formula:

$$\sum_{k \in \mathbb{Z}} f(k) = \sum_{k \in \mathbb{Z}} \hat{f}(k),$$

where $\hat{f}(x) = \int_{\mathbb{R}} f(y) e^{-2\pi i xy} \, dy$.

Exercise 9.7 Let G be a first countable compact group and $H \subset G$ a closed subgroup. Let $f \in C(G)^2$ with $\operatorname{supp}(f) \cap G.H = \emptyset$, where $G.H$ is the union of all conjugates gHg^{-1} of H. Show that

$$\sum_{\pi \in \widehat{G}} \dim(V_\pi^H) \operatorname{tr} \pi(f) = 0.$$

Chapter 10
The Heisenberg Group

In this chapter we prove the Stone-von Neumann Theorem, which gives a full characterization of the unitary dual of the Heisenberg group $\mathcal{H}$. We then apply the trace formula to describe the spectral decomposition of $L^2(\Lambda \backslash \mathcal{H})$, where Λ is the standard integer lattice in $\mathcal{H}$.

10.1 Definition

The Heisenberg group $\mathcal{H}$ is defined to be the group of real upper triangular 3×3 matrices with ones on the diagonal:

$$\mathcal{H} \stackrel{\text{def}}{=} \left\{ \left(\begin{array}{ccc} 1 & x & z \\ & 1 & y \\ & & 1 \end{array} \right) \middle| x, y, z \in \mathbb{R} \right\}.$$

It can also be identified with $\mathbb{R}^3$, where the group law is given by

$$(a, b, c)(x, y, z) \stackrel{\text{def}}{=} (a + x, b + y, c + z + ay).$$

The inverse of (a, b, c) is

$$(a, b, c)^{-1} = (-a, -b, ab - c).$$

The center of $\mathcal{H}$ is $Z(\mathcal{H}) = \{(0, 0, z) \mid z \in \mathbb{R}\}$, and the projection to the first two coordinates induces an isomorphism

$$\mathcal{H}/Z(\mathcal{H}) \cong \mathbb{R}^2.$$

An easy calculation ([Dei05] Chap. 12) shows that $\mathcal{H}$ is unimodular and that a Haar integral on $\mathcal{H}$ is given by

$$\int_{\mathcal{H}} f(h) \, dh \stackrel{\text{def}}{=} \int_{\mathbb{R}} \int_{\mathbb{R}} \int_{\mathbb{R}} f(a, b, c) \, da \, db \, dc, \quad f \in C_c(\mathcal{H}).$$

We will use this Haar measure on $\mathcal{H}$ for all computations in the sequel.

A. Deitmar, S. Echterhoff, *Principles of Harmonic Analysis*, Universitext, DOI 10.1007/978-3-319-05792-7_10, © Springer International Publishing Switzerland 2014

10.2   The Unitary Dual

In this section we are going to describe the unitary dual $\widehat{\mathcal{H}}$ of the Heisenberg group $\mathcal{H}$.

Let $\widehat{\mathcal{H}}_0$ denote the subset of $\widehat{\mathcal{H}}$ consisting of all classes $\pi \in \widehat{\mathcal{H}}$ such that $\pi(h) = 1$ whenever h lies in the center $Z(\mathcal{H})$ of $\mathcal{H}$. Since $\mathcal{H}/Z(\mathcal{H}) \cong \mathbb{R}^2$, it follows that

$$\widehat{\mathcal{H}}_0 = \widehat{\mathcal{H}/Z(\mathcal{H})} \cong \widehat{\mathbb{R}^2} \cong \left(\widehat{\mathbb{R}}\right)^2,$$

and the latter can be identified with $\mathbb{R}^2$ in the following explicit way. Let $(a, b) \in \mathbb{R}^2$ and define a character

$$\chi_{a,b} : \quad \begin{aligned} H &\rightarrow \quad \mathbb{T}, \\ (x, y, z) &\mapsto \quad e^{2\pi i(ax+by)}. \end{aligned}$$

The identification is given by $(a, b) \mapsto \chi_{a,b}$. In particular, it follows that all representations in $\widehat{\mathcal{H}}_0$ are one-dimensional. This observation indicates the importance of the behavior of the center under a representation.

As a consequence of the Lemma of Schur 6.1.7, for each $\pi \in \widehat{\mathcal{H}}$ there is a character $\chi_\pi : Z(\mathcal{H}) \rightarrow \mathbb{T}$ with $\pi(z) = \chi_\pi(z)\mathrm{Id}$ for every $z \in Z(\mathcal{H})$. This character χ_π is called the *central character* of the representation π.

For every character $\chi \neq 1$ of $Z(\mathcal{H})$, we will now construct an irreducible unitary representation of the Heisenberg group that has χ for its central character. So let $t \neq 0$ be a real number and consider the central character

$$\chi_t(0, 0, c) = e^{2\pi ict}.$$

For $(a, b, c) \in \mathcal{H}$ we define the operator $\pi_t(a, b, c)$ on $L^2(\mathbb{R})$ by

$$\pi_t(a, b, c)\phi(x) \overset{\text{def}}{=} e^{2\pi i(bx+c)t}\phi(x + a).$$

It is straightforward that π_t is a unitary representation.

Recall from Exercise 3.17 that the Schwartz space $\mathcal{S}(\mathbb{R}^n)$ consists of all C^∞-functions $f : \mathbb{R}^n \rightarrow \mathbb{C}$ such that $x^\alpha \partial^\beta f$ is bounded for all multi-indices $\alpha, \beta \in \mathbb{N}_0^n$. We have $\mathcal{S}(\mathbb{R}^n) \subseteq L^p(\mathbb{R}^n)$ for every $p \geq 1$ and $\mathcal{S}(\mathbb{R}^n)$ is stable under Fourier transform.

Theorem 10.2.1 (Stone-von Neumann). *For $t \neq 0$ the unitary representation π_t is irreducible. Every irreducible unitary representation of $\mathcal{H}$ with central character χ_t is isomorphic to π_t. It follows that*

$$\widehat{\mathcal{H}} = \mathbb{R}^2 \cup \{\pi_t : t \neq 0\}.$$

Proof We show irreducibility first. Fix $t \neq 0$, and let $V \subset L^2(\mathbb{R})$ be a closed non-zero subspace that is invariant under the set of operators $\pi_t(\mathcal{H})$. If $\phi \in V$, then so is the function $\pi_t(-a,0,0)\phi(x) = \phi(x-a)$. As V is closed, it therefore contains $\psi * \phi(x) = \int_{\mathbb{R}} \psi(a)\phi(x-a)\,da$ for $\psi \in \mathcal{S} = \mathcal{S}(\mathbb{R})$. These convolution products are smooth functions, so V contains a smooth function $\phi \neq 0$. One has $\pi_t(0,-b,0)\phi(x) = e^{-2\pi i btx}\phi(x) \in V$. By integration it follows that for $\psi \in \mathcal{S}$ one has that $\hat{\psi}(tx)\phi(x)$ lies in V. The set of possible functions $\hat{\psi}(tx)$ contains all smooth functions of compact support, as the Fourier transform is a bijection on the space of Schwartz functions to itself. Choose an open interval I, in which ϕ has no zero. It follows that $C_c^\infty(I) \subset V$. Translating, taking sums, and using a partition of unity argument gives $C_c^\infty(\mathbb{R}) \subset V$. This space is dense in $L^2(\mathbb{R})$, so $V = L^2(\mathbb{R})$, which means that π_t is irreducible.

For the second assertion of the theorem, let π be an irreducible unitary representation with central character χ_t. We want to show that π is isomorphic to π_t. For notational ease, let's first reduce to the case $t = 1$. Consider the map $\theta_t : \mathcal{H} \to \mathcal{H}$ given by $\theta_t(a,b,c) = (a, bt, ct)$. A calculation shows that θ_t is an automorphism of $\mathcal{H}$ with $\chi_1 \circ \theta_t = \chi_t$. So we conclude that π has central character χ_t if and only if $\pi \circ \theta_t^{-1}$ has central character χ_1. It therefore suffices to show the uniqueness for $t = 1$.

Let (η, V_η) be an irreducible unitary representation of $\mathcal{H}$ with central character χ_1. Let Γ be the subgroup of $\mathcal{H}$ consisting of all $(0,0,k)$ with $k \in \mathbb{Z}$. Then Γ lies in the center and $\chi_1(\Gamma) = \{1\}$. So the representation η factors over the quotient group $B \stackrel{\text{def}}{=} \mathcal{H}/\Gamma$, which is homeomorphic to the space $\mathbb{R}^2 \times \mathbb{R}/\mathbb{Z}$. For $\phi \in \mathcal{S}(\mathbb{R}^2)$, let $\phi_B(a,b,c) = \phi(a,b)e^{-2\pi i c}$. Then $\phi(a,b) = \phi_B(a,b,0)$. For $\phi, \psi \in \mathcal{S}(\mathbb{R}^2)$, one computes

$$\phi_B * \psi_B(a,b,c) = \int_B \phi_B(h)\psi_B(h^{-1}(a,b,c))\,dh$$

$$= \int_{\mathbb{R}^2}\int_{\mathbb{R}/\mathbb{Z}} \phi(x,y)e^{-2\pi i z}\psi_B(a-x, b-y, c-z+xy-xb)\,dx\,dy\,dz$$

$$= \int_{\mathbb{R}^2}\int_{\mathbb{R}/\mathbb{Z}} \phi(x,y)e^{-2\pi i z}\psi(a-x, b-y)e^{-2\pi i(c-z+xy-xb)}\,dx\,dy\,dz$$

$$= e^{-2\pi i c}\int_{\mathbb{R}^2} \phi(x,y)\psi(a-x,b-y)e^{2\pi i x(b-y)}\,dx\,dy.$$

Inspired by this, we define a new, non-commutative convolution product on $\mathcal{S}(\mathbb{R}^2)$ by

$$\phi *_B \psi(a,b) \stackrel{\text{def}}{=} \int_{\mathbb{R}^2} \phi(x,y)\psi(a-x,b-y)e^{2\pi i x(b-y)}\,dx\,dy.$$

For $\phi \in \mathcal{S}(\mathbb{R}^2)$ we define $\eta(\phi) = \eta(\phi_B)$. This equals

$$\eta(\phi) = \int_{\mathbb{R}^2} \phi(a,b)\eta(a,b,0)\,da\,db.$$

By construction of the convolution product one gets $\eta(\phi *_B \psi) = \eta(\phi)\eta(\psi)$, so η is an algebra homomorphism of the algebra $\mathcal{A} = (\mathcal{S}(\mathbb{R}^2), *_B)$. One computes that

$(\phi_B)^* = (\phi^*)_B$, where the involution on the left is the usual involution of functions g on the group B, given by $g^*(b) = \overline{g(b^{-1})}$. The involution on the algebra $\mathcal{A} = \mathcal{S}(\mathbb{R}^2)$ on the other hand is defined by $\phi^*(a,b) = \overline{\phi(-a,-b)}e^{2\pi i ab}$. So η is a homomorphism of *-algebras.

Lemma 10.2.2 *The *-algebra homomorphism* $\eta : \mathcal{A} \to \mathcal{B}(V_\eta)$ *is injective.*

Proof Assume $\eta(\phi) = 0$. Then, for $h = (a,b,0) \in B$ and $v, w \in V_\eta$ we have

$$0 = \langle \eta(h)\eta(\phi)\eta(h^{-1})v, w \rangle$$

$$= \int_{\mathbb{R}^2} \langle \eta(x,y,0)v, w \rangle \phi(x,y) e^{2\pi i (ay - bx)} \, dx \, dy$$

The continuous, rapidly decreasing function $\langle \eta(x,y,0)v, w \rangle \phi(x,y)$ therefore has zero Fourier transform, hence is zero. Varying v and w, this leads to ϕ being zero. □

In the special case $\eta = \pi_1$ the lemma has the consequence that relations among the $\pi_1(\phi)$ for $\phi \in \mathcal{A}$ already hold in $\mathcal{A}$. We will make use of this principle for special elements as follows. For $F, G \in \mathcal{S}(\mathbb{R})$ let $\tilde{\phi}_{F,G}(a,b) = \overline{F(a+b)}G(b)$, and $\phi_{F,G}(a,b) = \mathcal{F}_2\tilde{\phi}_{F,G}(a,b) \in \mathcal{S}(\mathbb{R}^2)$, where $\mathcal{F}_2$ denotes taking Fourier transform in the second variable. An application of the properties of the Fourier transform yields

$$\pi_1(\phi_{F,G})f(x) = \int_{\mathbb{R}^2} \phi_{F,G}(a,b)\pi_1(a,b,0)f(x) \, da \, db$$

$$= \int_{\mathbb{R}^2} \phi_{F,G}(a,b)f(x+a)e^{2\pi i bx} \, da \, db$$

$$= \int_{\mathbb{R}^2} \mathcal{F}_2\tilde{\phi}_{F,G}(a-x,b)f(a)e^{2\pi i bx} \, da \, db$$

$$= \int_{\mathbb{R}} \tilde{\phi}_{F,G}(a-x,x)f(a) \, da$$

$$= \int_{\mathbb{R}} \overline{F(a)}G(x)f(a) \, da = \langle f, F \rangle G(x).$$

In particular, if the norm of F is one, then $\pi_1(\phi_{F,F})$ is the orthogonal projection onto the one dimensional space $\mathbb{C}F$.

For $h \in B = \mathcal{H}/\Gamma$ and $\phi \in \mathcal{S}(\mathbb{R}^2)$ define $L_h\phi$ by the formula $(L_h\phi)_B = L_h(\phi_B)$. Explicitly, one computes that $L_{(x,y,z)}\phi(a,b) = \phi(a-x, b-y)e^{2\pi i (bx-xy+z)}$. For $\phi, \psi \in \mathcal{A}$ we also write $\phi\psi$ for the convolution product $\phi *_B \psi$.

Lemma 10.2.3 *In the algebra* $\mathcal{A}$, *one has the relations*

$$\phi_{F,G}\phi_{H,L} = \langle L, F \rangle \phi_{H,G} \quad and \quad \phi_{F,G}^* = \phi_{G,F}.$$

For $h \in B$ *one has* $L_h\phi_{F,G} = \phi_{F, \pi_1(h)G}$.

Proof Using the formula $\pi_1(\phi_{F,G})f = \langle f, F\rangle G$ one computes for $f \in L^2(\mathbb{R})$,

$$\pi_1(\phi_{F,G}\phi_{H,L})f = \pi_1(\phi_{F,G})\pi_1(\phi_{H,L})f$$
$$= \langle \pi_1(\phi_{H,L})f, F\rangle G$$
$$= \langle\langle f, H\rangle L, F\rangle G = \langle L, F\rangle\langle f, H\rangle G.$$

This implies the first claim as π_1 is injective. The second follows from

$$\langle \pi_1(\phi_{F,G})f, h\rangle = \langle f, F\rangle\langle G, h\rangle = \langle f, \pi_1(\phi_{G,F})h\rangle.$$

For the last claim compute

$$\pi_1(L_h\phi_{F,G})f = \pi_1(h)\pi_1(\phi_{F,G})f = \pi_1(h)\langle f, F\rangle G$$
$$= \langle f, F\rangle\pi_1(h)G = \pi_1(\phi_{F,\pi_1(h)G})f.$$

The lemma follows. $\square$

As $\mathcal{S}(\mathbb{R})$ is dense in $L^2(\mathbb{R})$, by the orthonormalizing scheme one can find a sequence F_j in $\mathcal{S}(\mathbb{R})$ that forms an orthonormal base of $L^2(\mathbb{R})$. For $j, k \in \mathbb{N}$ write $\phi_{j,k}$ for ϕ_{F_j,F_k}. Lemma 10.2.3 implies that $\phi_{j,k}\phi_{s,t} = \delta_{j,t}\phi_{s,k}$ (Kronecker-delta), and $\phi_{j,j}^2 = \phi_{j,j} = \phi_{j,j}^*$.

Let now (η, V_η) again denote an irreducible unitary representation with central character χ_1. By integration, η gives an algebra homomorphism $\mathcal{A} \to \mathcal{B}(V_\eta)$.

Lemma 10.2.4 *The $\eta(\phi_{j,j})_{j\in\mathbb{N}}$ are non-zero projections. They are pairwise orthogonal. Write V_j for the image of $\eta(\phi_{j,j})$. Then $\eta(\phi_{j,k})$ is an isometry from V_j to V_k and it annihilates every V_l for $l \neq j$.*

Proof Since $\phi_{j,j}^2 = \phi_{j,j} = \phi_{j,j}^*$, the same holds for $\eta(\phi_{j,j})$ hence the latter are projections. They are non-zero by Lemma 10.2.2. For $v \in V_j$ one computes $\langle \eta(\phi_{j,k})v, \eta(\phi_{j,k})v\rangle = \langle \eta(\phi_{k,j}\phi_{j,k})v, w\rangle = \langle \eta(\phi_{j,j})v, w\rangle = \langle v, v\rangle$. The claim follows. $\square$

Now choose $v_1 \in V_1$ of norm one and set $v_j = \eta(\phi_{1,j})v_1$. Then (v_j) is an orthonormal system in V_η. Define an isometry $T : L^2(\mathbb{R}) \to V_\eta$ by mapping F_j to v_j. Then

$$\eta(h)T(F_j) = \eta(h)v_j = \eta(h)\eta(\phi_{j,j})v_j$$
$$= \eta(L_h\phi_{j,j})v_j = \eta(\phi_{F_j,\pi_1(h)F_j})v_j.$$

By the fact that (F_j) is an orthonormal basis, one concludes

$$\pi_1(h)F_j = \sum_k \langle \pi_1(h)F_j, F_k\rangle F_k.$$

So that by Lemma 10.2.3

$$\eta(h)T(F_j) = \sum_k \langle \pi_1(h)F_j, F_k \rangle \eta(\phi_{F_j,F_k})v_j$$

$$= \sum_k \langle \pi_1(h)F_j, F_k \rangle v_k = T(\pi_1(h)F_j).$$

Hence T is an $\mathcal{H}$-homomorphism onto a non-zero closed subspace of V_η. As η is irreducible, T must be surjective, so T is unitary and η is equivalent to π_1. This finishes the proof of the Theorem of Stone and von Neumann.   □

10.3   The Plancherel Theorem for $\mathcal{H}$

We have an identification $\mathcal{H} \cong \mathbb{R}^3$. We interpret $f \in \mathcal{S}(\mathbb{R}^3)$ as a function on $\mathcal{H}$, and we write $\mathcal{S}(\mathcal{H})$ for this space of functions.

Theorem 10.3.1 (Plancherel Theorem). *Let $f \in \mathcal{S}(\mathcal{H})$. For every $t \in \mathbb{R}^\times$ the operator $\pi_t(f)$ is a Hilbert-Schmidt operator, and we have*

$$\int_{\mathbb{R}^\times} \|\pi_t(f)\|_{HS}^2 |t| \, dt = \int_{\mathcal{H}} |f(h)|^2 \, dh.$$

It follows that the Plancherel measure of $\mathcal{H}$ in the sense of Theorem 8.5.3 equals $|t| \, dt$ and that the set of one dimensional representations of $\mathcal{H}$ has Plancherel measure zero.

Proof   This is proved in [Dei05], Chap. 12.   □

10.4   The Standard Lattice

Let Λ be the set of all $(a, b, c) \in \mathcal{H}$, where a, b, c are integers. The multiplication and inversion keep this set stable, so Λ is a discrete subgroup.

Lemma 10.4.1 Λ *is a uniform lattice in* $\mathcal{H}$.

Proof   Let $K = [0, 1] \times [0, 1] \times [0, 1] \subset \mathcal{H}$. Then K is a compact subset. We claim that the projection map $K \to \mathcal{H}/\Lambda$ is surjective. This means we have to show that for every $h \in \mathcal{H}$ there exists $\lambda \in \Lambda$ such that $h\lambda \in K$. Let $h = (x, y, z) \in \mathcal{H}$, and let $\lambda = (a, b, c) \in \Lambda$. Then $h\lambda = (x + a, y + b, z + c + xb)$. So we can find $a, b \in \mathbb{Z}$ such that $x + a$ and $y + b$ lie in the unit interval. So we may assume as well that $x, y \in [0, 1]$. Assuming that, we only consider such λ, which are central, i.e., of the form $\lambda = (0, 0, c)$ and it becomes clear that we can find λ such that $h\lambda \in K$.   □

By Theorem 9.2.2 we conclude that as $\mathcal{H}$-representation one has the decomposition

$$L^2(\Lambda\backslash\mathcal{H}) \cong \bigoplus_{\pi\in\widehat{\mathcal{H}}} N(\pi)\pi,$$

with finite multiplicities $N(\pi)$.

We want to apply the trace formula to compute the numbers $N(\pi)$. For this recall the characters of $\mathcal{H}$ given by $\chi_{r,s}(a,b,c) = e^{2\pi i(ra+sb)}$, where $r,s \in \mathbb{R}$.

Theorem 10.4.2 *For a character $\chi_{r,s}$ of $\mathcal{H}$ the multiplicity $N(\chi_{r,s})$ in $L^2(\Lambda\backslash\mathcal{H})$ is equal to one if r and s are both integers. Otherwise it is zero. For $t \in \mathbb{R}$, $t \neq 0$, the multiplicity $N(\pi_t)$ is $|t|$ if t is an integer and zero otherwise. So the decomposition of the representation $\left(L^2(\Lambda\backslash\mathcal{H}), R\right)$ reads*

$$R \cong \bigoplus_{r,s\in\mathbb{Z}} \chi_{r,s} \oplus \bigoplus_{\substack{k\in\mathbb{Z}\\k\neq0}} |k|\pi_k.$$

Proof We first consider the subspace H of $L^2(\Lambda\backslash\mathcal{H})$, which is invariant under the action of the center $Z \subset \mathcal{H}$. This is isomorphic to $L^2\left(\mathbb{Z}^2\backslash\mathbb{R}^2\right)$ as $\Lambda\backslash\mathcal{H}/Z \cong \mathbb{Z}^2\backslash\mathbb{R}^2$ and the representation of $\mathcal{H}$ on H factors through the representation of $\mathbb{R}^2$ on $L^2\left(\mathbb{Z}^2\backslash\mathbb{R}^2\right)$. By the theory of Fourier series this gives us the value of $N\left(\chi_{r,s}\right)$ as in the theorem. The difficult part is to determine $N(\pi_t)$.

Let $h \in C_c^\infty(\mathcal{H})$ and set $f = h * h^*$. Then the trace formula says,

$$\operatorname{tr} R(f) = \sum_{\pi\in\widehat{\mathcal{H}}} N(\pi)\operatorname{tr}\pi(f) = \sum_{[\lambda]} \operatorname{vol}\left(\Lambda_\lambda\backslash\mathcal{H}_\lambda\right) \mathcal{O}_\lambda(f),$$

where the sum is taken over all conjugacy classes $[\lambda]$ in Λ, Λ_λ and $\mathcal{H}_\lambda$ denote the centralizers of λ in Λ and $\mathcal{H}$, respectively, and $\mathcal{O}_\lambda(f) = \int_{\mathcal{H}_\lambda\backslash\mathcal{H}} f(x^{-1}\lambda x)\,dx$. First consider $\pi \in \widehat{\mathcal{H}}$ with trivial central character, say $\pi = \chi_{r,s}$. Then

$$\operatorname{tr}\pi(f) = \chi_{r,s}(f) = \int_{\mathbb{R}}\int_{\mathbb{R}}\int_{\mathbb{R}} f(a,b,c)e^{2\pi i(ar+bs)}\,da\,db\,dc$$

$$= \mathcal{F}_1\mathcal{F}_2\mathcal{F}_3 f(-r,-s,0),$$

where $\mathcal{F}_i$ denotes Fourier transform on $\mathbb{R}$ applied to the i-th component of $\mathcal{H}$. Next the center Z_Λ of Λ, the set of all $(0,0,k)$ for $k \in \mathbb{Z}$, is contained in the center of $\mathcal{H}$ and therefore acts trivially on $L^2(\Lambda\backslash\mathcal{H})$. This implies that for $\pi \in \widehat{\mathcal{H}}$ one has that $N(\pi) \neq 0$ implies $\pi(Z_\Lambda) = \{1\}$. Therefore, $N(\pi_t) \neq 0$ implies that $t \in \mathbb{Z}$. For $t = k \in \mathbb{Z}$ one computes

$$\pi_k(f)\phi(x) = \int_{\mathbb{R}}\int_{\mathbb{R}}\left(\int_{\mathbb{R}} f(a,b,c)e^{2\pi ikc}\,dc\right)e^{2\pi ikbx}\phi(a+x)\,da\,db$$

$$= \int_{\mathbb{R}} \mathcal{F}_2\mathcal{F}_3 f(a-x,-kx,-k)\phi(a)\,da$$

So $\pi_k(f)$ is an integral operator with kernel

$$k(x, y) = \mathcal{F}_2\mathcal{F}_3 f(y - x, -kx, -k).$$

Analogously, the kernel of $\pi_k(h)$ is $\mathcal{F}_2\mathcal{F}_3 h(y - x, -kx, -k)$. The latter kernel is in $\mathcal{S}(\mathbb{R}^2)$ and therefore is admissible. By Proposition 9.3.1 and Fourier inversion, we have

$$\operatorname{tr} \pi_k(f) = \int_{\mathbb{R}} \mathcal{F}_2\mathcal{F}_3 f(0, -kx, -k)\, dx = \frac{1}{|k|}\mathcal{F}_3 f(0, 0, -k).$$

Together we get

$$\operatorname{tr} R(f) = \sum_{r,s \in \mathbb{Z}} \mathcal{F}f(r, s, 0) + \sum_{k \in \mathbb{Z} \setminus \{0\}} \frac{N(\pi_k)}{|k|} \mathcal{F}_3 f(0, 0, k),$$

where $\mathcal{F}f = \mathcal{F}_1\mathcal{F}_2\mathcal{F}_3 f$.

Next we consider the geometric side of the trace formula. Let $\lambda \in \Lambda$. First assume that λ lies in the center of Λ. Then $\lambda = (0, 0, k)$ for some $k \in \mathbb{Z}$, and the centralizer $\mathcal{H}_\lambda$ equals $\mathcal{H}$, so that the orbital integral equals $\mathcal{O}_\lambda(f) = f(\lambda) = f(0, 0, k)$. Next let $\lambda = (r, s, t) \in \Lambda = \mathbb{Z}^3$ with $(r, s) \neq (0, 0)$. Let $h = (a, b, c) \in \mathcal{H}$, then $h\lambda h^{-1} = (r, s, t + as - br)$. So h lies in the centralizer $\mathcal{H}_\lambda$ if and only if $as = br$. It follows that $\mathcal{H}_\lambda$ is the subspace generated by the vectors $(r, s, 0)$ and $(0, 0, c)$ as a subspace of $\mathbb{R}^3$. On $\mathcal{H}_\lambda$ we pick the Haar measure that comes from the euclidean length in $\mathbb{R}^3$; then the volume of $\Lambda_\lambda \backslash \mathcal{H}_\lambda$ is the minimal value $\sqrt{a^2 + b^2}$ where $a, b \in \mathbb{Z}$, not both zero, and $as = br$. This minimum is taken in $a = \frac{r}{\gcd(r,s)}$ and $b = \frac{s}{\gcd(r,s)}$, where $\gcd(r, s)$ denotes the greatest common divisor of r and s. Therefore,

$$\operatorname{vol}(\Lambda_\lambda \backslash \mathcal{H}_\lambda) = \frac{\sqrt{r^2 + s^2}}{\gcd(r, s)}.$$

The orbital integral equals

$$\mathcal{O}_\lambda(f) = \int_{U^\perp} f(r, s, t + xs - yr)\, d(x, y),$$

where U is the subspace of all $\binom{a}{b} \in \mathbb{R}^2$ with $as - br = 0$. Then $U = \mathbb{R}\binom{r}{s}$ and the orthogonal space is spanned by the norm one vector $\frac{1}{\sqrt{r^2+s^2}}\binom{s}{-r}$. So we get

$$\mathcal{O}_\lambda(f) = \int_{\mathbb{R}} f\left(r, s, t + x\frac{s^2 + r^2}{\sqrt{r^2 + s^2}}\right) dx = \frac{1}{\sqrt{r^2 + s^2}}\mathcal{F}_3 f(r, s, 0).$$

The number of conjugacy classes with given $(r, s, *)$ equals $\gcd(r, s)$, hence

$$\operatorname{tr} R(f) = \sum_{k \in \mathbb{Z}} f(0, 0, k) + \sum_{(r,s) \neq (0,0)} \mathcal{F}_3 f(r, s, 0).$$

Using the Poisson summation formula in the first two variables we get

$$\sum_{(r,s)\in\mathbb{Z}^2} \mathcal{F}_3 f(r,s,0) = \sum_{(r,s)\in\mathbb{Z}^2} \mathcal{F} f(r,s,0)$$

with $\mathcal{F} = \mathcal{F}_1 \mathcal{F}_2 \mathcal{F}_3$ and using the formula in the third variable gives

$$\sum_{k\in\mathbb{Z}} f(0,0,k) = \sum_{k\in\mathbb{Z}} \mathcal{F}_3 f(0,0,k).$$

Combining this with the two expressions for tr $R(f)$ we conclude

$$\sum_{k\neq 0} \frac{N(\pi_k)}{|k|} \mathcal{F}_3 f(0,0,k) = \sum_{k\neq 0} \mathcal{F}_3 f(0,0,k).$$

By Lemma 9.3.5 this implies $N(\pi_k) = |k|$. The theorem is proven. $\qquad\square$

10.5 Exercises and Notes

Exercise 10.1 Let $\mathcal{H}$ be the Heisenberg group, and let $\mathcal{Z}$ be its center. Show that every normal subgroup of $\mathcal{H}$ lies in the center $\mathcal{Z}$ or contains the center.

Exercise 10.2 On $\mathbb{R}^2$ consider the bilinear form

$$b(v,w) \stackrel{\text{def}}{=} v^t \begin{pmatrix} 0 & 1 \\ -1 & 0 \end{pmatrix} w.$$

Let L be the group $\mathbb{R}^2 \times \mathbb{R}$ with the multiplication $(v,t)(w,s) \stackrel{\text{def}}{=} (v + w, s + t + \frac{1}{2}b(v,w))$. Show: The map $\psi : \mathcal{H} \to L$, defined by $\psi(a,b,c) = ((a,b)^t, c - \frac{1}{2}ab)$ is a continuous group isomorphism with continuous inverse.

Exercise 10.3 Let L be defined as in Exercise 10.2. The group $G = SL_2(\mathbb{R})$ acts on L via $g(v,t) = (gv,t)$. Show that G acts by group homomorphisms, which fix the center of L point-wise. Conclude that for every $g \in G$ one has $\pi_1 \circ g \cong \pi_1$. Conclude that for every $g \in G$ there is a $T(g) \in U(L^2(\mathbb{R}))$, which is unique up to scalar multiplication, such that $\pi_1(gl) = T(g)\pi_1(l)T(g)^{-1}$, where we consider π_1 as a representation of L via Exercise 10.2. Show that the map $g \mapsto T(g)$ is a group homomorphism $G \to U(L^2(\mathbb{R}))/\mathbb{T}$.

Exercise 10.4 Let $\text{Aut}(\mathcal{H})$ be the group of all continuous group automorphisms of $\mathcal{H}$. Every $\phi \in \text{Aut}(\mathcal{H})$ preserves the center $\mathcal{Z}$, hence induces an element of $\text{Aut}(\mathcal{H}/\mathcal{Z}) = \text{Aut}(\mathbb{R}^2) = GL_2(\mathbb{R})$. Show that the ensuing map $\text{Aut}(\mathcal{H}) \to GL_2(\mathbb{R})$ is surjective.

Exercise 10.5 For $h \in \mathcal{H}$ let $\phi_h \in \text{Aut}(\mathcal{H})$ be defined by $\phi_h(x) = hxh^{-1}$. Every such automorphism is called an *inner automorphism*. Let $\text{Inn}(\mathcal{H})$ be the group of all inner automorphisms. Show that $\text{Inn}(\mathcal{H}) \cong \mathbb{R}^2$, and that, together with the last exercise, one gets an exact sequence

$$1 \to \mathbb{R}^2 \to \text{Aut}(\mathcal{H}) \to \text{GL}_2(\mathbb{R}) \to 1.$$

Exercise 10.6 Let $N \subset \text{GL}_n(\mathbb{R})$ be the group of all upper triangular matrices with ones on the diagonal. Show that $\Gamma = N \cap \text{GL}_n(\mathbb{Z})$ is a uniform lattice in N.

Exercise 10.7 The $(2n + 1)$-dimensional Heisenberg group $\mathcal{H}_n$ is defined as the group of all matrices of the form $\left(\begin{smallmatrix} 1 & x^t & z \\ & 1 & y \\ & & 1 \end{smallmatrix} \right)$ in $\text{GL}_{n+2}(\mathbb{R})$ where $x, y \in \mathbb{R}^n$ and $z \in \mathbb{R}$. Determine its unitary dual along the lines of the Stone-von Neumann Theorem.

Exercise 10.8 Let L be as in Exercise 10.2. Let $\Lambda \subset \mathbb{R}^2$ be a lattice with $b(\Lambda \times \Lambda) \subset 2\mathbb{Z}$. Show that $\Lambda \times \Lambda \times \mathbb{Z}$ is a uniform lattice in L. Use the proof of Theorem 10.4.2 to find the decomposition of $L^2(\Lambda \backslash L)$.

Notes

As shown in Exercise 10.3, the Stone-von Neumann Theorem yields a group homo-morphism $\text{SL}_2(\mathbb{R}) \to U(L^2(\mathbb{R}))/\mathbb{T}$. There is a unique non-trivial covering group $\text{SL}_2^2(\mathbb{R})$ of degree 2. On this group the homomorphism induces a proper unitary representation $\text{SL}_2^2(\mathbb{R}) \to U(L^2(\mathbb{R}))$. This representation is known as the *Weil representation*. It is used to explain the behavior of theta-series, in particular with respect to lifting of automorphic forms [How79, LV80, Wei64].

Chapter 11
$SL_2(\mathbb{R})$

The group $SL_2(\mathbb{R})$ is the simplest case of a so called reductive Lie group. Harmonic analysis on these groups turns out to be more complex then the previous cases of abelian, compact, or nilpotent groups. On the other hand, the applications are more rewarding. For example, via the theory of automorphic forms, in particular the Langlands program, harmonic analysis on reductive groups has become vital for number theory. In this chapter we prove an explicit Plancherel Theorem for functions in the Hecke algebra of the group $G = SL_2(\mathbb{R})$. We apply the trace formula to a uniform lattice and as an application derive the analytic continuation of the Selberg zeta function.

11.1 The Upper Half Plane

Let $G = SL_2(\mathbb{R})$ denote the *special linear group of degree 2*, i.e.

$$SL_2(\mathbb{R}) = \left\{ \begin{pmatrix} a & b \\ c & d \end{pmatrix} \in M_2(\mathbb{R}) : ad - bc = 1 \right\}.$$

The locally compact group $SL_2(\mathbb{R})$ acts on the *upper half plane*

$$\mathbb{H} = \{z \in \mathbb{C} : \mathrm{Im}(z) > 0\}$$

by *linear fractionals*, i.e., for $g = \begin{pmatrix} a & b \\ c & d \end{pmatrix} \in SL_2(\mathbb{R})$ and for $z \in \mathbb{H}$ one defines

$$gz = \frac{az + b}{cz + d}.$$

To see that this is well-defined one has to show that $cz + d \neq 0$. If $c = 0$ then $d \neq 0$ and so the claim follows. If $c \neq 0$ then $\mathrm{Im}(cz + d) = c\mathrm{Im}(z) \neq 0$. Next one has to show that gz lies in $\mathbb{H}$ if z does and that $(gh)z = g(hz)$ for $g, h \in SL_2(\mathbb{R})$. The latter is an easy computation, for the former we will now derive an explicit formula for the imaginary part of gz. Multiplying numerator and denominator by $c\bar{z} + d$ one gets $gz = \frac{ac|z|^2 + bd + 2bc\mathrm{Re}(z) + z}{|cz+d|^2}$, so in particular,

$$\text{Im}(gz) = \frac{\text{Im}(z)}{|cz + d|^2},$$

which is strictly positive if $\text{Im}(z)$ is. Note that the action of the central element $-1 \in SL_2(\mathbb{R})$ is trivial.

If $g = \left(\begin{smallmatrix} a & b \\ c & d \end{smallmatrix}\right) \in G$ stabilizes the point $i \in \mathbb{H}$, then $\frac{ai+b}{ci+d} = i$, or $ai + b = -c + di$, which implies $a = d$ and $b = -c$. So the stabilizer of the point $i \in \mathbb{H}$ is the *rotation group*:

$$K = SO(2) = \left\{ \begin{pmatrix} a & -b \\ b & a \end{pmatrix} : a, b \in \mathbb{R},\ a^2 + b^2 = 1 \right\},$$

which also can be described as the group of all matrices of the form

$$\begin{pmatrix} \cos t & -\sin t \\ \sin t & \cos t \end{pmatrix} \quad \text{for} \quad t \in \mathbb{R}.$$

The operation of G on $\mathbb{H}$ is transitive, as for $z = x + iy \in \mathbb{H}$ one has

$$z = \begin{pmatrix} \sqrt{y} & \frac{x}{\sqrt{y}} \\ 0 & \frac{1}{\sqrt{y}} \end{pmatrix} i.$$

It follows that via the map $G/K \to \mathbb{H}$, given by $gK \mapsto gi$, the upper half plane $\mathbb{H}$ can be identified with the quotient G/K.

Theorem 11.1.1 (Iwasawa Decomposition). *Let A be the group of all diagonal matrices in G with positive entries. Let N be the group of all matrices of the form $\left(\begin{smallmatrix} 1 & s \\ 0 & 1 \end{smallmatrix}\right)$ for $s \in \mathbb{R}$. Then one has $G = ANK$. More precisely, the map*

$$\psi : A \times N \times K \to G,$$

$$(a, n, k) \mapsto ank$$

is a homeomorphism.

Proof Let $g \in G$, and let $gi = x + yi$. Then, with

$$a = \begin{pmatrix} \sqrt{y} & \\ & 1/\sqrt{y} \end{pmatrix} \quad \text{and} \quad n = \begin{pmatrix} 1 & x/y \\ & 1 \end{pmatrix},$$

one has $gi = ani$ and so $g^{-1}an$ lies in K, which means that there exists $k \in K$ with $g = ank$. Using the explicit formula for gz above in the case $z = i$, one constructs the inverse map to ψ as follows. Let $\phi : G \to A \times N \times K$ be given by $\phi(g) = (\underline{a}(g), \underline{n}(g), \underline{k}(g))$, where

$$\underline{a}\begin{pmatrix} a & b \\ c & d \end{pmatrix} = \begin{pmatrix} \frac{1}{\sqrt{c^2 + d^2}} & \\ & \sqrt{c^2 + d^2} \end{pmatrix},$$

$$\underline{n}\begin{pmatrix} a & b \\ c & d \end{pmatrix} = \begin{pmatrix} 1 & ac + bd \\ & 1 \end{pmatrix},$$

$$k \begin{pmatrix} a & b \\ c & d \end{pmatrix} = \frac{1}{\sqrt{c^2 + d^2}} \begin{pmatrix} d & -c \\ c & d \end{pmatrix}.$$

A straightforward computation shows that $\phi\psi = \mathrm{Id}$ and $\psi\phi = \mathrm{Id}$. □

For $g \in \mathrm{SL}_2(\mathbb{R})$ we shall use throughout this chapter the notation $\underline{a}(g)$, $\underline{n}(g)$, and $\underline{k}(g)$ as explained above. Moreover, for $x, t, \theta \in \mathbb{R}$, we shall write

$$a_t \stackrel{\mathrm{def}}{=} \begin{pmatrix} e^t & \\ & e^{-t} \end{pmatrix} \in A$$

$$n_x \stackrel{\mathrm{def}}{=} \begin{pmatrix} 1 & x \\ & 1 \end{pmatrix} \in N$$

$$k_\theta \stackrel{\mathrm{def}}{=} \begin{pmatrix} \cos\theta & -\sin\theta \\ \sin\theta & \cos\theta \end{pmatrix} \in K.$$

A function $f : G \to \mathbb{C}$ is called *smooth* if the map $\mathbb{R}^3 \to \mathbb{C}$ given by

$$(t, x, \theta) \mapsto f(a_t n_x k_\theta)$$

is infinitely differentiable. We denote the space of smooth functions by $C^\infty(G)$. The space of smooth functions of compact support is denoted by $C_c^\infty(G)$.

Theorem 11.1.2 *The group $G = \mathrm{SL}_2(\mathbb{R})$ is unimodular.*

Proof Let $\phi : G \to \mathbb{R}_+^\times$ be a continuous group homomorphism. We show that $\phi \equiv 1$. First note that $\phi(K) = 1$ as K is compact. As ϕ restricted to A is a continuous group homomorphism, there exists $x \in \mathbb{R}$ such that $\phi(a_t) = e^{tx}$ for every $t \in \mathbb{R}$. Let $w = \begin{pmatrix} & 1 \\ -1 & \end{pmatrix}$, then $wa_t w^{-1} = a_{-t}$, and therefore $e^{tx} = \phi(a_t) = \phi(wa_t w^{-1}) = e^{-tx}$ for every $t \in \mathbb{R}$, which implies $x = 0$ and so $\phi(A) = 1$. Similarly, $\phi(n_x) = e^{rx}$ for some $r \in \mathbb{R}$. As we have $a_t n_x a_t^{-1} = n_{e^{2t}x}$ it follows $e^{rs} = e^{re^{2t}s}$ for every $t \in \mathbb{R}$, which implies $r = 0$, so $\phi(N) = 1$ and by the Iwasawa decomposition, we conclude $\phi \equiv 1$. □

We write $\underline{t}(g)$ for the unique $t \in \mathbb{R}$ with $\underline{a}(g) = a_t$, i.e., one has $\underline{a}(g) = a_{\underline{t}(g)}$.

Theorem 11.1.3 *For any given Haar measures on three of the four groups G, A, N, K, there is a unique Haar measure on the fourth such that for $f \in L^1(G)$ the decomposition formula*

$$\int_G f(x)\, dx = \int_A \int_N \int_K f(ank)\, dk\, dn\, da$$

holds. For $\phi \in L^1(K)$ and $x \in G$ one has

$$\int_K \phi(k)\, dk = \int_K \phi(\underline{k}(kx))\, e^{2\underline{t}(kx)}\, dk.$$

From now on we normalize Haar measures as follows. On K we normalize the volume to be one. On A we choose the measure $2dt$, where $t = \underline{t}(a)$, and on N we choose $\int_{\mathbb{R}} f(n_s)\,ds$. The factor 2 is put there to match the usual invariant measure $\frac{dx\,dy}{y^2}$ on the upper half plane.

Proof Let $B = AN$, the subgroup of G consisting of all upper triangular matrices with positive diagonal entries. Then an easy computation shows that $db = da\,dn$ is a Haar measure on B and that B is not unimodular. Indeed, one has $\Delta_B(a_x n) = e^{-2x}$, which follows from the equation $a_t n_x a_s n_y = a_{t+s} n_{y+e^{-2s}x}$. Let $\underline{b} : G \to B$ be the projection $\underline{b}(g) = \underline{a}(g)\underline{n}(g)$. The map $B \to G/K \cong \mathbb{H}$ mapping b to bK is a B-equivariant homeomorphism. Any G-invariant measure on G/K gives a Haar measure on B and as both these types of measures are unique up to scaling one gets that every B-invariant measure on G/K is already G-invariant. So the formula $\int_G f(x)\,dx = \int_{G/K} \int_K f(xk)\,dk\,dx$ leads to $\int_G f(x)\,dx = \int_B \int_K f(bk)\,dk\,db$. Since $db = da\,dn$, the integral formula follows.

For the second assertion let $\phi \in L^1(K)$. Let $\eta \in L^1(B)$ and set $g(bk) = \eta(b)\phi(k)$. Then g lies in $L^1(G)$. As G is unimodular, for $y \in G$ one has

$$
\int_B \int_K \eta(b)\phi(k)\,dk\,db = \int_G g(y)\,dy = \int_G g(yx)\,dy
$$

$$
= \int_G \eta(\underline{b}(yx))\phi(\underline{k}(yx))\,dy
$$

$$
= \int_B \int_K \eta(\underline{b}(bkx))\phi(\underline{k}(kx))\,dk\,db
$$

$$
= \int_B \int_K \eta(b\underline{b}(kx))\phi(\underline{k}(kx))\,dk\,db
$$

$$
= \int_B \int_K \eta(b)\Delta_B(\underline{b}(kx))^{-1}\phi(\underline{k}(kx))\,dk\,db
$$

$$
= \int_B \int_K \eta(b)e^{2\underline{t}(kx)}\phi(\underline{k}(kx))\,dk\,db,
$$

where we used the facts that $\underline{k}(bg) = \underline{k}(g)$ for all $b \in B$, $g \in G$ and $\underline{t}(\underline{b}(kx)) = \underline{t}(kx)$ for all $b \in B, k \in K$, and $x \in G$. Varying η, we get the claim of the theorem. $\qquad\square$

Hyperbolic Geometry

Let $g = \left(\begin{smallmatrix} a & b \\ c & d \end{smallmatrix}\right)$ be in $G = \mathrm{SL}_2(\mathbb{R})$. For $z \in \mathbb{H}$ one gets

$$
\frac{d}{dz}gz = \frac{d}{dz}\frac{az+b}{cz+d} = \frac{a(cz+d) - c(az+b)}{(cz+d)^2} = \frac{1}{(cz+d)^2}.
$$

Since on the other hand, $\text{Im}(gz) = \frac{\text{Im}(z)}{|cz+d|^2}$, we get $\left|\frac{d}{dz}gz\right| = \frac{\text{Im}(gz)}{\text{Im}(z)}$, or

$$\frac{\left|\frac{d}{dz}gz\right|}{\text{Im}(gz)} = \frac{\left|\frac{d}{dz}z\right|}{\text{Im}(z)}.$$

That is to say, the Riemannian metric $\frac{dx^2+dy^2}{y^2}$ is invariant under the group action of G on $\mathbb{H}$. For a continuously differentiable path $p : [0, 1] \to \mathbb{H}$ we get the induced *hyperbolic length* defined by

$$L(p) = \int_0^1 \frac{|p'(t)|}{\text{Im}(p(t))}\, dt.$$

Then it follows that $L(p) = L(g \circ p)$ for every $g \in G$, i.e., the length is G-invariant. The *hyperbolic distance* of two points $z, w \in \mathbb{H}$ is defined by

$$\rho(z, w) = \inf_p L(p),$$

where the infimum is extended over all paths p with $p(0) = z$ and $p(1) = w$.

Lemma 11.1.4 *For any two point $z, w \in \mathbb{H}$ there exists $g \in G$ such that $gz = i$ and $gw = yi$ for some $y \geq 1$.*

Proof As we have seen in the beginning of this chapter, the group action of G on $\mathbb{H}$ is transitive, hence there exists $h \in G$ with $hz = i$. We next apply an element $k \in K$ so that $g = kh$ does the job. For this we have to show that for any given $z \in \mathbb{H}$ there exists $k \in K$ such that $kz = yi$ for some $y \geq 1$. The map $\theta \mapsto k_\theta z$ is continuous, for $\theta = 0$ we have $k_\theta z = z$ and for $\theta = \pi/2$ we have $k_\theta = \left(\begin{smallmatrix} 0 & -1 \\ 1 & 0 \end{smallmatrix}\right)$ so that $k_{\pi/2}z = -1/z$. Hence the real parts of z and $k_{\pi/2}z$ have opposite sign, by the intermediate value theorem there exists $k \in K$ such that $\text{Re}(kz) = 0$. If now $kz = yi$ with $y < 1$, then we replace k with $k_{\pi/2}k$ to finish the proof   □

Lemma 11.1.5 *The hyperbolic distance is a metric on $\mathbb{H}$. It is G-invariant, i.e., $\rho(gz, gw) = \rho(z, w)$ holds for all $z, w \in \mathbb{H}$, $g \in G$. For $z, w \in \mathbb{H}$ one has*

$$\rho(z, w) = \log \frac{|z - \overline{w}| + |z - w|}{|z - \overline{w}| - |z - w|},$$

and

$$2\cosh \rho(z, w) = 2 + \frac{|z - w|^2}{\text{Im}(z)\text{Im}(w)}.$$

Proof The G-invariance follows from the invariance of the length. The axioms of a metric are immediate from the definition. For the explicit formulae, we start with the special case $z = i$ and $w = yi$ for $y \geq 1$. For any path p with $p(0) = i$ and $p(1) = yi$ we get

$$L(p) = \int_0^1 \sqrt{\mathrm{Re}(p'(t))^2 + \mathrm{Im}(p'(t))^2} \, \frac{dt}{\mathrm{Im}(p(t))}.$$

This is minimized by the path $p(t) = ity$, since for any path $p = \mathrm{Re}(p) + i\,\mathrm{Im}(p)$ the path $i\,\mathrm{Im}(p)$ will also connect the points i and yi. So one gets $\rho(i, yi) = \log y$, which also equals the right hand side of the first assertion in this case. Next the equivalence of the first and second formula are easy, as is the G-invariance of the right hand side of the second formula, which then concludes the proof. $\square$

11.2   The Hecke Algebra

Let A^+ be the subset of A consisting of all diagonal matrices with entries e^t, e^{-t}, where $t > 0$. Let $\overline{A^+} = A^+ \cup \{1\}$ be its closure in G.

Theorem 11.2.1 (Cartan Decomposition). *The group G can be written as $G = K\overline{A^+}K$, i.e. every $x \in G$ is of the form $x = k_1 a k_2$ with $a \in \overline{A^+}$, $k_1, k_2 \in K$. The element a is uniquely determined by x. If $a \neq 1$, which means that $x \notin K$, then also k_1 and k_2 are uniquely determined up to sign, i.e., if $k_1 a k_2 = k_1' a k_2'$, then either $(k_1, k_2) = (k_1', k_2')$ or $(k_1, k_2) = (-k_1', -k_2')$.*

For $f \in L^1(G)$ we have the integral formula

$$\int_G f(x)\,dx = 2\pi \int_K \int_0^\infty \int_K f(ka_t l) \left(e^{2t} - e^{-2t} \right) dk\,dt\,dl.$$

Proof For $x \in G$ the matrix xx^t is symmetric and positive definite. As it has determinant one, it follows from linear algebra, that there exists $k \in K$ and $t \geq 0$, such that $kxx^t k^t$ is the diagonal matrix with entries e^{2t}, e^{-2t}. For two elements $x, x_1 \in G$ the condition $xx^t = x_1 x_1^t$ is equivalent to $1 = x^{-1}(x_1 x_1^t)(x^t)^{-1} = (x^{-1}x_1)(x^{-1}x_1)^t$. The last is equivalent to $x^{-1}x_1 \in K$. So there is $k' \in K$ with $x = k^{-1}a_t k'$.

This shows existence of the decomposition. For the uniqueness note that e^{2t} is the larger of the two eigenvalues of xx^t and thus determined by x. For the uniqueness of k_1, k_2 assume that $a \in A^+$ and $k_1 a k_2 = l_1 a l_2$ with $k_1, k_2, l_1, l_2 \in K$. Then one has $ak_2 l_2^{-1} = k_1^{-1} l_1 a$. But the equation $ak = k'a$ with $k, k' \in K$ implies $k = k' = \pm 1$ as we show now. Let $a = a_t, k = \left(\begin{smallmatrix} a & -b \\ b & a \end{smallmatrix} \right), k' = \left(\begin{smallmatrix} c & -d \\ d & c \end{smallmatrix} \right)$. Then

$$\begin{pmatrix} e^t a & -e^t b \\ e^{-t}b & e^{-t}a \end{pmatrix} = \begin{pmatrix} e^t & \\ & e^{-t} \end{pmatrix} \begin{pmatrix} a & -b \\ b & a \end{pmatrix} = ak = k'a$$

$$= \begin{pmatrix} c & -d \\ d & c \end{pmatrix} \begin{pmatrix} e^t & \\ & e^{-t} \end{pmatrix} = \begin{pmatrix} e^t c & -e^{-t}d \\ e^t d & e^{-t}c \end{pmatrix}$$

Consider the norm of the first column of this matrix to get

$$e^{2t} = e^{2t}(c^2 + d^2) = e^{2t}a^2 + e^{-2t}b^2 = e^{2t}a^2 + e^{-2t}(1 - a^2),$$

or $e^{2t}(1 - a^2) = e^{-2t}(1 - a^2)$, which implies $a = \pm 1$ and therefore $b = 0$. But then also $d = 0$ and the claim follows. So this means $k_2 l_2^{-1} = k_1^{-1} l_1 = \pm 1$ and therefore the claimed uniqueness up to sign.

Let $M = \{\pm 1\} \subseteq K$. In order to verify the integral formula, consider the map $\phi : K/M \times A^+ \to AN \smallsetminus \{1\}$ defined by

$$\phi(kM, a) = \underline{a}(ka)\underline{n}(ka).$$

Lemma 11.2.2 *The map ϕ is a C^1 diffeomorphisms. In the coordinates $\mathbb{R}/\pi\mathbb{Z} \times \mathbb{R}_{>0} \ni (\theta, s) \mapsto (k_\theta M, a_s)$ on $K/M \times A$ and $(t, x) \mapsto a_t n_x$ on AN one has for the differential matrix*

$$|\det(D\phi)(\theta, s)| = |e^{2s} - e^{-2s}|.$$

Proof A computation shows that

$$\phi(k_\theta, a_s) = \underline{a}(k_\theta a_s)\underline{n}(k_\theta, a_s) = a_t n_x,$$

where

$$t = -\frac{1}{2} \log \left(e^{2s} \sin^2 \theta + e^{-2s} \cos^2 \theta \right)$$
$$x = \left(e^{2s} - e^{-2s} \right) \sin \theta \cos \theta$$

According to the Cartan decomposition, the map $K/M \times A^+ \to (G \smallsetminus K)/K$ is bijective. By the Iwasawa decomposition, the map $AN \to G/K$ is bijective as well, hence ϕ is bijective.

The map ϕ is continuously differentiable. Once we have shown the claimed formula for the differential, it follows that the differential matrix is invertible and so the inverse function is continuously differentiable as well. We have

$$\det(D\phi)(\theta, s) = \det \begin{pmatrix} \frac{\partial t}{\partial \theta} & \frac{\partial t}{\partial s} \\ \frac{\partial x}{\partial \theta} & \frac{\partial x}{\partial s} \end{pmatrix} = \frac{\partial t}{\partial \theta} \frac{\partial x}{\partial s} - \frac{\partial t}{\partial s} \frac{\partial x}{\partial \theta}.$$

From this one gets the lemma by a computation $\qquad \square$

The transformation formula for the variables $(x, t) = \phi(\theta, s)$ shows

$$\int_G f(g) \, dg = 2 \int_\mathbb{R} \int_\mathbb{R} \int_K f(a_t n_y l) \, dl \, dy \, dt$$

$$= 2 \int_0^\pi \int_0^\infty \int_K f(k_\theta a_s l) \left(e^{2s} - e^{-2s} \right) dl \, ds \, d\theta$$

$$= \int_0^{2\pi} \int_0^\infty \int_K f(k_\theta a_s l) \left(e^{2s} - e^{-2s} \right) dl \, ds \, d\theta$$

$$= 2\pi \int_K \int_0^\infty \int_K f(k a_s l) \left(e^{2s} - e^{-2s} \right) dl \, ds \, dk,$$

for every $f \in L^1(G)$, where the transition from the integral over $[0, \pi]$ to the integral over $[0, 2\pi]$ in the middle equation is justified by the fact that $k_{\theta+\pi} a_s = k_\theta a_s m$ with $m = \pm 1 \in M$ for all $\theta \in \mathbb{R}$ and $s > 0$. This finishes the proof of the theorem.   □

Corollary 11.2.3 *The map*

$$K\backslash G/K \to [2, \infty), \qquad x \mapsto \mathrm{tr}\,(x^t x)$$

is a bijection.

Proof The map is a bijection when restricted to $\overline{A^+}$, so the corollary follows from the theorem.   □

Definition A function f on G is said to be *K-bi-invariant* if it factors through $K\backslash G/K$. We define the *Hecke algebra* $\mathcal{H}$ of G to be the set of K-bi-invariant functions f on G, which are in $L^1(G)$. So we can characterize $\mathcal{H}$ as the space of all $f \in L^1(G)$ with $L_k f = f = R_k f$ for every $k \in K$, where $L_k f(x) = f(k^{-1}x)$ and $R_k f(x) = f(xk)$. We know that for $f, g \in L^1(G)$,

$$L_k(f * g) = (L_k f) * g \quad \text{and} \quad R_k(f * g) = f * (R_k g).$$

We conclude that $\mathcal{H}$ is a convolution subalgebra of $L^1(G)$. Further, $\mathcal{H}$ is stable under the involution $f^*(x) = \overline{f(x^{-1})}$, so $\mathcal{H}$ is a *-subalgebra of $L^1(G)$.

Theorem 11.2.4

(a) *The Hecke algebra $\mathcal{H}$ is commutative.*

(b) *For every irreducible unitary representation π of G the space of K-invariants,*

$$V_\pi^K = \{v \in V_\pi : \pi(k)v = v \,\forall k \in K\}$$

   is zero or one dimensional.

(c) *For every irreducible representation π of G and for every $f \in \mathcal{H}$ we have $\pi(f) = P_K \pi(f) P_K$, where $P_K : V_\pi \to V_\pi^K$ denotes the orthogonal projection.*

Proof For $x \in G$ the Cartan decomposition implies that $KxK = Kx^{-1}K$, as this is the case for $x \in A$, since conjugating $x \in A$ with $\left(\begin{smallmatrix} & -1 \\ 1 & \end{smallmatrix}\right) \in K$ gives x^{-1}. This implies that for every $f \in \mathcal{H}$ one has $f(x^{-1}) = f(x)$. For general $f \in L^1(G)$ let $f^\vee(x) = f(x^{-1})$, then $(f * g)^\vee = g^\vee * f^\vee$ for all $f, g \in L^1(G)$. For $f, g \in \mathcal{H}$, one has $f^\vee = f$ and likewise for g and $f * g$, so that

$$f * g = (f * g)^\vee = g^\vee * f^\vee = g * f.$$

So $\mathcal{H}$ is commutative, which proves (a).

For (b) assume $V_\pi^K \neq 0$. The Hecke algebra acts on V_π^K. We show that V_π^K is irreducible under $\mathcal{H}$, so let $U \subset V_\pi^K$ a closed, $\mathcal{H}$-stable subspace. We show that $U = 0$ or $U = V_\pi^K$. For this assume $U \neq 0$, then, as π is irreducible, one has

$\overline{\pi(L^1(G))U} = V_\pi$. Let $P_K : V_\pi \to V_\pi^K$ be the orthogonal projection. Then $P_K v = \int_K \pi(k)v\,dk$ for $v \in V_\pi$ (see Proposition 7.3.3), as we normalize the Haar measure on K to have volume one. For $f \in L^1(G)$ let

$$\tilde{f}(x) = \int_K \int_K f(kxl)\,dk\,dl \in \mathcal{H}.$$

It follows that $P_K \pi(f) P_K = \pi(\tilde{f})$. Let $u \in U$ and $f \in L^1(G)$. Then

$$P_K \pi(f)u = P_K \pi(f) P_K u = \pi(\tilde{f})u \in U.$$

So we conclude that $U = P_K V_\pi = V_\pi^K$ and thus V_π^K is irreducible. Finally, to see that every irreducible *-representation $\eta : \mathcal{H} \to \mathcal{B}(V_\eta)$ on a Hilbert space V_η is one-dimensional, observe that for each $f \in \mathcal{H}$ the operator $\eta(f)$ commutes with the self-adjoint irreducible set $\eta(\mathcal{H}) \subset \mathcal{B}(V_\eta)$, since $\mathcal{H}$ is commutative. Thus $\eta(\mathcal{H}) \subset \mathbb{C}\mathrm{Id}$ by Schur's Lemma (Theorem 5.1.6). As η is irreducible, it must be one dimensional.

For (c) observe that $\tilde{f} = f$ for every $f \in \mathcal{H}$. Thus it follows from the above computations that $\pi(f) = P_K \pi(f) P_K$ for every $f \in \mathcal{H}$. $\qquad\square$

Let $\widehat{G}_K$ be the set of all $\pi \in \widehat{G}$ such that the space V_π^K of K-invariants is non-zero. We will now give a list of the $\pi \in \widehat{G}_K$. For $\lambda \in \mathbb{C}$ let V_λ be the Hilbert space of all functions $\phi : G \to \mathbb{C}$ with firstly, $\phi(ma_tnx) = e^{t(2\lambda+1)}\phi(x)$ for $m = \pm 1 \in G, a_t \in A, n \in N$ and $x \in G$. By the Iwasawa decomposition, such ϕ is uniquely determined by its restriction to K. We secondly insist that $\phi|_K$ be in $L^2(K)$. We equip V_λ with the inner product of $L^2(K)$. The group G acts on this space by $\pi_\lambda(y)\phi(x) = \phi(xy)$. Note that the restriction to the subgroup K of the representation π_λ is the induced representation $\mathrm{Ind}_M^K(1)$ as in Sect. 7.4. The Frobenius reciprocity (Theorem 7.4.1) implies that

$$\pi_\lambda|_K \cong \bigoplus_{l \in \mathbb{Z}} \varepsilon_{2l},$$

where for $l \in \mathbb{Z}$ the character ε_l on K is defined by

$$\varepsilon_l \begin{pmatrix} \cos\theta & -\sin\theta \\ \sin\theta & \cos\theta \end{pmatrix} = e^{il\theta}.$$

Proposition 11.2.5 If $\lambda \in i\mathbb{R}$, then the representation π_λ is unitary.

Proof The map $\phi \mapsto \phi|_K$ yields an isomorphism of Hilbert spaces, $V_\lambda \cong L^2(\bar{K})$, where $\bar{K} = K/\pm 1$. The representation π_λ can, on $L^2(\bar{K})$, be written as $\pi_\lambda(y)\phi(k) = e^{\underline{t}(ky)(2\lambda+1)}\phi(\underline{k}(ky))$. To see this, recall that $t = \underline{t}(ky)$ is the unique real number such that $a_t = \underline{a}(ky)$ in the Iwasawa decomposition $ky = \underline{a}(ky)\underline{n}(ky)\underline{k}(ky)$, and therefore

$$\pi_\lambda(y)\phi(k) = \phi(ky) = \phi\left(\underline{a}_{t(ky)}\underline{n}(ky)\underline{k}(ky)\right) = e^{\underline{t}(ky)(2\lambda+1)}\phi(\underline{k}(ky)).$$

It follows that $|\pi_\lambda(y)\phi(k)|^2 = e^{\underline{t}(ky)(4\mathrm{Re}(\lambda)+2)}|\phi(\underline{k}(ky))|^2$. By the second assertion of Theorem 11.1.3, one sees that π_λ is indeed unitary if $\lambda \in i\mathbb{R}$. $\qquad\square$

Definition For a general representation (π, V_π) of G we let $V_{\pi,K}$ denote the space of all *K-finite vectors*, i.e., the space of all vectors $v \in V_\pi$ such that $\pi(K)v$ spans a finite dimensional space in V_π. The vector space $V_{\pi,K}$ is in general not stable under G, but is always stable under K. Since V_π has a decomposition $V_\pi = \widehat{\bigoplus}_{i \in I} U_i$, where U_i is an irreducible (hence finite-dimensional) K-representation, it follows that $V_{\pi,K}$ is dense in V_π.

The representations π_λ for $\lambda \in i\mathbb{R}$ are called the *unitary principal series* representations. One can show that π_λ is irreducible and unitarily equivalent to $\pi_{-\lambda}$ if $\lambda \in i\mathbb{R}$. These are the only equivalences that occur. One can show that for $0 < \lambda < 1/2$ there is an inner product on the space $V_{\lambda,K}$ such that the completion of $V_{\lambda,K}$ with respect to this inner product is the space of a unitary representation of G. By abuse of notation, this representation is again denoted (π_λ, V_λ). These are called the *complementary series* representations. The set $\widehat{G}_K$ consists of

- the trivial representation,
- the unitary principal series representations π_{ir}, where $r \geq 0$, and
- the complementary series π_λ for $0 < \lambda < 1/2$.

No two members of this list are equivalent. The proofs of these facts can be found in [Kna01], Chapter II.

Note that the one dimensional space V_λ^K is spanned by the element p_λ with

$$p_\lambda(mank) = e^{\ell(a)(2\lambda+1)}.$$

By Corollary 11.2.3 there exists for every $f \in \mathcal{H}$ a unique function ϕ_f on $[0, \infty)$ such that

$$f(x) = \phi_f\left(\text{tr}\,(x^t x) - 2\right).$$

Consider the special case $x \in AN$, say $x = a_t n_s$, then $\text{tr}\,(x^t x) = (s^2 + 1)e^{2t} + e^{-2t}$. For $f \in \mathcal{H}$ there exists a function h_f such that

$$\pi_{ir}(f)p_{ir} = h_f(r)p_{ir}.$$

The function h_f is called the *eigenvalue function* of f. Here ir can vary in $i\mathbb{R} \cup (0, 1/2)$. Since $p_{ir}(1) = 1$, we can compute $h_f(r)$ as follows

$$h_f(r) = \pi_{ir}(f)p_{ir}(1) = \int_G f(x)p_{ir}(x)\,dx$$

Lemma 11.2.6 *The map $f \mapsto h_f$ is injective on $\mathcal{H}$. We have $h_f(r) = \text{tr}\,\pi_{ir}(f)$, and for $f, g \in \mathcal{H}$ the formula*

$$h_{f*g} = h_f h_g$$

holds. The function h_f can be computed via the following integral transformations. First set

$$q_f(x) = A(\phi_f)(x) \overset{\text{def}}{=} \int_\mathbb{R} \phi_f\left(x + s^2\right) ds, \qquad x \geq 0.$$

The map $\phi \mapsto A(\phi)$ is called the Abel transform. *Next define*

$$g_f(u) \overset{\text{def}}{=} q_f \left(e^u + e^{-u} - 2 \right), \qquad u \in \mathbb{R}.$$

Then one has

$$h_f = \int_{\mathbb{R}} g_f(u) e^{iru} \, du.$$

Proof For the injectivity take an $f \in \mathcal{H}$ with $h_f = 0$. Then $\pi(f) = 0$ for every $\pi \in \hat{G}$. By the Plancherel Theorem the representation $(R, L^2(G))$ is a direct integral of irreducible representations and so it follows that $R(f) = 0$. In particular it follows that $g * f = 0$ for every $g \in C_c(G)$. Letting g run through a Dirac net, it follows $f = 0$.

The equation $h_f(r) = \operatorname{tr} \pi_{ir}(f)$ is a consequences of Theorem 11.2.4 and $h_{f*g} = h_f h_g$ follows from $\pi(f * g) = \pi(f)\pi(g)$ for all $f, g \in L^1(G)$.

Using Iwasawa coordinates and the K-invariance of f, we compute

$$h_f(r) = \int_{AN} f(an) e^{L(a)(2ir+1)} \, da \, dn$$

$$= 2 \int_{-\infty}^{\infty} \int_{-\infty}^{\infty} \phi_f \left(e^{2t} + e^{-2t} + s^2 - 2 \right) e^{2tir} \, ds \, dt,$$

where we used the transformation $s \mapsto e^{-t}s$. As $q_f(x) = A\phi_f(x)$ and g is even, we have

$$h_f(r) = 2 \int_{-\infty}^{\infty} q_f \left(e^{2t} + e^{-2t} - 2 \right) e^{2tir} \, dt = \int_{\mathbb{R}} g_f(u) e^{iru} \, du. \qquad \square$$

Definition Let $\mathcal{S}_{[0,\infty)}$ be the space of all infinitely differentiable functions ϕ on $[0, \infty)$ such that the function $x^n \phi^{(m)}(x)$ is bounded for all $m, n \geq 0$.

Lemma 11.2.7 *The Abel transform is invertible in the following sense: Let ϕ be continuously differentiable on $[0, \infty)$ such that*

$$|\phi \left(x + s^2 \right)|, |s\phi' \left(x + s^2 \right)| \leq g(s)$$

for some $g \in L^1([0, \infty))$, then $q = A(\phi)$ is continuously differentiable and

$$\phi = \frac{-1}{\pi} A(q').$$

Moreover, the Abel transform maps $\mathcal{S}_{[0,\infty)}$ to itself and defines a bijection $A : \mathcal{S}_{[0,\infty)} \to \mathcal{S}_{[0,\infty)}$.

Proof We first show that for any ϕ satisfying the conditions we have $\lim_{x\to\infty}\phi(x) = 0$. To see this, let $h(s) = s\phi'(s^2)$. Then h is integrable on $[0,\infty]$. It follows that

$$\phi(y) - \phi(0) = \int_0^y \frac{h(\sqrt{t})}{\sqrt{t}}\, dt = 2\int_0^{\sqrt{y}} h(u)\, du.$$

Letting $y \to \infty$, we see that $\lim_{x\to\infty}\phi(x)$ exists and since $\phi(x + s^2)$ is integrable, this limit is zero.

Next by the theorem of dominated convergence one sees that q is continuously differentiable and that $q' = A(\phi')$. Using polar coordinates, we compute

$$-\frac{1}{\pi}\int_{\mathbb{R}}\int_{\mathbb{R}} \phi'\left(x + s^2 + t^2\right) ds\, dt = -2\int_0^\infty r\phi'\left(x + r^2\right) dr$$
$$= -\phi\left(x + r^2\right)\big|_0^\infty = \phi(x).$$

It is easy to see that the Abel transform as well as its inverse map $\mathcal{S}_{[0,\infty)}$ to itself. The lemma follows. $\qquad\qquad\square$

Lemma 11.2.8 *Let E be the space of all entire functions h such that $x^n h^{(m)}(x + ki)$ is bounded in $x \in \mathbb{R}$ for all $m, n \geq 0$ and every $k \in \mathbb{R}$. Let F be the space of all smooth functions g on $\mathbb{R}$ such that $(e^u + e^{-u})^n g^{(m)}(u)$ is bounded for all $m, n \geq 0$. Then the Fourier transform*

$$\Phi(h)(u) \overset{\mathrm{def}}{=} \frac{1}{2\pi}\int_{\mathbb{R}} h(r)e^{-iru}\, dr, \qquad h \in E,$$

defines a linear bijection $\Phi : E \to F$. Its inverse is given by

$$\Phi^{-1}(g)(r) = \int_{\mathbb{R}} g(u)e^{iru}\, du.$$

The map Φ maps the subspace of even functions E^{ev} in E to the space of even functions F^{ev} in F.

Proof By some simple estimates, the space F can also be characterized as the space of all smooth g such that $e^{-ku}g^{(m)}(u)$ is bounded for every $k \in \mathbb{R}$ and every $m \geq 0$.

The space F is a subspace of the Schwartz space $\mathcal{S}(\mathbb{R})$ be definition. By the identity theorem of holomorphic functions, the restriction $h \mapsto h|_{\mathbb{R}}$ is an injection of E into $\mathcal{S}(\mathbb{R})$. As the Fourier transform is a bijection on $\mathcal{S}(\mathbb{R})$, it suffices to show that it maps E to F and vice versa.

For $h \in E$ let $g = \Phi(h)$. With $k \in \mathbb{R}$, and $m \geq 0$ compute

$$e^{-ku}g^{(m)}(u) = e^{-ku}\frac{1}{2\pi i^m}\int_{\mathbb{R}} h(r)r^m e^{-iru}\, dr$$

$$= \frac{1}{2\pi i^m} \int_{\mathbb{R}} h(r) r^m e^{-i(r-ik)u} \, dr$$

$$= \frac{1}{2\pi i^m} \int_{\mathbb{R}} h(r+ik)(r+ik)^m e^{-iru} \, dr.$$

The latter is the Fourier transform of a Schwartz function and hence a bounded function in u. It follows that g lies in F.

For the converse, let $g \in F$. Then the Fourier integral

$$h(r) = \int_{\mathbb{R}} g(u) e^{iru} \, du$$

converges for every $r \in \mathbb{C}$, so h extends to a unique entire function. Further, for $m, n \geq 0$ and $k \in \mathbb{R}$ we have

$$x^n h^{(m)}(x + ik) = x^n i^m \int_{\mathbb{R}} u^m g(u) e^{-ku} e^{ixu} \, du.$$

The latter function is bounded in $x \in \mathbb{R}$. So h lies in E as claimed. The last assertion is clear as the Fourier transform preserves evenness. $\qquad\square$

Recall the definition of the function h_f for $f \in \mathcal{H}$ as given preceding Lemma 11.2.6.

Proposition 11.2.9 *Let $\mathcal{HS}$ be the space of all smooth functions f on G of the form $f(x) = \phi(\operatorname{tr}(x^t x) - 2)$ for some $\phi \in S_{[0,\infty)}$. Then $\mathcal{HS}$ is a subalgebra of the Hecke algebra $\mathcal{H}$ and the map $\Psi : f \mapsto h_f$ is a bijection onto the space E^{ev}.*

For a given $h \in E^{\mathrm{ev}}$ the function $f = \Psi^{-1}(h)$ is computed as follows. First one defines the even function

$$g(u) = \frac{1}{2\pi} \int_{\mathbb{R}} h(r) e^{-iru} \, dr.$$

Then $q : [0, \infty) \to \mathbb{C}$ is defined to be the unique function with $g(u) = q(e^u + e^{-u} - 2)$. Further one sets $\phi = -\frac{1}{\pi} A(q')$. Then

$$f(x) = \phi\left(\operatorname{tr}(x^t x) - 2\right).$$

Proof First note that that the map $q \mapsto g$ with $g(u) = q(e^u + e^{-u} - 2)$ is a bijection between $S_{[0,\infty)}$ and the space F^{ev}. Finally, Lemma 11.2.7 and 11.2.8 give the claim. $\qquad\square$

11.3   An Explicit Plancherel Theorem

The Plancherel Theorem says that there exists a measure μ on $\widehat{G}$ such that for $g \in L^1(G) \cap L^2(G)$ one has

$$\|g\|_2^2 = \int_{\widehat{G}} \|\pi(g)\|_{HS}\, d\mu(\pi).$$

The techniques developed so far allow us as a side-result, to give an explicit measure on $\widehat{G}_K$, for which this equation holds with $f \in \mathcal{H}_{\text{sym}}$. Any such computation is called an *Explicit Plancherel Theorem*.

Theorem 11.3.1   *For every* $g \in \mathcal{H}_{\text{sym}}$ *one has*

$$\|g\|_2^2 = \frac{1}{4\pi} \int_{\mathbb{R}} \|\pi_{ir}(g)\|_{HS}^2\, r \tanh(\pi r)\, dr.$$

Moreover, for every $f \in \mathcal{H}_{\text{sym}}$ *one has*

$$f(1) = \frac{1}{4\pi} \int_{\mathbb{R}} \operatorname{tr}(\pi_{ir}(f))\, r \tanh(\pi r)\, dr.$$

Proof   We show the second assertion first. Let $h = h_f$, $\phi = \phi_f$ and $g = g_f$ be as in the discussion at the end of the previous section. Recall in particular that

$$A\phi\left(e^u + e^{-u} - 2\right) = g(u) = \frac{1}{2\pi} \int_{\mathbb{R}} h(r)e^{iru}\, dr$$

(since h is even), from which it follows that $g'(u) = \frac{i}{2\pi} \int_{\mathbb{R}} r h(r)e^{iru}\, dr$. Using this and Lemma 11.2.7 we compute

$$f(1) = \phi(0) = -\frac{1}{\pi} \int_{\mathbb{R}} (A\phi)'(x^2)\, dx.$$

As $g_f(u) = A\phi(e^u + e^{-u} - 2) = A\phi((e^{u/2} - e^{-u/2})^2)$, we get $g'(u) = (A\phi)'((e^{u/2} - e^{-u/2})^2)(e^u - e^{-u})$. Putting $x = e^{u/2} - e^{-u/2}$ in the above integral, we get

$$f(1) = -\frac{1}{2\pi} \int_{\mathbb{R}} \frac{g'(u)}{e^{u/2} - e^{-u/2}}\, du$$

$$= -\frac{i}{4\pi^2} \int_{\mathbb{R}} \int_{\mathbb{R}} r h(r) \frac{e^{iru}}{e^{u/2} - e^{-u/2}}\, dr\, du.$$

As h is even, the latter equals

$$-\frac{i}{8\pi^2} \int_{\mathbb{R}} r h(r) \int_{\mathbb{R}} \frac{e^{iru} - e^{-iru}}{e^{u/2} - e^{-u/2}}\, du\, dr.$$

The first step of the following computation is justified by the fact that the integrand is even. We compute

$$-\frac{i}{8\pi^2}\int_{\mathbb{R}}\frac{e^{iru}-e^{-iru}}{e^{u/2}-e^{-u/2}}\,du = \frac{1}{4\pi^2 i}\int_0^\infty\frac{e^{iru}-e^{-iru}}{e^{u/2}-e^{-u/2}}\,du$$

$$= \frac{1}{4\pi^2 i}\int_0^\infty e^{-u/2}\frac{e^{iru}-e^{-iru}}{1-e^{-u}}\,du$$

$$= \frac{1}{4\pi^2 i}\int_0^\infty e^{-u/2}\left(e^{iru}-e^{-iru}\right)\sum_{n=0}^\infty e^{-nu}\,du$$

$$= \frac{1}{4\pi^2 i}\sum_{n=0}^\infty\int_0^\infty e^{-u(n+\frac{1}{2}-ir)}\,du - \int_0^\infty e^{-u(n+\frac{1}{2}+ir)}\,du$$

$$= \frac{1}{4\pi^2 i}\sum_{n=0}^\infty\frac{1}{n+\frac{1}{2}-ir}-\frac{1}{n+\frac{1}{2}+ir}.$$

For this latter expression we temporarily write $\psi(r)$. Then

$$\psi\left(i\left(r+\frac{1}{2}\right)\right) = \frac{1}{4\pi^2 i}\sum_{n=0}^\infty\frac{1}{n+1+r}-\frac{1}{n-r} = \frac{1}{4\pi i}\cot(\pi r).$$

The last step is the well known Mittag-Leffler expansion of the cotangent function. We conclude

$$\psi(r) = \frac{1}{4\pi i}\tan(\pi i r) = \frac{1}{4\pi}\tanh(\pi r).$$

The second assertion of the theorem follows. For the first, put $f = g * g^*$ and apply the theorem to this f. Then, on the one hand, $f(1) = g * g^*(1) = \|g\|_2^2$, and on the other, for $\pi \in \widehat{G}$,

$$\operatorname{tr}\pi(f) = \operatorname{tr}\pi(g)\pi(g)^* = \|\pi(g)\|_{\mathrm{HS}}^2.$$

This implies the theorem. □

11.4 The Trace Formula

For $g \in \mathrm{SL}_2(\mathbb{R})$ the two eigenvalues in $\mathbb{C}$ must be inverse to each other as the determinant is one. Since g is a real matrix, its characteristic polynomial is real, and so the eigenvalues are either both real, or complex conjugates of each other. Let $g \neq \pm 1$. There are three cases.

- g is in $\mathrm{SL}_2(\mathbb{C})$ conjugate to a diagonal matrix with entries $\varepsilon, \bar\varepsilon$ for some $\varepsilon \in \mathbb{C}$ of absolute value one. In this case, g is called an *elliptic element* of G; or

- g is conjugate to $\pm\left(\begin{smallmatrix}1 & 1\\ & 1\end{smallmatrix}\right)$. In this case g is called a *parabolic element*; or

- g is conjugate to a diagonal matrix with entries $t, 1/t$ for some $t \in \mathbb{R}$, in which case g is called a *hyperbolic element*.

Let g be elliptic, say g is conjugate to $\left(\begin{smallmatrix} a+bi & \\ & a-bi \end{smallmatrix}\right)$. As an element of G, the element g is conjugate to some $\left(\begin{smallmatrix} a & -b \\ b & a \end{smallmatrix}\right)$ in K. This implies that g has a unique fixed point in $\mathbb{H}$.

Proposition 11.4.1 *A uniform lattice $\Gamma \subset G$ contains no parabolic elements.*

Proof Consider the map $\eta : \mathbb{H} \to [0, \infty)$ given by

$$\eta(z) = \inf\{\rho(z, \gamma z) : \gamma \in \Gamma, \ \gamma \neq \pm 1, \ \gamma \text{ not elliptic}\}.$$

It is easy to see that the map η is continuous. Further it is Γ-invariant and therefore it constitutes a continuous function $\Gamma\backslash\mathbb{H} \to (0, \infty)$. Since $\Gamma\backslash\mathbb{H} \cong \Gamma\backslash G/K$ is compact, the function η attains its minimum, hence there exists $\theta > 0$ such that $\eta(z) \geq \theta$ for all $z \in \mathbb{H}$. Now *assume* that Γ contains a parabolic element, say $p = g \left(\begin{smallmatrix} 1 & 1 \\ & 1 \end{smallmatrix}\right) g^{-1} \in \Gamma$ for some $g \in G$. Then for $y > 1$ we have

$$\rho(g(yi), pg(yi)) = \rho(g(yi), g(yi + 1)) = \rho(yi, yi + 1)$$

and the latter tends to zero as $y \to \infty$, which follows from

$$\rho(yi, yi + 1) \leq \int_0^1 |p'(t)| \frac{dt}{\text{Im}(p(t))} = \frac{1}{y},$$

where $p(t) = yi + t$. We therefore have a *contradiction!* Hence Γ does not contain any parabolic element. $\square$

For a hyperbolic element g, with eigenvalues $\lambda, 1/\lambda$ for $|\lambda| > 1$, define the *length* of g as $l(g) = 2 \log |\lambda|$.

Let $\Gamma \subset G$ be a uniform lattice. For convenience we will assume that Γ contains no elliptic elements. Then Γ consists, besides ± 1, of hyperbolic elements only. We call such a group a *hyperbolic lattice*. In [Bea95], there are given many examples of uniform lattices in G without elliptic elements. For instance, every Riemannian manifold of genus $g \geq 2$ is a quotient of the upper half plane by a hyperbolic lattice in G.

So let Γ be a hyperbolic lattice in G. Let $r_0 = \frac{i}{2}$, and let $(r_j)_{j\geq 1}$ be a sequence in $\mathbb{C}$ such that $ir_j \in i\mathbb{R} \cup (0, \frac{1}{2})$ with the property that π_{ir_j} is isomorphic to a subrepresentation of $(R, L^2(\Gamma\backslash G))$ and the value $r = r_j$ is repeated in the sequence as often as $N_\Gamma(\pi_{ir})$ times, i.e, as often as π_{ir_j} appears in the decomposition of R. Let $f \in \mathcal{H}$ such that the operator $R(f)$ is of trace class, and define $\phi = \phi_f, g = g_f$ and $h = h_f$ as in the previous two sections (See Lemma 11.2.6). Recall that $f(x) = \phi(\text{tr}\,(x^t x) - 2)$, $g(u) = A\phi(e^u + e^{-u} - 2)$, where A denotes the Abel transform, and $h(r) = \int_\mathbb{R} g(u)e^{iru}\,du$. Recall from Lemma 11.2.6 that $h(r) = \text{tr}\,\pi_{ir}(f)$ for every $ir \in i\mathbb{R} \cup (0, \frac{1}{2})$. Moreover, it follows from Theorem 11.2.4 that $\text{tr}\,\pi(f) = 0$ for all $\pi \in \widehat{G} \smallsetminus \widehat{G}_K$. We therefore get $\text{tr}\,R(f) = \sum_{j=0}^\infty h(r_j)$. Suppose that the trace formula of Theorem 9.3.2 is valid for the function f. Then

$$\sum_{j=0}^{\infty} h(r_j) = \sum_{[\gamma]} \mathrm{vol}(\Gamma_\gamma \backslash G_\gamma) \mathcal{O}_\gamma(f).$$

Recall the hyperbolic tangent function $\tanh(x) = \frac{e^x - e^{-x}}{e^x + e^{-x}}$.

An element $\gamma \in \Gamma \backslash \{1\}$ is called *primitve*, if it is not a power in Γ, i.e., if the equation $\gamma = \sigma^n$ with $n \in \mathbb{N}$ and $\sigma \in \Gamma$ implies $n = 1$.

Lemma 11.4.2 *If Γ is a torsion free uniform lattice, every element γ of $\Gamma \backslash \{1\}$ is a positive power of a uniquely determined primitive element γ_0. This element generates the centralizer Γ_γ of γ in Γ. We call it the primitive element underlying γ.*

Proof Let $\gamma \in \Gamma \backslash \{1\}$. By assumption, γ is hyperbolic. Replacing Γ with a conjugate group we may assume that γ is the diagonal matrix with entries e^t, e^{-t} for some $t > 0$, as the other case of $\gamma = -\mathrm{diag}(e^t, e^{-t})$ gives the same result. Then the centralizer G_γ of γ in G equals $\pm A$, the group of all diagonal matrices in G and $\Gamma_\gamma = \Gamma \cap \pm A$. As $-1 \notin \Gamma$, since Γ is torsion-free, it follows that there is $\gamma_0 \in \Gamma$ such that the centralizer Γ_γ in Γ is equal to $\langle \gamma_0 \rangle$. Replacing γ_0 by γ_0^{-1} if necessary, we can assume that $\gamma = \gamma_0^n$ for some $n \in \mathbb{N}$. It follows that γ_0 is primitive.   □

Theorem 11.4.3 *Assume that Γ is a torsion free uniform lattice in $\mathrm{SL}(2, \mathbb{R})$. Let $\varepsilon > 0$, and let h be a holomorphic function on the strip $\{|\mathrm{Im}(z)| < \frac{1}{2} + \varepsilon\}$. Suppose that h is even, i.e., $h(-z) = h(z)$ for every z, and that $h(z) = O(|z|^{-2-\varepsilon})$ as $|z|$ tends to infinity. Let $g(u) = \frac{1}{2\pi} \int_{\mathbb{R}} h(r) e^{-iru} \, dr$. Then one has*

$$\sum_{j=0}^{\infty} h(r_j) = \frac{\mathrm{vol}(\Gamma \backslash G)}{4\pi} \int_{\mathbb{R}} r h(r) \tanh(\pi r) \, dr$$

$$+ \sum_{[\gamma] \neq 1} \frac{l(\gamma_0)}{e^{l(\gamma)/2} - e^{-l(\gamma)/2}} g(l(\gamma)),$$

where for $\gamma \neq 1$, γ_0 is the primitive element underlying γ.

Proof We start with functions $f \in \mathcal{H}_{\mathrm{sym}}$, for which the trace formula holds and then we extend the range of the trace formula up to the level of the theorem. So let $f \in \mathcal{H}_{\mathrm{sym}}$ such that the trace formula is valid for f. For instance, $f \in C_c^\infty(G)^2 = C_c^\infty(G) * C_c^\infty(G)$ will suffice. At first we consider the class $[\gamma]$ with $\gamma = 1$. Then

$$\mathrm{vol}(\Gamma_\gamma \backslash G_\gamma) \mathcal{O}_\gamma(f) = \mathrm{vol}(\Gamma \backslash G) f(1).$$

Theorem 11.3.1 tells us that

$$f(1) = \frac{1}{4\pi} \int_{\mathbb{R}} \mathrm{tr}\,(\pi_{ir}(f)) r \tanh(\pi r) \, dr = \frac{1}{4\pi} \int_{\mathbb{R}} h(r) r \tanh(\pi r) \, dr.$$

Next let γ be an element of Γ with $\gamma \neq 1$ and recall that this implies $G_\gamma = \pm A$. If $\gamma_0 = \mathrm{diag}(e^t, e^{-t})$ and if we identify A with $\mathbb{R}$ via the exponential map, the group

$\Gamma_\gamma = \langle \gamma_0 \rangle$ corresponds to the subgroup $t\mathbb{Z}$. It follows that $\mathrm{vol}(\Gamma_\gamma \backslash G_\gamma) = 2t = l(\gamma_0)$, where the factor 2 is due to the normalization of Haar measure on A.

By the Iwasawa decomposition, the set $G_\gamma \backslash G$ can be identified with $NK / \pm 1$. As f is K-bi-invariant and $f(x) = \phi(\mathrm{tr}\,(x^t x) - 2)$ for every $x \in G$, the orbital integral $\mathcal{O}_\gamma(f)$ equals

$$\int_{\mathbb{R}} f(n_s^{-1} \gamma n_s)\, ds = \int_{\mathbb{R}} \phi\left(e^{2t} + e^{-2t} + s^2(e^t - e^{-t})^2 - 2\right)\, ds,$$

so that

$$\mathcal{O}_\gamma(f) = \frac{1}{e^t - e^{-t}} A\phi\left(e^{2t} + e^{-2t} - 2\right)$$

$$= \frac{1}{e^t - e^{-t}} g(2t) = \frac{1}{e^{l(\gamma)/2} - e^{-l(\gamma)/2}} g(l(\gamma)).$$

By the general trace formula as stated before the theorem, we see that the theorem holds if $f \in \mathcal{H}_{\mathrm{sym}}$ is admissible for the trace formula.

We now derive the trace formula for the special case of the *heat kernel*. Let

$$h(r) = h_t(r) = e^{-(\frac{1}{4}+r^2)t}.$$

Note that $h_t \in E$ and so Proposition 11.2.9 applies. One gets $g(u) = \frac{e^{-t/4}}{\sqrt{4\pi t}} e^{-\frac{u^2}{4t}}$. Let $f_t = \Psi^{-1}(h_t)$ with $\Psi : \mathcal{H}_{\mathrm{sym}} \to E^{\mathrm{ev}}$ as in Proposition 11.2.9. Recall from Sect. 9.2 the definition of the space $C_{\mathrm{unif}}(G)$ of uniformly integrable functions on G.

Proposition 11.4.4 *The function f_t lies in $C_{\mathrm{unif}}(G)^2$, so the trace formula is valid for f.*

Proof Note that $h_t = h_{t/2}^2$, which means $f_t = f_{t/2} * f_{t/2}$ and so, in order to show that the trace formula is valid for f_t, it suffices to show that $f_t \in C_{\mathrm{unif}}(G)$ for every $t > 0$, as it then follows that $f_t \in C_{\mathrm{unif}}(G)^2$. Let $r > 0$ and define

$$U(r) \overset{\mathrm{def}}{=} K\{a_s : 0 \le s < r\}K \subset G.$$

Then $U(r)$ is an open neighborhood of the unit. Note that $U(r) = \{x \in G : \mathrm{tr}\,(x^t x) < e^{2r} + e^{-2r}\}$ and the boundary satisfies

$$\partial U(r) = \{x \in G : \mathrm{tr}\,(x^t x) = e^{2r} + e^{-2r}\} = Ka_r K.$$

Lemma 11.4.5 *For $0 < r < \frac{s}{2}$ we have*

$$U(r)a_s U(r) \subset U(s + 2r) \backslash U(s - 2r).$$

Proof As $U(r)$ is invariant under K-multiplication from both sides, it suffices to show everything modulo K-multiplication on both sides. Suppose that for $0 \le y < r$

and $k \in K$ we can show that $a_y k a_x$ and $a_x k a_y$ both lie in $U(x+r) \setminus U(x-r)$. Then, modulo K-multiplication one has $a_y k a_s = a_t$ for $s - r < t < s + r$. Iterating the argument with t taking the part of s, one gets for $0 \le y' < r$,

$$a_y k a_s k' a_{y'} = a_t k'' a_{y'} \in U(t+r) \setminus U(t-r) \subset U(s+2r) \setminus U(s-2r).$$

So it suffices to show that for $k \in K$ one has $a_y k a_s, a_s k a_y \in U(s+r) \setminus U(s-r)$ for $0 \le y < r$ and arbitrary s. For $x \in G$ let $T(x) = \text{tr}\,(x^t x)$. Note that

$$T \begin{pmatrix} a & b \\ c & d \end{pmatrix} = a^2 + b^2 + c^2 + d^2.$$

We have to show that

$$e^{2(s-r)} + e^{2(r-s)} < T(a_y k a_s) < e^{2(s+r)} + e^{-2(s+r)}.$$

Now any $k \in K$ can be written as $k = \begin{pmatrix} a & -b \\ b & a \end{pmatrix}$ for some $a, b \in \mathbb{R}$ with $a^2 + b^2 = 1$. Then

$$T(a_y k a_s) = T \begin{pmatrix} e^{y+s} a & -e^{y-s} b \\ e^{s-y} b & e^{-(y+s)} a \end{pmatrix}$$

$$= e^{2(y+s)} + e^{-2(y+s)} + b^2 (e^{2(y-s)} + e^{2(s-y)} - e^{2(y+s)} - e^{-2(y+s)}).$$

Here we have used $a^2 = 1 - b^2$. Now $b \in [-1, 1]$ and the above is a quadratic polynomial in b, which takes its extremal values at the zero of its derivative, i.e., at $b = 0$ or at $b = \pm 1$. In both cases we get the claim.   □

The proof of the proposition now proceeds as follows: One notes that the function ϕ_t with $f_t(x) = \phi_t (\text{tr}\, x^t x - 2)$ is monotonically decreasing. This follows from $\phi_t = -\frac{1}{\pi} A(q'_t)$. Hence Lemma 11.4.5 implies that $(f_t)_{U(r)}(a_s) \le f_t(a_{s-2r})$ for $0 \le r < s/2$. (Recall the notation $f_U(x) = \sup |f(UxU)|$.) Therefore it suffices to show that for any $r \ge 0$,

$$\int_{\{x \in G : T(x) \ge 2r\}} \phi_t(\text{tr}\,(x^t x) - 2 - 2r)\, dx \; < \; \infty.$$

For this we use the integration formula of the Cartan decomposition in Theorem 11.2.1, which shows that the integral equals

$$2\pi \int_{e^{2x} + e^{-2x} - 2 > 2r} \phi_t \left(e^{2x} + e^{-2x} - 2 - 2r \right) \left(e^{2x} - e^{-2x} \right) dx.$$

Substituting $u = e^{2x} + e^{-2x}$ this becomes

$$\pi \int_{u > 2r+2} \phi_t(u - 2 - 2r)\, du = \pi \int_0^\infty \phi_t(x)\, dx.$$

As $h_t \in E^{\text{ev}}$, the function ϕ_t lies in $S_{[0,\infty)}$ and so this integral is indeed finite.   □

The trace formula for the function f_t says

$$\sum_{j=0}^{\infty} e^{-\left(\frac{1}{4}+r_j^2\right)t} = \frac{\text{vol}(\Gamma\backslash G)}{4\pi} \int_{\mathbb{R}} re^{-\left(\frac{1}{4}+r^2\right)t} \tanh(\pi r)\, dr$$

$$+ \sum_{[\gamma]\neq 1} \frac{l(\gamma_0)}{e^{l(\gamma)/2} - e^{-l(\gamma)/2}} \frac{1}{2\pi} \int_{\mathbb{R}} e^{-(r^2+1/4)t} e^{irl(\gamma)}\, dr,$$

where we used the equation $g_t(u) = \frac{1}{2\pi}\int_{\mathbb{R}} h_t(r)e^{iru}\,dr$, which follows from inverse Fourier transform and the fact that h_t is even. Let $\mu(t)$ denote either side of this equation. For a complex number s with $\text{Re}(s^2) < -\frac{1}{4}$ let

$$\alpha(s) \stackrel{\text{def}}{=} \int_1^{\infty} \mu(t)e^{t\left(s^2+\frac{1}{4}\right)}\, dt.$$

By realizing μ via the left hand side of the trace formula gives

$$\alpha(s) = \sum_{j=0}^{\infty} \frac{e^{s^2-r_j^2}}{r_j^2 - s^2}$$

and using the right hand side of the trace formula gives

$$\alpha(s) = \frac{\text{vol}(\Gamma\backslash G)}{4\pi} \int_{\mathbb{R}} r \frac{e^{s^2-r^2}}{r^2 - s^2} \tanh(\pi r)\, dr$$

$$+ \sum_{[\gamma]\neq[1]} \frac{l(\gamma_0)}{e^{l(\gamma)/2} - e^{-l(\gamma)/2}} \frac{1}{2\pi} \int_{\mathbb{R}} \frac{e^{s^2-r^2}}{r^2 - s^2} e^{irl(\gamma)}\, dr.$$

Now take h as in the assumptions of the theorem, but with the stronger growth condition $h(z) = O(\exp(-a|z|^4))$ for some $a > 0$ and $|\text{Im}(z)| < \frac{1}{2} + \varepsilon$. For $T > 0$, let R_T denote the positively oriented rectangle with vertices $\pm T \pm i\frac{\varepsilon+1}{2}$. By the Residue Theorem we can compute

$$\frac{1}{2\pi i} \int_{R_T} \frac{e^{s^2-r^2}}{r^2 - s^2} sh(s)\, ds = \frac{1}{2}(h(r) + h(-r)) = h(r)$$

whenever r lies in the interior of the rectangle, and 0 else. For $T \to \infty$ this converges to $h(r)$ for every $r \in \mathbb{R} \cup i(0,\frac{1}{2})$. Thus, using the realization $\alpha(s) = \sum_{j=0}^{\infty} \frac{e^{s^2-r_j^2}}{r_j^2-s^2}$ it follows that that $\frac{1}{2\pi i}\int_{R_T} \alpha(s)sh(s)\,ds$ converges to the right hand side of Theorem 11.4.3 if $T \to \infty$. On the other hand, using the realization of $\alpha(s)$ given by the left hand side of the trace formula and interchanging the order of integration, which is justified by the growth condition on h, shows that $\frac{1}{2\pi i}\int_{R_T} \alpha(s)sh(s)\,ds$ equals

$$\frac{\text{vol}(\Gamma\backslash G)}{4\pi} \int_{-T}^{T} rh(r)\tanh(\pi r)\, dr + \sum_{[\gamma]\neq[1]} \frac{l(\gamma_0)}{e^{l(\gamma)/2} - e^{-l(\gamma)/2}} \frac{1}{2\pi} \int_{-T}^{T} h(r)e^{irl(\gamma)}\, dr.$$

This converges to the left hand side of Theorem 11.4.3 if $T \to \infty$ since $g(l(\gamma)) = \frac{1}{2\pi} \int_{\mathbb{R}} h(r) e^{irl(\gamma)} dr$.

This proves the theorem for h satisfying the stronger growth condition. For arbitrary h, let $a > 0$ and set $h_a(z) = h(z) \exp(-az^4)$. Then the function h_a satisfies the stronger growth condition for $|\mathrm{Im}(z)| < \frac{1}{2} + \varepsilon$ and the limit $a \to 0$, using Lebesgue's convergence theorem for the integrals, gives the claim. □

11.5 Weyl's Asymptotic Law

In the proof of the trace formula, we have used the "heat kernel" $h_t(r) = e^{-(\frac{1}{4}+r^2)t}$. The reason for this being called so is the following. The *Laplace operator* for hyperbolic geometry on $\mathbb{H}$,

$$\Delta = -y^2 \left(\left(\frac{\partial}{\partial x}\right)^2 + \left(\frac{\partial}{\partial y}\right)^2 \right),$$

is invariant under G, i.e., $\Delta L_g = L_g \Delta$ for every $g \in G$. Therefore Δ defines a differential operator on the quotient $\Gamma \backslash \mathbb{H}$, which we denote by the same letter. It can be shown that its eigenvalues are $\lambda_j = (\frac{1}{4} + r_j^2)$ for $j \geq 0$. Since this requires additional arguments from Lie theory and is not essential for our purposes, we will not give the proof, but only mention the fact as an explanation for the terminology. The interested reader may consult Helgason's book [Hel01].

The hyperbolic *heat operator* is $e^{-t\Delta}$ for $t > 0$. This is an integral operator whose kernel $k_t(z, w)$ describes the amount of heat flowing in time t from point z to point w. Therefore

$$\sum_{j=0}^{\infty} e^{-(\frac{1}{4}+r_j^2)t} = \sum_{j=0}^{\infty} e^{-t\lambda_j} = \mathrm{tr}\, e^{-t\Delta}$$

is the *heat trace* on $\Gamma \backslash \mathbb{H}$.

Proposition 11.5.1 *As $t \to 0$, one has*

$$t \sum_{j=0}^{\infty} e^{-(\frac{1}{4}+r_j^2)t} \to \frac{\mathrm{vol}(\Gamma \backslash \mathbb{H})}{4\pi}.$$

Proof As $g(u) = \frac{e^{-t/4}}{\sqrt{4\pi t}} e^{-\frac{u^2}{4t}}$, the trace formula for the heat kernel gives

$$t \sum_{j=0}^{\infty} e^{-t(\frac{1}{4}+r_j^2)} = t \frac{\mathrm{vol}(\Gamma \backslash \mathbb{H})}{4\pi} \int_{-\infty}^{\infty} r e^{-(\frac{1}{4}+r^2)t} \tanh(\pi r) \, dr$$

$$+ t \sum_{[\gamma] \neq 1} \frac{l(\gamma_0)}{e^{l(\gamma)/2} - e^{-l(\gamma)/2}} \frac{e^{-\frac{t}{4} - \frac{l(\gamma)^2}{4t}}}{\sqrt{4\pi t}}.$$

Substituting r with $r/\sqrt{t}$ shows that the first summand equals

$$\frac{\mathrm{vol}(\Gamma\backslash\mathbb{H})}{4\pi}e^{-t/4}\int_{\mathbb{R}}re^{-r^2}\tanh\left(\pi\frac{r}{\sqrt{t}}\right)dr.$$

The integral equals $\int_0^\infty 2re^{-r^2}\tanh\left(\pi\frac{r}{\sqrt{t}}\right)dr$. As $t\to 0$, the tanh-term tends to 1 monotonically from below; therefore the integral tends to

$$\int_0^\infty 2re^{-r^2}\,dr = -e^{-r^2}\Big|_0^\infty = 1.$$

It remains to show that

$$\sqrt{t}\sum_{[\gamma]\neq 1}\frac{l(\gamma_0)}{e^{l(\gamma)/2}-e^{-l(\gamma)/2}}\frac{e^{-\frac{t}{4}-\frac{l(\gamma)^2}{4t}}}{\sqrt{4\pi}}$$

tends to zero as $t\to 0$. This is clear as the sum is finite for every $0<t<1$ and each summand tends to zero monotonically as soon as $t<l/2$, where l is the minimal length $l(\gamma)$ for $\gamma\in\Gamma\backslash\{1\}$. $\square$

We use this proposition to derive Weyl's asymptotic formula.

Theorem 11.5.2 *For $T>0$, let $N(T)$ be the number of eigenvalues $\lambda_j=\frac{1}{4}+r_j^2$ of Δ that are $\leq T$. Then, as $T\to\infty$, one has*

$$N(T)\sim\frac{\mathrm{vol}(\Gamma\backslash\mathbb{H})}{4\pi}T,$$

where the asymptotic equivalence $\sim$ means that the quotient of the two sides tends to 1, as $T\to\infty$.

Proof We need a lemma. Recall the definition of the Γ-function from Sect. 11.2.6.

Lemma 11.5.3 *Let μ be a Borel measure on $[0,\infty)$ such that*

$$\lim_{t\to 0}t\int_{[0,\infty)}e^{-t\lambda}\,d\mu(\lambda)=C$$

for some $C>0$. Then the following hold.

(a) *If f is a continuous function on $[0,1]$, then*

$$\lim_{t\to 0}t\int_{[0,\infty)}f(e^{-t\lambda})e^{-t\lambda}\,d\mu(\lambda)=C\int_0^\infty f(e^{-x})e^{-x}\,dx.$$

(b) *One has*

$$\lim_{t\to 0}t\int_{[0,\frac{1}{t}]}d\mu(\lambda)=C.$$

Proof (a) By The Stone-Weierstraß Theorem A.10.1, the set of polynomials is dense in $C([0,1])$. We first show that it suffices to prove the lemma for polynomials in the role of f. So let $f_n \to f$ be a convergent sequence in $C([0,1])$ and assume the lemma holds for each f_n. We have to show that it holds for f as well. Let $\varepsilon > 0$. Then there exists n_0 such that $\| f_n - f \|_{[0,1]} < \varepsilon$ for every $n \geq n_0$. For such n one gets

$$\left| t \int_{[0,\infty)} \left(f_n(e^{-t\lambda}) - f(e^{-t\lambda}) \right) e^{-t\lambda}\, d\mu(\lambda) \right| < \varepsilon t \int_{[0,\infty)} e^{-t\lambda}\, d\mu(\lambda),$$

and the latter tends to εC as $t \to 0$.

On the other hand,

$$\left| C \int_0^\infty \left(f_n(e^{-x}) - f(e^{-x}) \right) e^{-x}\, dx \right| < \varepsilon C.$$

So it suffices to prove the lemma for a polynomial and indeed for $f(x) = x^n$, in which case it comes down to

$$\lim_{t \to 0} t \int_{[0,\infty)} e^{-t(n+1)\lambda}\, d\mu = (n+1)^{-1} \lim_{t \to 0} t \int_{[0,\infty)} e^{-t\lambda}\, d\mu(\lambda)$$

$$= \frac{C}{(n+1)}$$

$$= C \int_0^\infty e^{-(n+1)t}\, dt.$$

Now for (b). Consider any continuous function $f \geq 0$ on the interval such that $f(x) = \frac{1}{x}$ for $x \geq e^{-1}$. Then

$$t \int_{[0,\frac{1}{t}]} f\left(e^{-t\lambda}\right) e^{-t\lambda}\, d\mu(\lambda) = t \int_{[0,\frac{1}{t}]} d\mu(\lambda),$$

so that for the limit superior we have the bound

$$\limsup_{t \to 0} t \int_{[0,\frac{1}{t}]} d\mu(\lambda) \leq \lim_{t \to 0} t \int_{[0,\infty)} f(e^{-t\lambda}) e^{-t\lambda}\, d\mu(\lambda)$$

$$= C \int_0^\infty f(e^{-x}) e^{-x}\, dx$$

$$= C + C \int_1^\infty f(e^{-x}) e^{-x}\, dx.$$

As the last integral can be chosen arbitrarily small, by using the Monotone Convergence Theorem we get that the limit superior in question is $\leq C$. Similarly, by choosing $f(x)$ to vanish for $x \leq e^{-1}$ and satisfy $0 \leq f(x) \leq 1/x$ one gets

$$\liminf_{t \to 0} t \int_{[0,\frac{1}{t}]} d\mu(\lambda) \geq C. \qquad \square$$

To get the theorem, we apply part (b) of the last lemma to the measure $\mu = \sum_{j=0}^{\infty} \delta_{\lambda_j}$. Indeed, substituting $T = \frac{1}{t}$, the left hand side of the above equation becomes $\lim_{T \to \infty} \frac{1}{T} N(T)$, while, by the proposition, we have $C = \lim_{t \to 0} t \sum_{j=0}^{\infty} e^{-t\lambda_j} = \frac{\mathrm{vol}(\Gamma \backslash \mathbb{H})}{4\pi}$. $\qquad\square$

11.6   The Selberg Zeta Function

As in the previous sections, let Γ be a torsion free hyperbolic uniform lattice in SL$(2, \mathbb{R})$. The compact surface $\Gamma \backslash \mathbb{H}$ is homeomorphic to a 2-sphere with a finite number of handles attached. The number of handles g is ≥ 2. It is called the *genus* of the surface $\Gamma \backslash \mathbb{H}$ (See [Bea95]).

The Selberg zeta function for Γ is defined for $s \in \mathbb{C}$ with $\mathrm{Re}(s) > 1$ as

$$Z(s) = \prod_{\gamma} \prod_{k \geq 0} \left(1 - e^{-(s+k)l(\gamma)}\right),$$

where the first product runs over all primitive hyperbolic conjugacy classes in Γ.

Theorem 11.6.1 *The product $Z(s)$ converges for $\mathrm{Re}(s) > 1$ and the function $Z(s)$ extends to an entire function with the following zeros. For $k \in \mathbb{N}$ the number $s = -k$ is a zero of multiplicity $2(g-1)(2k+1)$, where g is the genus of $\Gamma \backslash \mathbb{H}$. For every $j \geq 0$ the numbers*

$$\frac{1}{2} + ir_j, \quad \text{and} \quad \frac{1}{2} - ir_j$$

are zeros of $Z(s)$ of multiplicity equal to the multiplicity $N_\Gamma(\pi_{ir_j})$. These are all zeros.

Proof Let $a, b \in \mathbb{C}$ with real part $> \frac{1}{2}$. Then the function

$$h(r) = \frac{1}{a^2 + r^2} - \frac{1}{b^2 + r^2}$$

satisfies the conditions of the trace formula of Theorem 11.4.3. One computes (Exercise 11.5),

$$g(u) = \frac{1}{2\pi} \int_{\mathbb{R}} h(r) e^{-iru} \, dr = \frac{e^{-a|u|}}{2a} - \frac{e^{-b|u|}}{2b}.$$

We compute, formally at first,

$$\frac{Z'}{Z}(s) = \partial_s \left(\log \left(\prod_{\gamma_0} \prod_{k \geq 0} \left(1 - e^{-(s+k)l(\gamma_0)}\right) \right) \right)$$

$$= \partial_s \left(-\sum_{\gamma_0} \sum_{k \geq 0} \sum_{n=1}^{\infty} \frac{e^{-n(s+k)l(\gamma_0)}}{n} \right)$$

$$= \sum_{\gamma_0} \sum_{k \geq 0} \sum_{n=1}^{\infty} e^{-n(s+k)l(\gamma_0)} l(\gamma_0).$$

If γ_0 runs over all primitive classes, then $\gamma = \gamma_0^n$ will run over all classes $\neq 1$. Using $l(\gamma_0^n) = nl(\gamma_0)$ we get

$$\frac{Z'}{Z}(s) = \sum_{\gamma} \sum_{k \geq 0} e^{-(s+k)l(\gamma)} l(\gamma_0)$$

$$= \sum_{\gamma} e^{-sl(\gamma)} \frac{l(\gamma_0)}{1 - e^{-l(\gamma)}}$$

$$= \sum_{\gamma} e^{-(s-\frac{1}{2})l(\gamma)} \frac{l(\gamma_0)}{e^{l(\gamma)/2} - e^{-l(\gamma)/2}}.$$

Up to this point we have ignored questions of convergence. To deal with these, note that the geometric side of the trace formula for our function h equals $\frac{\text{vol}(\Gamma \backslash G)}{4\pi} \int_{\mathbb{R}} rh(r) \tanh(\pi r) \, dr$ plus

$$\frac{1}{2} \sum_{[\gamma] \neq 1} \left(\frac{e^{-al(\gamma)}}{a} - \frac{e^{-bl(\gamma)}}{b} \right) \frac{l(\gamma_0)}{e^{l(\gamma)/2} - e^{-l(\gamma)/2}}.$$

By the trace formula, the latter sum converges absolutely for all complex numbers a, b with $\text{Re}(a), \text{Re}(b) > \frac{1}{2}$. In the special case $b = 2a > 1$ all summands are positive and the estimate

$$\frac{e^{-al(\gamma)}}{a} - \frac{e^{-2al(\gamma)}}{2a} > \frac{e^{-al(\gamma)}}{a} - \frac{e^{-al(\gamma)}}{2a} = \frac{1}{2} \frac{e^{-al(\gamma)}}{a}$$

shows that the series

$$\sum_{[\gamma] \neq 1} e^{-al(\gamma)} \frac{l(\gamma_0)}{e^{l(\gamma)/2} - e^{-l(\gamma)/2}} = \frac{Z'}{Z}(a + \frac{1}{2})$$

converges locally uniformly absolutely for $\text{Re}(a) > \frac{1}{2}$. To be precise, for every $a_0 > \frac{1}{2}$ consider the open set $U = \{\text{Re}(a) > a_0\}$. For $a \in U$ and every $\gamma \in \Gamma \backslash \{1\}$ one has $|e^{-al(\gamma)}| = e^{-\text{Re}(a)l(\gamma)} < e^{-a_0 l(\gamma)}$. This shows locally uniform absolute convergence of the logarithmic derivative

$$\frac{Z'}{Z}(s) = \sum_{\gamma_0} \sum_{k \geq 0} \sum_{n=1}^{\infty} e^{-n(s+k)l(\gamma_0)} l(\gamma_0)$$

for $\text{Re}(s) > 1$. By direct comparison we conclude the absolute locally uniform convergence of the series $-\sum_{\gamma_0} \sum_{k \geq 0} \sum_{n=1}^{\infty} \frac{e^{-n(s+k)l(\gamma_0)}}{n}$, which is the logarithm of Z. This implies the locally uniform convergence of the product $Z(s)$ in the region $\{\text{Re}(s) > 1\}$.

The geometric side of the trace formula for h equals

$$\frac{\mathrm{vol}(\Gamma\backslash G)}{4\pi}\int_{\mathbb{R}} rh(r)\tanh(\pi r)\,dr + \frac{1}{2a}\frac{Z'}{Z}\left(a+\frac{1}{2}\right) - \frac{1}{2b}\frac{Z'}{Z}\left(b+\frac{1}{2}\right).$$

The spectral side is

$$\sum_{j=0}^{\infty}\frac{1}{2a}\left(\frac{1}{a+ir_j}+\frac{1}{a-ir_j}\right) - \frac{1}{2b}\left(\frac{1}{b+ir_j}+\frac{1}{b-ir_j}\right).$$

The trace formula implies that this series converges for complex numbers a, b with $\mathrm{Re}(a), \mathrm{Re}(b) > \frac{1}{2}$. Being a Mittag-Leffler series, it converges for all $a, b \in \mathbb{C}$, which are not one of the poles $\pm ir_j$, and it represents a meromorphic function in, say $a \in \mathbb{C}$ with simple poles at the $\pm ir_j$ of residue $1/2a$ times the multiplicity of $\pm ir_j$.

We want to evaluate the integral

$$\int_{\mathbb{R}} rh(r)\tanh(\pi r)\,dr.$$

The Mittag-Leffler series of $\tanh$ equals

$$\tanh(\pi z) = \frac{1}{\pi}\sum_{n=0}^{\infty}\frac{1}{z+i(n+\frac{1}{2})}+\frac{1}{z-i(n+\frac{1}{2})},$$

where the sum converges absolutely locally uniformly outside the set of poles $i(\frac{1}{2}+\mathbb{Z})$. For $n \in \mathbb{N}$ the path γ_n consisting of the interval $[-n, n]$ and the half-circle in $\{\mathrm{Im} z > 0\}$ around zero of radius n will not pass through a pole. Note that the function $\tanh(\pi r)$ is periodic, i.e, $\tanh(\pi(r+2i)) = \tanh(\pi r)$. Further, it is globally bounded on any set of the form $\{z \in \mathbb{C} : |z - i(k+1/2)| \geq \varepsilon\ \forall k \in \mathbb{Z}\}$ for any $\varepsilon > 0$. As $rh(r)$ is decreasing to the power r^{-3}, it follows that the integral $\int_{\gamma_n} rh(r)\tanh(\pi r)\,dr$ converges to the integral in question. By the residue theorem we conclude

$$\int_{\mathbb{R}} rh(r)\tanh(\pi r)\,dr = 2\pi i\sum_{z:\mathrm{Im} z>0}\mathrm{res}_{r=z}(rh(r)\tanh(\pi r)).$$

We have

$$rh(r) = \frac{1}{2}\left(\frac{1}{r+ia}+\frac{1}{r-ia}\right) - \frac{1}{2}\left(\frac{1}{r+ib}+\frac{1}{r-ib}\right).$$

We will assume $a \neq b$, both in $\mathbb{C}\backslash(\frac{1}{2}+\mathbb{Z})$. Then the poles of $rh(r)$ and of $\tanh(\pi r)$ are disjoint and we conclude that the integral equals

$$\pi i\tanh(\pi ia) - \pi i\tanh(\pi ib) + 2i\sum_{n=0}^{\infty}\frac{1}{2}\left(\frac{1}{i(n+\frac{1}{2})+ia}+\frac{1}{i(n+\frac{1}{2})-ia}\right) - (\ldots),$$

where the dots indicate the same term for b instead of a. Plugging in the Mittag-Leffler series of tanh, one shows that the integral equals

$$2 \sum_{n=0}^{\infty} \left(\frac{1}{a + \frac{1}{2} + n} - \frac{1}{b + \frac{1}{2} + n} \right).$$

From hyperbolic geometry (see [Bea95] Theorem 10.4.3) we take

Lemma 11.6.2 *The positive number* $\frac{\mathrm{vol}(\Gamma \backslash G)}{4\pi}$ *is an integer. More precisely, it is equal to* $g - 1$, *where* $g \geq 2$ *is the genus of the compact Riemann surface* $\Gamma \backslash \mathbb{H}$.

After a change of variables $a \mapsto a + \frac{1}{2}$ and the same for b, comparing the two sides of the trace formula tells us that

$$\frac{Z'}{Z} \left(a + \frac{1}{2} \right) = \frac{a}{b} \frac{Z'}{Z} \left(b + \frac{1}{2} \right) + 4a(1 - g) \sum_{n=0}^{\infty} \left(\frac{1}{a + \frac{1}{2} + n} - \frac{1}{b + \frac{1}{2} + n} \right)$$

$$+ \sum_{j=0}^{\infty} \frac{1}{a + ir_j} + \frac{1}{a - ir_j} - \frac{a}{b} \frac{1}{b + ir_j} + \frac{a}{b} \frac{1}{b - ir_j}.$$

Fixing an appropriate b, this extends to a meromorphic function on $\mathbb{C}$ with simple poles at $a = -n$ and $a = \frac{1}{2} \pm ir_j$. It follows that Z extends to an entire function on $\mathbb{C}$ and by a theorem from Complex Analysis (see [Rud87], Theorem 10.43) it follows that the poles of $\frac{Z'}{Z}$ are precisely the zeros of Z with multiplicity the respective residues. These are $2(2n + 1)(g - 1)$ for $a = -n$ and 1 in all other cases.   □

We define the *Ruelle zeta function* of Γ as the infinite product

$$R(s) = \prod_{[\gamma]} (1 - e^{-sl})^{\gamma}.$$

Corollary 11.6.3 *The product defining the Ruelle zeta function converges for* $\mathrm{Re}(s) > 1$ *and the so defined Ruelle zeta function extends to a meromorphic function on* $\mathbb{C}$. *Its poles and zeros all lie in the union of* $\mathbb{R}$ *with the two vertical lines* $\mathrm{Re}(s) = \frac{1}{2}$ *and* $\mathrm{Re}(s) = -\frac{1}{2}$. *One has*

$$R(s) \doteq \frac{Z(s)}{Z(s + 1)}.$$

Proof The correlation between the Ruelle and the Selberg zeta function is immediate from the Euler product. The rest of the Corollary follows from this and Theorem 11.6.1.   □

Note that, as $r_j \in i \left[-\frac{1}{2}, \frac{1}{2} \right] \cup \mathbb{R}$, the Selberg zeta function satisfies a weak form of the Riemann hypothesis, as its zeros in the critical strip $\{ 0 < \mathrm{Re}(s) < 1 \}$ are all in the set $\{ \mathrm{Re}(s) = \frac{1}{2} \}$ with the possible exception of finitely many zeros in the interval $[0, 1]$.

Note further, that one has a simple zero at $s = 1$ and no other poles or zeros in $\{\mathrm{Re}(s) \geq 1\}$. This information, together with the product expansion, suffices to use standard machinery from analytic number theory as in [Cha68] to derive the following theorem.

Theorem 11.6.4 (Prime Geodesic Theorem). *For $x > 0$ let $\pi(x)$ be the number of hyperbolic conjugacy classes $[\gamma]$ in Γ with $l(\gamma) \leq x$. Then, as $x \to \infty$,*

$$\pi(x) \sim \frac{e^{2x}}{2x}.$$

11.7 Exercises and Notes

Exercise 11.1 Show that $\int_{\mathbb{R}} \frac{e^{-u/2} \sin(ru)}{1+e^{-u}} \, du = \pi \tanh(\pi r)$.

(Hint: Write $\sin(ru) = \frac{1}{2i}(e^{iru} - e^{-iru})$ and thus decompose the integral into the sum of two integrals, each of which can be computed by the residue theorem.)

Exercise 11.2 Show that $g \in G = \mathrm{SL}_2(\mathbb{R})$ is

$$\text{hyperbolic} \Leftrightarrow |\mathrm{tr}(g)| > 2,$$
$$\text{parabolic} \Leftrightarrow |\mathrm{tr}(g)| = 2,$$
$$\text{elliptic} \Leftrightarrow |\mathrm{tr}(g)| < 2.$$

Exercise 11.3 Show that a circle or a line in $\mathbb{C}$ is described by the equation $Az\bar{z} + Bz + \underline{B}\underline{z} + C = 0$, where $A, C \in \mathbb{R}$. Show that the linear fractional $z \mapsto \frac{az+b}{cz+d}$ for $\left(\begin{smallmatrix} a & b \\ c & d \end{smallmatrix}\right) \in \mathrm{GL}_2(\mathbb{C})$ maps circles and lines to circles and lines.

Exercise 11.4 Let $A \in M_n(\mathbb{C})$. Show that

$$\det(\exp(A)) = \exp(\mathrm{tr}(A)).$$

Exercise 11.5 Let $a \in \mathbb{C}$ with $\mathrm{Re}(a) > 0$. Show that $\int_{\mathbb{R}} e^{-a|u|} e^{iru} \, dr = \frac{2a}{a^2+r^2}$.

Exercise 11.6 Let G be a locally compact group and K a compact subgroup. The *Hecke algebra* $\mathcal{H} = L^1(K \backslash G / K)$ is defined to be the space of all L^1-functions on G which are invariant under right and left translations from K. Show that $\mathcal{H}$ is an algebra under convolution. Show that for every $(\pi, V_\pi) \in \widehat{G}$ the space V_π^K of K-invariants is either zero, or an irreducible $\mathcal{H}$-module in the sense that it does not contain a closed $\mathcal{H}$-stable subspace.

Exercise 11.7 Continue the notation of the last exercise. The pair (G, K) is called a *Gelfand pair* if $\mathcal{H}$ is commutative. Show that if (G, K) is a Gelfand pair, then for every $(\pi, V_\pi) \in \widehat{G}$ the space V_π^K is at most one dimensional.

Exercise 11.8 Keep the notation of Exercise 11.6. Suppose that there is a continuous map $G \to G$, $x \mapsto x^c$ such that $(xy)^c = y^c x^c$ and $(x^c)^c = x$ as well as $x^c \in KxK$ for every $x \in G$. Show that G is unimodular and that (G, K) is a Gelfand pair.

(Hint: Let μ be the Haar measure on G and set $\mu^c(A) = \mu(A^c)$. Show that μ^c is a right Haar measure and that $\int_G f(x) \, d\mu(x) = \int_G f(x) \, d\mu^c(x)$ holds for every $f \in \mathcal{H}$. Consider the equation $\int_G f(xy) \, d\mu(x) = \Delta(y^{-1}) \int_G f(x) \, d\mu(x)$ for $f \in \mathcal{H}$ and make the integrand on the right hand side K-bi-invariant.)

Exercise 11.9 Let $f \in \mathcal{H}$ with $\operatorname{tr}(\pi_{ir}(f)\pi_{ir}(x)) = 0$ for every $x \in G$ and every $r \in \mathbb{R}$. Show that $f \equiv 0$.

Exercise 11.10 Let Δ denote the hyperbolic Laplace operator. Show that the function $z \mapsto \operatorname{Im}(z)^s$ for $s \in \mathbb{C}$ is an eigenfunction of Δ of eigenvalue $s(1 - s)$.

Exercise 11.11 Read and understand the proof of the prime number theorem in [Cha68]. Apply the same methods to give a proof of Theorem 11.6.4.

Notes

The Selberg zeta function has been introduced in Selberg's original paper on the trace formula [Sel56]. It has fascinated mathematicians from the beginning as its relation to the trace formula is similar to the relation of the Riemann zeta function to the Poisson summation formula and, as we have seen, a weak form of the Riemann hypothesis can be proved for the Selberg zeta function. However, Selberg's zeta continues to live in a world separate from Riemann's, and although many tried, no one has found a bridge between these worlds yet.

The name *Prime Geodesic Theorem* for Theorem 11.6.4 is derived from the following geometric facts. On the upper half plane $\mathbb{H}$ there is a Riemannian metric given by $\frac{dx^2 + dy^2}{y^2}$, which is left stable by the action of the group G, in other words, G acts by isometries. If $\Gamma \subset G$ is a torsion-free discrete cocompact subgroup, the quotient $\Gamma \backslash \mathbb{H}$ will inherit the metric and thus become a Riemannian manifold, the projection $\mathbb{H} \to \Gamma \backslash \mathbb{H}$ is a covering. A closed geodesic c in $\Gamma \backslash \mathbb{H}$ is covered by geodesics of infinite lengths in $\mathbb{H}$ and any such geodesic is being closed by an element $\gamma \in \Gamma$, which is uniquely determined up to conjugacy. The map $c \mapsto \gamma$ sets up a bijection between closed geodesics and primitive conjugacy classes in Γ. The number $l(\gamma)$ is just the length of the geodesic c. So indeed, Theorem 11.6.4 gives an asymptotic of lengths of closed geodesics. This theorem has been generalized several times, the most general version being a theorem of Margulis [KH95], which gives a similar asymptotic for compact manifolds of strictly negative curvature.

Chapter 12
Wavelets

In this chapter we will give an introduction to the theory of wavelets and wavelet transforms from the viewpoint of Harmonic Analysis. We will not be able to cover all theoretical aspects of wavelet theory, but we shall at least give a first introduction into this fascinating field. A much more complete coverage is given in the Lecture Notes [Füh05], which we recommend for further reading.

12.1 First Ideas

Let (X, μ) be any measure space. Roughly speaking, a wavelet is an element $\phi \in L^2(X, \mu)$ together with a family of transformations $t \mapsto \pi_t \phi$ indexed by the elements of some other measure space (T, ν). The corresponding *wavelet transform* is a map

$$\Phi : L^2(X, \mu) \to L^2(T, \nu),$$

which maps a function ψ to the function $\Phi(\psi)$, which is defined by

$$\Phi(\psi)(t) = \langle \psi, \pi_t \phi \rangle .$$

One usually considers only cases where the wavelet transform is injective and allows some reconstruction formula for a given ψ out of the transformed data. In many interesting cases, the measure space T is a locally compact group equipped with Haar measure and the transformation $t \mapsto \pi_t$ is a unitary representation of T on $L^2(X, \mu)$, so it is evident that Harmonic Analysis is entering the picture. Although there are many important cases that do not fit directly into this setting (like most discretized versions), we shall restrict ourselves to this setting below. The basic example is given by the continuous wavelet transform of the real line:

Example 12.1.1 Recall the *ax + b-group* G consisting of all matrices of the form $\begin{pmatrix} a & b \\ 0 & 1 \end{pmatrix}$ with $a \in \mathbb{R}^* = \mathbb{R} \smallsetminus \{0\}$ and $b \in \mathbb{R}$. It is the group of affine transformations $(a, b) \mapsto T_{a,b}$ on the real line given by $T_{a,b}(x) = ax + b$. A short computation shows that left Haar measure on G is given by the formula

A. Deitmar, S. Echterhoff, *Principles of Harmonic Analysis,* Universitext,
DOI 10.1007/978-3-319-05792-7_12, © Springer International Publishing Switzerland 2014

$$\int_G f(a,b)\, d(a,b) = \int_{\mathbb{R}^*} \int_{\mathbb{R}} f(a,b)\, db\, \frac{da}{|a|^2}$$

and that $\Delta : (a,b) \mapsto \frac{1}{|a|}$ is the modular function on G.

The action of G on $\mathbb{R}$ induces a unitary representation $(a,b) \mapsto \pi(a,b)$ of the $ax + b$-group on $L^2(\mathbb{R})$ via

$$(\pi(a,b)\phi)(x) = \frac{1}{\sqrt{|a|}} \phi\left(T_{(a,b)}^{-1}(x)\right) = \frac{1}{\sqrt{|a|}} \phi\left(\frac{x-b}{a}\right).$$

Now fix an element $\phi \in L^2(\mathbb{R})$. Then for each $\psi \in L^2(\mathbb{R})$ we obtain a bounded continuous function $W_\phi(\psi) : G \to \mathbb{C}$ given by

$$(a,b) \mapsto W_\phi(\psi)(a,b) = \langle \psi, \pi(a,b)\phi \rangle.$$

In order to get useful transformations in this way, we have to find conditions on ϕ, which guarantee that the transform $\psi \mapsto W_\phi(\psi)$ is injective with image in $L^2(G)$ and allows for a suitable inversion formula.

Of course, the optimal solution to the above stated problem would be to find functions $\phi \in L^2(\mathbb{R})$ such that the resulting transform $\psi \mapsto W_\phi(\psi)$ is an isometry from $L^2(\mathbb{R})$ into $L^2(G)$. In that case the reconstruction formula would be given by the adjoint operator $W_\phi^* : L^2(G) \to L^2(\mathbb{R})$, since for every $\psi, \eta \in L^2(\mathbb{R})$ we would have

$$\langle \psi, \eta \rangle = \langle W_\phi(\psi), W_\phi(\eta) \rangle = \langle W_\phi^* W_\phi(\psi), \eta \rangle,$$

and hence

$$\psi = W_\phi^* W_\phi(\psi),$$

as elements in $L^2(\mathbb{R})$.

Before we give specific answers to the questions asked in the above example, we want to put the problem into a more general framework: Suppose that G is an arbitrary locally compact group and that (π, V_π) is a unitary representation of G on some Hilbert space V_π. Given any fixed vector $\xi \in V_\pi$, we obtain a corresponding transformation

$$W_\xi : V_\pi \to C(G),$$

which sends $\eta \in V_\pi$ to the matrix coefficient $x \mapsto \langle \eta, \pi(x)\xi \rangle$. We call this the (generalized) *continuous wavelet transform* corresponding to π and ξ. Note that this transformation is injective if and only if ξ is a cyclic vector for π.

If W_ξ mapped V_π into the Hilbert space $L^2(G)$, we would be able to use Hilbert space methods. But in general the function $x \mapsto \langle \eta, \pi(x)\xi \rangle$ may not be an L^2-function. So, in order to be able to use inner products, we have to restrict W_ξ to the linear space

$$D_\xi = \{ \eta \in V_\pi : W_\xi(\eta) \in L^2(G) \}.$$

We then get a linear operator $W_\xi : D_\xi \to L^2(G)$. Up to now we have always considered operators from Banach spaces to Banach spaces. The current example shows that sometimes linear operators occur naturally, which are only defined on linear subspaces of Banach spaces. So let V, W be Banach spaces, and let $D \subset V$ be a linear subspace. A linear map $T : D \to W$ will loosely be called a *linear operator from V to W with domain D*.

Let D be a linear subspace of the Banach space V.

- A linear operator $T : D \to W$ is called a *closed operator* if its *graph*

$$\mathcal{G}(T) = \{(\eta, T\eta) : \eta \in D\}$$

  is a closed subset of $V \times W$.

- A linear operator $T : D \to W$ is called a *bounded operator* if its norm

$$\|T\| = \sup_{\substack{v \in D \\ \|v\|=1}} \|Tv\|$$

  is finite.

- A linear operator $T : D \to W$ is called *densely defined* if D is dense in V.

Note (Exercise 12.1) that if T is densely defined and bounded, it extends to a unique bounded operator $T : V \to W$.

Lemma 12.1.2 *The operator* $W_\xi : D_\xi \to L^2(G)$ *is closed.*

Proof Suppose that $(\eta_n)_{n \in \mathbb{N}}$ is a sequence in V_π such that $\eta_n \to \eta$ in V_π and $W_\xi(\eta_n) \to \phi$ in $L^2(G)$. Passing to a subsequence (Theorem B.4.3) if necessary, we may assume in addition that $W_\xi(\eta_n) \to \phi$ point-wise almost everywhere. But for every $x \in G$ we have:

$$|W_\xi(\eta)(x) - W_\xi(\eta_n)(x)| = |\langle \eta - \eta_n, \pi(x)\xi \rangle|$$
$$\leq \|\eta - \eta_n\| \|\xi\|,$$

which converges to 0. This implies that $\phi = W_\xi(\eta)$ almost everywhere. $\qquad\square$

Definition A vector $\xi \in V_\pi$ is called *square integrable* if $D_\xi = V_\pi$, i.e., if $W_\xi(\eta) : x \mapsto \langle \eta, \pi(x)\xi \rangle$ is square integrable on G for every $\eta \in V_\pi$. We put

$$D_\pi \stackrel{\text{def}}{=} \{\xi \in V_\pi : \xi \text{ is square integrable}\}.$$

It is easily checked that D_π is a $\pi(G)$-invariant linear subspace of V_π, and it follows from the above lemma and the closed graph theorem (C.1.6) that $W_\xi : V_\pi \to L^2(G)$ is bounded for every $\xi \in D_\pi$. The representation (π, V_π) is called a *square integrable representation*, if D_π is dense in V_π. A square integrable vector ξ is called an *admissible vector* if $W_\xi : V_\pi \to L^2(G)$ is isometric. Having fixed notation as above, the general problems to be solved are now as follows:

1. Under what conditions does a unitary representation (π, V_π) have any nontrivial square integrable vectors, or when is π a square integrable representation?

2. Suppose the unitary representation (π, V_π) is square integrable. Characterize all admissible vectors in V_π (if they exist).

3. Given an admissible vector ξ for the representation (π, V_π). Find an explicit reconstruction formula for the corresponding continuous wavelet transform W_ξ : $V_\pi \to L^2(G)$.

As explained in the above example, if ξ is an admissible vector for (π, V_π), then the reconstruction operator for the wavelet transform $W_\xi : V_\pi \to L^2(G)$ is given by the adjoint $W_\xi^* : L^2(G) \to V_\pi$. So Problem 3 reduces to the problem of finding an explicit formula for W_ξ^*. In the next section we shall give a general answer to the first two problems in case where the representation (π, V_π) is irreducible. This case covers many of the known examples and applies in particular to the continuous wavelet transform of the real line as discussed in Example 12.1.1 above.

In what follows next, we shall present some more or less easy observations concerning the above stated problems in general. Since the wavelet transformation $W_\xi : V_\pi \to C_b(G)$ is injective if and only if ξ is a cyclic vector for π, we should restrict our attention to cyclic representations and cyclic vectors thereof. Then the following easy lemma shows that we are really only interested in square integrable representations.

Lemma 12.1.3 *Suppose that $\xi \in V_\pi$ is a cyclic square integrable vector of the unitary representation (π, V_π). Then π is a square integrable representation.*

Proof If $\xi \in D_\pi$ is cyclic then D_π contains the dense subspace span$\{\pi(x)\xi : x \in G\}$ of V_π. □

We now give a description of the adjoint operator $W_\xi^* : L^2(G) \to V_\pi$:

Lemma 12.1.4 *Suppose that $\xi \in V_\pi$ is a square integrable vector for the unitary representation (π, V_π). Then the adjoint operator $W_\xi^* : L^2(G) \to V_\pi$ of the wavelet transform W_ξ is given by the formula*

$$W_\xi^*(\phi) = \int_G \phi(x)\pi(x)\xi \, dx.$$

Proof This integral is to be understood in the weak sense, i.e., it signifies the unique vector such that

$$\left\langle \int_G \phi(x)\pi(x)\xi \, dx, \eta \right\rangle = \int_G \phi(x) \langle \pi(x)\xi, \eta \rangle \, dx$$

for every $\eta \in H$. Note that it follows from the square integrability of ξ that the integral on the right hand side of this formula exists. For $\eta \in V_\pi$ and $\phi \in L^2(G)$ we have

$$\langle W_\xi^*(\phi), \eta \rangle = \langle \phi, W_\xi(\eta) \rangle$$

$$= \int_G \phi(x) \overline{W_\xi(\eta)(x)} \, dx$$

$$= \int_G \phi(x) \, \langle \pi(x)\xi, \eta \rangle \, dx.$$

□

As a consequence, we obtain the following general reconstruction formula for continuous wavelet transforms corresponding to admissible vectors $\xi \in V_\pi$:

Proposition 12.1.5 *Suppose that* $\xi \in V_\pi$ *is an admissible vector for the unitary representation* (π, V_π). *Then for all* $\eta \in V_\pi$ *we have*

$$\eta = W_\xi^* W_\xi \eta = \int_G \langle \eta, \pi(x)\xi \rangle \, \pi(x)\xi \, dx.$$

Moreover, the image $W_\xi(V_\pi) \subset L^2(G)$ *of the wavelet transform is a closed subspace of* $L^2(G)$ *with orthogonal projection* $P_\xi : L^2(G) \to W_\xi(V_\pi)$ *given by the formula*

$$\left(P_\xi \phi\right)(y) = \left(W_\xi W_\xi^*(\phi)\right)(y) = \int_G \phi(x) \, \langle \pi(x)\xi, \pi(y)\xi \rangle \, dx.$$

Proof It is a general fact for isometries $S : V \to W$ between Hilbert spaces V and W that S^*S is the identity on V and $SS^* : W \to S(V)$ is the projection onto the closed range $S(V) \subset W$ of S (we leave the verification as an easy exercise for the reader). Thus the proposition follows directly from the explicit description of W_ξ^* given in the previous lemma. □

Corollary 12.1.6 *Suppose that* $\xi \in V_\pi$ *is an admissible vector for the unitary representation* (π, V_π). *Then the orthogonal projection* $P_\xi : L^2(G) \to W_\xi(V_\pi)$ *is given by right convolution with* $W_\xi(\xi) \in L^2(G)$, *i.e.* $P_\xi(\phi) = \phi * W_\xi(\xi)$ *for every* $\phi \in L^2(G)$.

Proof By definition of W_ξ and convolution of functions on G, we have

$$\phi * W_\xi(\xi)(y) = \int_G \phi(x) \langle \xi, \pi(x^{-1}y)\xi \rangle \, dx$$

$$= \int_G \phi(x) \langle \pi(x)\xi, \pi(y)\xi \rangle \, dx$$

and the result follows from the previous proposition. □

We close this section by observing that every group has at least one square integrable representation:

Example 12.1.7. Consider the left regular representation of G. We claim that every $g \in C_c(G) \subset L^2(G)$ is square integrable. Since $C_c(G)$ is dense in $L^2(G)$ it then follows that $(L, L^2(G))$ is square integrable.

Consider the right regular representation R of G on $L^2(G)$ given by $R(y)\phi(x) = \sqrt{\Delta(y)}\phi(xy)$, where $\Delta = \Delta_G$ denotes the modular function on G. Put $\tilde{g}(y) = \Delta(y)^{-1/2}\overline{g(y)}$, and let $\phi \in L^2(G)$ be arbitrary. Consider the continuous function $x \mapsto \langle \phi, L(x)g \rangle$. We compute

$$
\begin{aligned}
\langle \phi, L(x)g \rangle &= \int_G \phi(y)\overline{g(x^{-1}y)}\,dy = \int_G \overline{g(y)}\phi(xy)\,dy \\
&= \int_G \tilde{g}(y)\sqrt{\Delta(y)}\phi(xy)\,dy \\
&= \int_G \tilde{g}(y)\,R(y)\phi(x)\,dy = R(\tilde{g})\phi(x).
\end{aligned}
$$

It follows that $x \mapsto \langle \phi, L(x)g \rangle$ is in $L^2(G)$ indeed.

12.2   Discrete Series Representations

In this section we want to characterize the irreducible square integrable representations of a locally compact group as precisely those irreducible representations, which are direct summands of the left regular representation $(L, L^2(G))$. Note that such representations are commonly called the *discrete series representations* of G. Aside of their relation to wavelet theory, as discussed here, they play a very important role in the general representation theory of reductive Lie groups, and a good deal of research work has been done to characterize the discrete series representations of certain locally compact groups. We start with some easy observations:

Suppose that (π, V_π) is a unitary representation, let $\xi \in V_\pi$ be any vector, and let $D_\xi \subset V_\pi$ be the domain of $W_\xi : D_\xi \to L^2(G)$. Then, if $y \in G$, we have

$$
\begin{aligned}
(W_\xi(\pi(y)\eta))(x) &= \langle \pi(y)\eta, \pi(x)\xi \rangle \\
&= \langle \eta, \pi(y^{-1}x)\xi \rangle \\
&= (L(y)W_\xi(\eta))(x),
\end{aligned}
$$

which proves that D_ξ is a $\pi(G)$-invariant subset of V_π. Moreover, we see that W_ξ intertwines the representation π restricted to D_ξ with the left regular representation $L : G \to L^2(G)$. Thus, if ξ is admissible it follows that the closed subspace $W_\xi(V_\pi) \subset L^2(G)$ is L-invariant and that W_ξ establishes a unitary equivalence between (π, V_π) and a subrepresentation of $(L, L^2(G))$. So we have shown

Proposition 12.2.1 *Suppose that ξ is an admissible vector for the unitary representation (π, V_π). Then (π, V_π) is equivalent to a subrepresentation of $(L, L^2(G))$.*

Definition In what follows we shall need the following generalization of Schur's Lemma for possibly unbounded intertwining operators. Let (π, V_π) and (ρ, V_ρ) be two representations of the locally compact group G. Let $D \subset V_\pi$ be a linear subspace. A linear operator $T : D \to V_\rho$ is called an *intertwining operator* if D is stable under π and

$$\rho(x)T = T\pi(x)$$

holds for every $x \in G$.

Proposition 12.2.2 *Let* (π, V_π) *and* (ρ, V_ρ) *be two unitary representations of* G *with* π *irreducible and let* $T : D \to V_\rho$ *with* $D \subset V_\pi$ *be a densely defined closed intertwining operator. Then* $D = V_\pi$ *and* $T = \mu S$ *for some* $0 \le \mu \in \mathbb{R}$ *and some isometry* $S : V_\pi \to V_\rho$.

Proof We may assume that $T \ne 0$. Denote by $\langle \cdot, \cdot \rangle_\pi$ and $\langle \cdot, \cdot \rangle_\rho$ the inner products on V_π and V_ρ, respectively. Since $\mathcal{G}(T)$ is a closed subspace of $V_\pi \times V_\rho$, we see that

$$\langle (\xi, T\xi), (\eta, T\eta) \rangle \overset{\text{def}}{=} \langle \xi, \eta \rangle_\pi + \langle T\xi, T\eta \rangle_\rho$$

defines an inner product on $\mathcal{G}(T)$, which turns $\mathcal{G}(T)$ into a Hilbert space. Let $P : \mathcal{G}(T) \to V_\pi$ denote the projection to the first factor, and let P^* be its adjoint operator. Then $PP^* : V_\pi \to V_\pi$ is a bounded intertwining operator for the irreducible representation π, which, by the classical version of Schur's Lemma (see Lemma 6.1.7), must be a multiple of the identity. Since $T \ne 0$ it follows $PP^* \ne 0$, and therefore $V_\pi = PP^*(V_\pi) \subset P(\mathcal{G}(T)) = D$, so T is everywhere defined. By the closed graph theorem (Theorem C.1.6) it must be bounded. Now we can apply Schur's Lemma to the positive intertwining operator T^*T of π to obtain some $\mu > 0$ such that $T^*T = \mu^2 \text{Id}_{V_\pi}$. Put $S = \frac{1}{\mu}T$. Then

$$\langle S\xi, S\eta \rangle_\rho = \langle S^*S\xi, \eta \rangle_\pi = \frac{1}{\mu^2} \langle T^*T\xi, \eta \rangle_\pi = \langle \xi, \eta \rangle_\pi,$$

which implies that S is an isometry. $\qquad\square$

As a first consequence, we obtain the following characterization of admissible vectors in irreducible representations:

Proposition 12.2.3 *Let* (π, V_π) *be an irreducible unitary representation, and let* $0 \ne \xi \in V_\pi$. *Then the following are equivalent.*

(a) *There exists* $0 \ne \eta \in V_\pi$ *such that* $W_\xi(\eta) \in L^2(G)$.

(b) ξ *is a square integrable vector.*

(c) *There exists* $0 < \lambda \in \mathbb{R}$ *such that* $\lambda\xi$ *is admissible.*

Moreover, if these assertions are true, then $\lambda = \frac{\|\xi\|}{\|W_\xi(\xi)\|}$.

Proof The implications (c) $\Rightarrow$ (b) and (b) $\Rightarrow$ (a) are trivial, so we only have to show that (a) implies (c). The discussion preceding Proposition 12.2.1 shows that D_ξ is a $\pi(G)$-invariant linear subspace of V_π, which is nonzero by condition (a). Thus, since π is irreducible, it must be dense in V_π. We further observed above that the densely defined linear operator

$$W_\xi : D_\xi \to L^2(G)$$

intertwines π with the left regular representation $(L, L^2(G))$ and Lemma 12.1.2 shows that W_ξ is a closed operator. Thus by Proposition 12.2.2 we know that $W_\xi = \mu S$ for some $0 \le \mu \in \mathbb{R}$ and some isometry $S : V_\pi \to L^2(G)$. Since $(W_\xi(\xi))(e) = \langle \xi, \xi \rangle \ne 0$, and since $x \mapsto \langle \xi, \pi(x)\xi \rangle$ is continuous, it follows that $W_\xi(\xi) \ne 0$, and W_ξ is not the zero operator. Thus $\mu \ne 0$. Applying $W_\xi = \mu S$ to ξ implies that $S = \frac{1}{\mu} W_\xi = W_{\lambda \xi}$ with $\lambda = \frac{1}{\mu} = \frac{\|\xi\|}{\|W_\xi(\xi)\|}$. $\qquad\qquad\square$

Corollary 12.2.4 *Let (π, V_π) be an irreducible representation of the locally compact group G. Then the following are equivalent:*

(a) *(π, V_π) is square integrable.*

(b) *There exists an admissible vector $\xi \in V_\pi$.*

(c) *(π, V_π) is a discrete series representation, i.e., it is equivalent to a subrepresentation of the left regular representation $(L, L^2(G))$.*

Proof The equivalence (a) $\Leftrightarrow$ (b) follows from Proposition 12.2.3 together with Lemma 12.1.3. The part (b) $\Rightarrow$ (c) follows from Proposition 12.2.1. Thus it only remains to show that (c) implies (a). Recall from Example 12.1.7 that every function $\phi \in C_c(G)$ is a square integrable element in $L^2(G)$ for the regular representation $(L, L^2(G))$. Suppose that $T : V_\pi \to L^2(G)$ is an isometric intertwining operator. Let $P : L^2(G) \to T(V_\pi)$ denote the orthogonal projection on the image $T(V_\pi) \subseteq L^2(G)$. Since $C_c(G)$ is dense in $L^2(G)$, the image $P(C_c(G))$ is dense in $T(V_\pi)$ and $D = \{\xi \in V_\pi : T\xi \in P(C_c(G))\}$ is dense in V_π. But all elements $\xi \in D$ are square integrable vectors for (π, V_π): Using the fact that $L(x)$ commutes with P for every $x \in G$ and that $P \circ T = T$ we obtain

$$\langle \eta, \pi(x)\xi \rangle = \langle T\eta, L(x)T\xi \rangle = \langle T\eta, L(x)P\phi \rangle = \langle T\eta, L(x)\phi \rangle,$$

where $\phi \in C_c(G)$ such that $T\xi = P\phi$. But $x \mapsto \langle T\eta, L(x)\phi \rangle$ is square integrable. $\qquad\qquad\square$

We now come to the main result of this section, which is mainly due to Duflo and Moore in [DM76]. But its importance to wavelet theory was first highlighted by Grossmann, Morlet and Paul in [GMP85]:

Theorem 12.2.5 *Let (π, V_π) be a discrete series representation of the locally compact group G, and let D_π be the set of square integrable vectors in V_π.*

(a) *There exists a closed densely defined operator $C_\pi : D_\pi \to V_\pi$ satisfying the orthogonality relation*

$$\langle C_\pi \xi', C_\pi \xi \rangle \langle \eta, \eta' \rangle = \langle W_\xi(\eta), W_{\xi'}(\eta') \rangle$$

for all $\xi, \xi' \in D_\pi$ and all $\eta, \eta' \in V_\pi$.

(b) *The operator $C_\pi : D_\pi \to V_\pi$ is injective and $\xi \in V_\pi$ is admissible if and only if $\xi \in D_\pi$ with $\|C_\pi \xi\| = 1$*

(c) *If, in addition, G is unimodular, then all vectors $\xi \in V_\pi$ are square-integrable, so $D_\pi = V_\pi$ and there exists a unique constant $c_\pi > 0$ such that C_π can be chosen equal to $c_\pi \mathrm{Id}_{V_\pi}$.*

Proof Choose any $\eta \in V_\pi$ with $\|\eta\| = 1$. Then, if C_π exists as in the theorem, we must have

$$\langle C_\pi \xi, C_\pi \xi' \rangle = \langle W_{\xi'}(\eta), W_\xi(\eta) \rangle.$$

Thus, in order to obtain the operator C_π, we have to check that

$$B_\eta(\xi, \xi') \stackrel{\text{def}}{=} \langle W_\xi(\eta), W_{\xi'}(\eta) \rangle$$

is a closed, positive definite, hermitian form on $D_\pi \times D_\pi$ as explained in the appendix preceding Theorem C.4.5. Linearity (resp. conjugate linearity) in the first (resp. second) variable follows easily from the equation

$$\langle W_{\xi'}(\eta), W_\xi(\eta) \rangle = \int_G \langle \eta, \pi(x)\xi' \rangle \langle \pi(x)\xi, \eta \rangle \, dx.$$

Positivity follows from the fact that every non-zero vector $\xi \in V_\pi$ is cyclic by irreducibility of π, and hence the function $x \mapsto \langle \eta, \pi(x)\xi \rangle$ does not vanish everywhere if $0 \neq \xi$. To see that $B_\eta : D_\pi \times D_\pi \to \mathbb{C}$ is closed, let $(\xi_n)_{n \in \mathbb{N}}$ be any sequence in D_π, which converges to some $\xi \in V_\pi$ and such that

$$B_\eta(\xi_n - \xi_m, \xi_n - \xi_m) \to 0 \quad \text{for} \quad n, m \to \infty.$$

Since

$$B_\eta(\xi_n - \xi_m, \xi_n - \xi_m) = \|W_{\xi_n - \xi_m}(\eta)\|_2^2$$
$$= \|W_{\xi_n}(\eta) - W_{\xi_m}(\eta)\|_2^2,$$

it follows that $(W_{\xi_n}(\eta))_{n \in \mathbb{N}}$ is a Cauchy-sequence in $L^2(G)$. Hence it converges to some $\phi \in L^2(G)$. By passing to a subsequence if necessary we may further assume that $(W_{\xi_n}(\eta))_{n \in \mathbb{N}}$ converges to ϕ point-wise almost everywhere. On the other hand, we have

$$|W_{\xi_n}(\eta)(x) - W_\xi(\eta)(x)| = |\langle \eta, \pi(x)(\xi_n - \xi) \rangle| \leq \|\xi_n - \xi\|$$

from which we may conclude that $\phi = W_\xi(\eta)$ in $L^2(G)$ and all requirements of Theorem C.4.5 are satisfied. We therefore obtain an operator $C_\pi : D_\pi \to V_\pi$ such that

$$\langle C_\pi\xi, C_\pi\xi'\rangle = \langle W_{\xi'}(\eta), W_\xi(\eta)\rangle.$$

We now check that $B_\eta = B_{\eta'}$ whenever η' is some other unit vector in V_π, hence proving that the constructed operator C_π does not depend on the choice of η. To see this, we simply use the fact that by Proposition 12.2.3 we have $W_\xi = \mu_\xi S$ with S an isometry and $\mu_\xi = \frac{\|W_\xi(\xi)\|_2}{\|\xi\|}$ whenever $0 \neq \xi \in D_\pi$. It then follows that

$$B_\eta(\xi, \xi) = \langle W_\xi(\eta), W_\xi(\eta)\rangle = \mu_\xi^2 \langle S\eta, S\eta\rangle = \mu_\xi^2.$$

Hence the expression $B_\eta(\xi, \xi)$ does not depend on the chosen unit vector η and, via polarization, neither does the hermitian form B_η. This shows that we have

$$\langle C_\pi\xi', C_\pi\xi\rangle \|\eta\|^2 = \langle W_\xi(\eta), W_{\xi'}(\eta)\rangle$$

for all $\xi, \xi' \in D_\pi$ and all $\eta \in V_\pi$. Now if $\langle C_\pi\xi', C_\pi\xi\rangle = 0$ it follows from Cauchy-Schwartz applied to the positive semi-definite hermitian form $(\eta, \eta') \mapsto \langle W_\xi(\eta), W_{\xi'}(\eta')\rangle$ that this form vanishes on $V_\pi \times V_\pi$ and if $\langle C_\pi(\xi'), C_\pi(\xi)\rangle \neq 0$ we can apply the polarization formula to the positive definite hermitian form

$$(\eta, \eta') \mapsto \frac{1}{\langle C_\pi\xi', C_\pi\xi\rangle} \langle W_\xi(\eta), W_{\xi'}(\eta')\rangle$$

to obtain the orthogonality relation of (a) for all $\xi, \xi' \in D_\pi$ and all $\eta, \eta' \in V_\pi$.

The injectivity of C_π follows from the orthogonality, since the inner product $\langle W_\xi(\eta), W_\xi(\eta)\rangle$ is > 0 if ξ and η are non-zero. Moreover, a vector $\xi \in V_\pi$ is admissible if and only if $\xi \in D_\pi$ and W_ξ is isometric. But the latter is equivalent to

$$\|C_\pi\xi\|^2 = \langle W_\xi(\eta), W_\xi(\eta)\rangle = 1$$

for all unit vectors $\eta \in V_\pi$. This finishes the proof of (b).

Assume now that G is unimodular. For $f \in L^2(G)$, the function $f^*(x) = \overline{f(x^{-1})}$ then lies in $L^2(G)$ again, and for $g \in L^2(G)$ one has $\langle f^*, g^*\rangle = \langle g, f\rangle$. For $\xi, \eta \in V_\pi$ one has $W_\xi(\eta)^* = W_\eta(\xi)$. Therefore, for $\xi, \xi', \eta, \eta' \in D_\pi$ one has

$$\langle C_\pi\xi', C_\pi\xi\rangle\langle \eta, \eta'\rangle = \langle W_\xi(\eta), W_{\xi'}(\eta')\rangle$$
$$= \langle W_{\eta'}(\xi'), W_\eta(\xi)\rangle$$
$$= \langle C_\pi\eta, C_\pi\eta'\rangle\langle\xi', \xi\rangle.$$

For fixed $\eta = \eta' \neq 0$, one sees that there exists a constant $d_\pi > 0$ such that $\langle\xi', \xi\rangle = d_\pi \langle C_\pi\xi', C_\pi\xi\rangle$. For arbitrary $\xi, \eta \in V_\pi$, we choose a sequence ξ_n that converges to ξ in V_π. The above implies that $\|W_{\xi_n}(\eta) - W_{\xi_m}(\eta)\|^2 = \|W_{\xi_n-\xi_m}(\eta)\|^2 =$

$\|C_\pi(\eta)\|^2 \|\xi_n - \xi_m\|^2$ for all $n, m \in \mathbb{N}$. Thus $W_{\xi_n}(\eta)$ is a Cauchy-sequence in $L^2(G)$, hence converges to an L^2-function f. Passing to a subsequence if necessary, we may assume that it converges point-wise almost everywhere to f. But the sequence $W_{\xi_n}(\eta)$ converges point-wise to the continuous function $W_\xi(\eta)$, hence $f = W_\xi(\eta)$ and the latter is in $L^2(G)$. As ξ was arbitrary, it follows $D_\pi = V_\pi$. The number $c_\pi = \sqrt{d_\pi}^{-1}$ will satisfy the claim. $\qquad\square$

Remark 12.2.6 Using functional calculus for unbounded operators one can strengthen the above theorem by showing that there exists a *unique positive* operator C_π, which satisfies the requirements of the theorem (See Remark C.4.6). Since we did not discuss functional calculus for unbounded operators and since this stronger result is not needed in what follows, we decided to state the theorem in the present weaker form.

Example 12.2.7 We should discuss the content of the theorem in case of a *compact* group G. In this case every bounded continuous function is square integrable, and hence we have $D_\pi = V_\pi$ for every irreducible representation (π, V_π). By Theorem 7.2.3, we know that every irreducible representation of the compact group G is finite dimensional. We claim that the operator C_π is then given by multiplication with the constant $\frac{1}{\sqrt{\dim V_\pi}}$, so in other words, $d_\pi = \dim(V_\pi)$. To see this, recall that it follows from the Peter-Weyl Theorem (Theorem 7.2.1) that for any vectors $\xi, \xi', \eta, \eta' \in V_\pi$ we have the orthogonality relation

$$
\begin{aligned}
\langle W_\xi(\eta), W_{\xi'}(\eta') \rangle &= \int_G \langle \eta, \pi(x)\xi \rangle \overline{\langle \eta', \pi(x)\xi' \rangle} \, dx \\
&= \frac{1}{d_\pi} \langle \xi, \xi' \rangle \langle \eta, \eta' \rangle,
\end{aligned}
$$

which follows easily from part (b) of Theorem 7.2.1 by choosing an orthonormal basis and writing all vectors as linear combinations of that basis. Thus we see that multiplication by $\frac{1}{\sqrt{d_\pi}}$ satisfies the requirements for the operator C_π in the theorem. This relation between C_π and the dimension d_π led to the study of the operator $K_\pi = (C_\pi)^{-2}$ by Duflo and Moore in [DM76] (with C_π being the positive operator of Remark 12.2.6), which they called the *formal dimension operator* of the square integrable irreducible representation π.

Definition If G is unimodular, then $C_\pi = c_\pi \mathrm{Id}_{V_\pi}$ for some positive number c_π, and we can define the *formal dimension* of the discrete series representation π to be the positive number $d_\pi = \frac{1}{c_\pi^2}$.

12.3 Examples of Wavelet Transforms

In this section we want to give several examples of continuous wavelet transforms where the results of the previous sections are used. But before we start we want to give a useful result concerning the construction of certain irreducible representations of semi-direct product groups. For notation, we say that a subset L of a locally compact space Y is *locally closed* if $L = O \cap C$ for some open subset O of Y and some closed subset C of Y. It is clear that all open and all closed sets are locally closed and one can show that the locally closed subsets are precisely those that are locally compact with respect to the subspace topology.

Lemma 12.3.1 *Let $L \subset Y$ be a locally closed subset of the locally compact space Y. Then for every function $f \in C_0(L)$ there exists $F \in C_0(Y)$ such that $f = F|_L$.*

Proof Let $L = O \cap C$ with O open and C closed in Y. By Tietze's Extension Theorem (Theorem A.8.3), every function in $C_0(C)$ is the restriction of some function in $C_0(Y)$. On the other hand, since $L = O \cap C$ is open in C, it follows that $C_0(L)$ is a subspace of $C_0(C)$. $\square$

The following proposition is a very special case of Mackey's Theory of irreducible representations for group extensions. For a more general treatment of this theory we refer to Folland's book [Fol95].

Proposition 12.3.2 *Suppose that $G = N \rtimes H$ is the semi-direct product of the abelian locally compact group N by the locally compact group H. Consider the action of H on the dual group $\widehat{N}$ of N given by $(h \cdot \chi)(n) = \chi(h^{-1}n)$. Suppose further that there exists $\chi \in \widehat{N}$ such that the map $H \to H(\chi)$ defined by $h \mapsto h \cdot \chi$ is a homeomorphism (this implies that $H(\chi)$ is locally closed in $\widehat{N}$). Then the representation $(\pi_\chi, L^2(H))$ defined by*

$$(\pi_\chi(n,h)\xi)(l) = \chi(l^{-1}n)\xi(h^{-1}l)$$

is irreducible.

Proof It is straightforward to check that $\pi = \pi_\chi$ is a unitary representation. To see that it is irreducible let $0 \neq V \subset L^2(H)$ be any closed invariant subspace, and let $\xi \in V$. Let $f \in L^1(N)$. We then get

$$(\pi|_N(f)\xi)(s) = \int_N f(n)(\pi(n,1)\xi)(l)\,dn$$

$$= \int_N f(n)\chi(l^{-1}n)\xi(l)\,dn = \hat{f}(l \cdot \chi)\xi(l).$$

Since the set of Fourier transforms $\{\hat{f} : f \in L^1(N)\}$ is dense in $C_0(\widehat{N})$, we see that $l \mapsto g(l \cdot \chi)\xi(l)$ lies in V for every $g \in C_0(\widehat{N})$. Since the orbit $H(\chi)$ is locally closed, every function in $C_0(H(\chi))$ can be realized as a restriction of some function in $C_0(\widehat{N})$. Thus, identifying H with $H(\chi)$ via $h \mapsto h \cdot \chi$ shows that $M_g\xi \in V$ for every $g \in C_0(H)$ and $\xi \in V$, where M_g denotes multiplication operator with g.

Suppose now that $f \in C_c(H)$. Integrating the restriction of π to H against f yields the operator

$$(\pi|_H(f)\xi)(l) = \int_H f(h)\xi(h^{-1}l)\,dh.$$

Thus we observe that V is invariant under all operators of the form $M_g\pi|_H(f)$ for all $f, g \in C_c(H)$. Computing

$$(M_g\pi|_H(f)\xi)(l) = \int_H g(l)f(h)\xi(h^{-1}l)\,dh = \int_H g(l)f(lh)\xi(h^{-1})\,dh$$

$$= \int_H \Delta_H(h^{-1})g(l)f(lh^{-1})\xi(h)\,dh,$$

we see that $M_g\pi|_H(f)$ is an integral operator on $L^2(H)$ with kernel $(h, l) \mapsto \Delta_H(h^{-1})g(l)f(lh^{-1})$ in $C_c(H \times H)$. We leave it as an exercise to show that every function in $L^2(H \times H)$ can be approximated in the L^2-norm by finite linear combinations of functions of the form $(h, l) \mapsto \Delta_H(h^{-1})g(l)f(lh^{-1})$ with $f, g \in C_c(H)$. This shows that V is invariant under all Hilbert-Schmidt operators on $L^2(H)$. Since $V \neq \{0\}$ this implies that $V = L^2(H)$ and π is irreducible.   □

We are now ready to treat our motivating example of the continuous wavelet transform of the real line:

Example 12.3.3 Recall from Example 12.1.1 the definition of the $ax + b$-group G and of the unitary representation $(\pi, L^2(\mathbb{R}))$ given by

$$(\pi(a, b)\phi)(x) = \frac{1}{\sqrt{|a|}}\phi\left(\frac{x-b}{a}\right).$$

We want to study the admissible vectors for π. For this we first observe that π is irreducible, and hence the results of the previous section apply. In order to apply the above proposition we want to realize π as a representation on $L^2(\mathbb{R}^*)$ with respect to Haar measure $\frac{1}{|x|}dx$. As a first step we apply the Plancherel isomorphism $\mathcal{F} : L^2(\mathbb{R}) \to L^2(\mathbb{R})$ given by $\phi \to \hat{\phi}$ for $\phi \in L^1(\mathbb{R}) \cap L^2(\mathbb{R})$, where we use the formula

$$\hat{\phi}(x) = \int_{\mathbb{R}} \phi(t)e^{-2\pi itx}\,dt.$$

for the Fourier Transform on $\mathbb{R}$. We then compute

$$(\pi(a, b)\phi)\hat{\,}(x) = \int_{\mathbb{R}} \frac{1}{\sqrt{|a|}}\phi(\frac{t-b}{a})e^{-2\pi itx}\,dt$$

$$= \sqrt{|a|}\int_{\mathbb{R}} \phi(u)e^{2\pi ix(au+b)}\,du$$

$$= \sqrt{|a|}e^{-2\pi ixb}\hat{\phi}(ax),$$

so that on the transformed side we get the representation $(\widehat{\pi}, L^2(\mathbb{R}))$

$$(\widehat{\pi}(a,b)\hat{\phi})(x) = \sqrt{|a|}e^{-2\pi i x b}\hat{\phi}(ax).$$

If we now identify $L^2(\mathbb{R})$ with $L^2(\mathbb{R}^*)$ via $g \mapsto (x \mapsto \sqrt{|x|}g(x))$, and then use the transform on $L^2(\mathbb{R}^*)$ given by $\eta \mapsto \check{\eta}$ with $\check{\eta}(x) = \eta(\frac{1}{x})$, we see that $\widehat{\pi}$ transforms to the representation (also called $\widehat{\pi}$) given by

$$(\widehat{\pi}(a,b)\xi)(x) = e^{-2\pi i b/x}\xi(a^{-1}x)$$

for $\xi \in L^2(\mathbb{R}^*)$. But this is precisely the representation π_χ of the semi-direct product $G = \mathbb{R} \rtimes \mathbb{R}^*$ corresponding to the character $\chi(x) = e^{-2\pi i x}$ of $\mathbb{R}$ as in the previous proposition. One easily checks that χ satisfies all conditions of that proposition. It follows that π is irreducible.

We now want to show that π is a discrete series representation, and we want to compute explicitly the Duflo-Moore operator $C_\pi : D_\pi \to L^2(\mathbb{R})$. Indeed, using the Plancherel isomorphism for the real line, we shall show the following:

- $\phi \in D_\pi$ if and only if $x \mapsto \frac{1}{\sqrt{|x|}}\hat{\phi}(x) \in L^2(\mathbb{R})$, and

- $C_\pi\phi = \mathcal{F}^{-1}(\frac{1}{\sqrt{|x|}}\hat{\phi})$, where $\mathcal{F}^{-1}$ denotes the inverse Plancherel isomorphism on $L^2(\mathbb{R})$.

As a result, it will follow that $\phi \in L^2(\mathbb{R})$ is admissible if and only if

$$1 = \|C_\pi\phi\|^2 = \int_{\mathbb{R}} \frac{|\hat{\phi}(x)|^2}{|x|}\,dx.$$

For the first item we directly compute the norm $\|W_\phi(\psi)\|_2$ for vectors $\phi, \psi \in L^2(\mathbb{R})$. In the following computations we write $\hat{\phi}$ also for the Plancherel transform on $L^2(\mathbb{R})$, although the Fourier integral itself might not always be defined. Using the Plancherel isomorphism and the above computation for $(\pi(b,a)\eta)\widehat{}$, we get

$$\|W_\phi(\psi)\|_2^2 = \int_G |\langle \psi, \pi(b,a)\eta \rangle|^2\,d(a,b)$$

$$= \int_G \left|\left\langle \hat{\psi}, (\pi(b,a)\eta)\widehat{}\right\rangle\right|^2\,d(a,b)$$

$$= \int_G \left|\int_{\mathbb{R}} \hat{\psi}(x)e^{2\pi i x b}\overline{\hat{\phi}(ax)}\,dx\right|^2 |a|\,d(a,b).$$

The interior integral is the Fourier integral for the function $F_a(x) = \hat{\psi}(x)\overline{\hat{\phi}(ax)}$ at the point $-b \in \mathbb{R}$. Thus, applying the formula for the Haar measure on G and the Plancherel isomorphism in the variable b the last term is equal to

$$\int_G |\hat{F}_a(-b)|^2 |a| \, d(a,b) = \int_{\mathbb{R}^*} \int_{\mathbb{R}} |\hat{F}_a(-b)|^2 \frac{1}{|a|} \, db \, da$$

$$= \int_{\mathbb{R}^*} \int_{\mathbb{R}} |\hat{\psi}(x)\overline{\hat{\phi}(ax)}|^2 \, dx \frac{1}{|a|} \, da$$

$$= \int_{\mathbb{R}} |\hat{\psi}(x)|^2 \left(\int_{\mathbb{R}^*} |\hat{\phi}(ax)|^2 \frac{1}{|a|} \, da \right) dx$$

$$= \|\hat{\psi}\|_2^2 \int_{\mathbb{R}^*} |\hat{\phi}(a)|^2 \frac{1}{|a|} \, da,$$

where the last two equations follow from Fubini and the translation invariance of Haar measure $\frac{1}{|a|} da$ on $\mathbb{R}^*$. In particular, $\|W_\phi(\psi)\|_2$ is finite if and only if $\int_{\mathbb{R}^*} |\hat{\phi}(a)|^2 \frac{1}{|a|} \, da$ is finite.

Now a similar (but a bit more tedious) computation shows that for two elements $\phi, \phi' \in D_\pi$ and $\psi \in L^2(G)$ with $\|\psi\|_2 = 1$ we get the equation

$$\langle W_{\phi'}(\psi), W_\phi(\psi) \rangle = \int_{\mathbb{R}^*} \hat{\phi}(a)\overline{\hat{\phi}'(a)} \frac{1}{|a|} \, da = \langle C_\pi \phi, C_\pi \phi' \rangle$$

with C_π as in the second item of our claim.

Notice that the Duflo-Moore operator C_π for the representation π is not a bounded operator since its domain is not all of $L^2(\mathbb{R})$. This reflects the fact that G is not unimodular.

If $\phi \in L^2(\mathbb{R})$ is an admissible vector, then by Proposition 12.1.5 the reconstruction formula for the continuous wavelet transform

$$W_\phi : L^2(\mathbb{R}) \to L^2(G)$$

is given by the adjoint operator $W_\phi^* : L^2(G) \to L^2(\mathbb{R})$, which is given by the formula

$$W_\phi^*(f) = \int_G f(a,b)\pi(a,b)\phi \, d(a,b),$$

where the integral has to be understood in the weak sense. However, in good cases (see Exercise 12.2 below) we obtain a point-wise reconstruction formula

$$\psi(x) = \int_{\mathbb{R}^*} \int_{\mathbb{R}} f(a,b) \frac{1}{\sqrt{|a|}} \phi\left(\frac{x-b}{a}\right) \frac{1}{|a|^2} \, db \, da$$

from the transform $f = W_\phi(\psi)$ of ψ.

We now proceed by describing some explicit admissible functions $\phi \in L^2(\mathbb{R})$ for the continuous wavelet transform of the real line.

Lemma 12.3.4 *Suppose that* $\phi \in L^2(\mathbb{R})$ *such that* $x \mapsto x\phi(x)$ *lies in* $L^1(\mathbb{R})$. *Then* ϕ *is a square integrable vector for the continuous wavelet transform on* $\mathbb{R}$ *if and only if* $\hat{\phi}(0) = \int_{\mathbb{R}} \phi(x) \, dx = 0$.

Proof We first note that the conditions of the lemma imply that $\phi \in L^1(\mathbb{R}) \cap L^2(\mathbb{R})$. Then $\hat{\phi}$ lies in $C_0(\mathbb{R}) \cap L^2(\mathbb{R})$, and the existence of $\int_{\mathbb{R}} |\hat{\phi}(x)|^2 \frac{1}{|x|} \, dx$ implies that $\hat{\phi}(0) = 0$.

Conversely, the condition $x\phi \in L^1(\mathbb{R})$ implies that $\hat{\phi}$ is continuously differentiable with $\frac{d}{dx}\hat{\phi} = -2\pi i \widehat{x\phi}(x)$ (see [Dei05, Theorem 3.3.1]). Let M be any upper bound of $|\hat{\phi}'|$ on the interval $[-1, 1]$. Then the mean value theorem implies that

$$|\hat{\phi}(x)| \leq M|x|$$

for every $x \in [-1, 1]$, which implies that

$$\int_{\mathbb{R}^*} |\hat{\phi}(x)|^2 \frac{1}{|x|} \, dx \leq \int_{|x|\leq 1} M^2 |x| \, dx + \int_{|x|\geq 1} |\hat{\phi}(x)|^2 \, dx$$

$$\leq M^2 + \|\phi\|^2. \qquad \square$$

Corollary 12.3.5 *Suppose that* $f \in S(\mathbb{R})$ *is a Schwartz-function. Then* f' *is a square integrable vector for the continuous wavelet transform.*

Proof By the computation rules for the Fourier transform as in [Dei05, Theorem 3.3.1], we have

$$\widehat{f'}(x) = 2\pi i x \hat{f}(x),$$

so that $\widehat{f'}(0) = 0$. $\qquad \square$

Example 12.3.6. To get an explicit example for an admissible vector ϕ for the continuous wavelet transform on the real line we look at the Gauss function $f(x) = e^{-\pi x^2}$. This function has the remarkable property that $\hat{f} = f$ (See [Dei05, Proposition 3.4.6]). By the above corollary we see that the derivatives of any positive order of f are square integrable vectors for the continuous wavelet transform. The first two of them are given by

$$f'(x) = -2\pi x e^{-\pi x^2}$$

$$f''(x) = 2\pi(2\pi x^2 - 1)e^{-\pi x^2}.$$

This yields even and odd square integrable functions. Their Fourier transforms are given by

$$\widehat{f'}(x) = 2\pi i x e^{-\pi x^2}$$

$$\widehat{f''}(x) = -4\pi^2 x^2 e^{-\pi x^2}.$$

We can use this to compute the integrals $\int_{\mathbb{R}_*} |\hat{\psi}(x)|^2 \frac{1}{|x|}\,dx$ for $\psi = f'$ and $\psi = f''$. In the first case we obtain the integral

$$8\pi^2 \int_0^\infty x e^{-2\pi x^2}\,dx = 2\pi,$$

as follows from the standard substitution $z = \pi x^2$. In the second case we arrive at the integral

$$32\pi^4 \int_0^\infty x^3 e^{-2\pi x^2}\,dx = 4\pi^2,$$

as follows from the same simple substitution plus a straightforward partial integration. Thus, scaling f' and f'' by the factors $-\frac{1}{\sqrt{2\pi}}$ and $-\frac{1}{2\pi}$, respectively, we obtain the admissible vectors

$$\phi_1(x) = \sqrt{2\pi}\, x e^{-\pi x^2} \quad \text{and}$$

$$\phi_2(x) = (1 - 2\pi x^2)e^{-\pi x^2}$$

for the continuous wavelet transform of the real line. The following figure shows the graph of the function ϕ_2, which is known under the name *Mexican hat*:

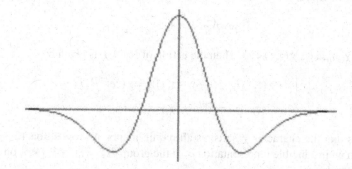

The big advantage of using wavelet transforms with functions like the Mexican hat lies in the fact that such transforms have much better local properties than the usual Fourier transform. Note that the wavelet transform is given by integration of a given function ψ on the real line with translates and dilates of the function ϕ. Now if ψ is a function with a large peak at the point $b \in \mathbb{R}$, say, then it is in general impossible to localize this peak in any local data of the Fourier transform. However, if we translate the Mexican hat by b, its peak will match the peak of the function ψ, and therefore this peak will have a strong effect at the wavelet transform $W_\phi(\psi)(a,b)$ if $\phi = \phi_2$ is the Mexican hat. In addition, this effect will be amplified by the dilations with the factors $a \in \mathbb{R}^*$. Of course, the possibility of having infinitely many choices for the admissible function (often called *mother wavelet* in the literature) gives the possibility to adjust the transformations to special needs.

Before we finish, we want to give one other example of a (generalized) continuous wavelet transform by using a variant of the Heisenberg group $\mathcal{H}$ as studied in Chap. 10.

Example 12.3.7 The transform we are studying here is known as the *windowed Fourier transform* or *Gabor transform* named after the 1971 physics Nobel prize winner Dennis Gabor, who was one of the first who systematically studied this transform. The idea behind the Gabor transform is to force the Fourier transform to be local by taking a window function $\phi \in L^2(\mathbb{R})$ and use this function to cut out the relevant pieces of a given function $\psi \in L^2(\mathbb{R})$ in its Fourier integral. To be more precise, if $\phi, \psi \in L^2(\mathbb{R})$ are given then we define the transform $W_\phi(\psi) : \mathbb{R} \times \mathbb{R} \to \mathbb{C}$ by the formula

$$W_\phi(\psi)(t, y) = \int_{\mathbb{R}} \psi(x)\overline{\phi(x - y)}e^{-2\pi itx}\, dx.$$

We shall show that this transform also fits in our general scheme. For this we consider the group $\mathcal{H}_r = \mathcal{H}/Z$, where $\mathcal{H}$ denotes the real Heisenberg group as introduced in Chap. 10, and Z denotes the standard lattice in the center of $\mathcal{H}$. The group may be realized as the semi-direct product $(\mathbb{T} \times \mathbb{R}) \rtimes \mathbb{R}$ with action of $y \in \mathbb{R}$ on a pair $(z, t) \in \mathbb{T} \times \mathbb{R}$ given by

$$y(z, t) = (ze^{-2\pi iyx}, t).$$

The dual group $\widehat{\mathbb{T} \times \mathbb{R}}$ of $\mathbb{T} \times \mathbb{R}$ is given by the characters $\chi_{(n,s)}$ indexed by $n \in \mathbb{Z}$ and $s \in \mathbb{R}$ such that

$$\chi_{(n,s)}(z, t) = z^n e^{-2\pi ist}.$$

Let $\chi = \chi_{(1,0)}$, i.e., $\chi(z, t) = z$. Then the action of $\mathbb{R}$ on χ is given by

$$y \cdot \chi(z, t) = \chi(-y \cdot (z, t)) = \chi(ze^{2\pi ity}, t)$$
$$= ze^{2\pi ity} = \chi_{(1, -y)}(z, t).$$

It follows that the character χ satisfies all requirements of Proposition 12.3.2, and we obtain an irreducible representation σ of the group $\mathcal{H}_r = (\mathbb{T} \times \mathbb{R}) \rtimes \mathbb{R}$ on $L^2(\mathbb{R})$ by the formula

$$\left(\sigma(z, t, y)\phi\right)(x) = ze^{2\pi itx}\phi(x - y).$$

The corresponding wavelet transform for a function $\psi \in L^2(\mathbb{R})$ is then given by

$$\left(W_\phi(\psi)\right)(z, t, y) = \bar{z} \int_{\mathbb{R}} \psi(x)e^{-2\pi itx}\overline{\phi(x - y)}\, dx.$$

The parameter $z \in \mathbb{T}$ obviously plays no important role in this transformation, and we observe that (up to this parameter) we end up with the Gabor transform.

As in the case of the ordinary Heisenberg group one easily checks that the Lebesgue measure on $\mathbb{T} \times \mathbb{R} \times \mathbb{R}$ serves as Haar measure on $\mathcal{H}_r$ and that $\mathcal{H}_r$ is unimodular. Thus it follows from Theorem 12.2.5 that either every vector $\phi \in L^2(\mathbb{R})$ is square integrable or σ has no square integrable vectors at all. Let's do the calculations: Since $|\bar{z}|^2 = 1$, applying the Plancherel formula to the function $F_y(x) = \psi(x)\overline{\phi(x - y)}$ and applying Fubini several times we get

$$\|W_\phi(\psi)\|_2^2 = \int_\mathbb{R} \int_\mathbb{R} \int_\mathbb{T} |W_\phi(\psi)(z,t,y)|^2 \, dz \, dt \, dy$$

$$= \int_\mathbb{R} \int_\mathbb{R} \left| \int_\mathbb{R} \psi(x)\overline{\phi(x-y)}e^{-2\pi itx} \, dx \right|^2 dt \, dy$$

$$= \int_\mathbb{R} \int_\mathbb{R} |\widehat{F_y}(t)|^2 \, dt \, dy = \int_\mathbb{R} \int_\mathbb{R} |F_y(x)|^2 \, dx \, dy$$

$$= \int_\mathbb{R} \int_\mathbb{R} |\psi(x)\overline{\phi(x-y)}|^2 \, dx \, dy = \|\psi\|_2^2 \|\phi\|_2^2,$$

which is always finite. So we see that σ is a discrete series representation. Since $\mathcal{H}_r$ is unimodular, the Duflo-Moore operator is given by multiplication with a positive scalar, which must be one since

$$\|C_\sigma \phi\|^2 = \|W_\phi(\psi)\|_2^2 = \|\phi\|_2^2$$

if $\|\psi\|_2 = 1$. Thus the unit vectors in $L^2(\mathbb{R})$ are precisely the admissible vectors and the formal dimension of the representation σ is one.

12.4   Exercises and Notes

Exercise 12.1. Show that a densely defined bounded operator $T : D \subset V \to W$ extends to a unique bounded operator $V \to W$.

(Use the methods of the proof of Lemma C.1.2.)

Exercise 12.2. Let $\phi \in L^2(\mathbb{R})$ be an admissible vector for the continuous wavelet transform of Example 12.1.1 that lies in $\mathcal{S}(\mathbb{R})$ (For example ϕ is the Mexican hat of Example 12.3.6). Show that the reconstruction formula for the wavelet transform $W_\phi(\psi)$ holds point-wise for every $\psi \in \mathcal{S}(\mathbb{R})$, i.e.

$$\psi(x) = \int_{\mathbb{R}^*} \int_\mathbb{R} W_\phi(\psi)(a,b) \frac{1}{\sqrt{|a|}} \phi\left(\frac{x-b}{a}\right) \frac{1}{|a|^2} \, db \, da$$

for every $x \in \mathbb{R}$.

Exercise 12.3. Let $\phi \in L^1(\mathbb{R}) \cap L^2(\mathbb{R})$ with compact support. Show that ϕ is a square integrable vector for the representation $(\pi, L^2(\mathbb{R}))$ underlying the continuous wavelet transform of Example 12.1.1 if and only if $\int_\mathbb{R} \phi(x) \, dx = 0$. In particular, the *Haar wavelet* $\phi : \mathbb{R} \to \mathbb{R}$ defined by

$$\phi(x) = \begin{cases} \frac{x}{|x|} & \text{for } |x| \le 1 \\ 0 & \text{for } |x| > 1 \end{cases}$$

is square integrable for π. Compute the scaling constant $\lambda > 0$ such that $\psi = \lambda\phi$ is admissible for π and give explicit formulas for the continuous wavelet transform corresponding to ψ and for its reconstruction.

Exercise 12.4. (a) Let G be any locally compact group, and let (π, V_π) be any discrete series representation of G. Let $C_\pi : V_\pi \supset D_\xi \to V_\pi$ be the corresponding Duflo-Moore operator, and let $\xi_1, \xi_2 \in D_\pi$ be two square integrable vectors for π such that $\langle C_\pi \xi_1, C_\pi \xi_2 \rangle \neq 0$. Show the following (mixed) reconstruction formula for the wavelet transform with respect to ξ_1:

$$\eta = \frac{1}{\langle C_\pi \xi_2, C_\pi \xi_1 \rangle} \int_G \langle \eta, \pi(x)\xi_1 \rangle \, \pi(x)\xi_2 \, dx.$$

(Hint: Use the orthogonality relations for C_π.)

(b) Assume now that G is the $ax + b$-group and that $(\pi, L^2(\mathbb{R}))$ is the representation corresponding to the continuous wavelet transform of the real line. Use the formula in (a) to give a reconstruction formula for the wavelet transform corresponding to the Mexican hat by using reconstruction corresponding to the Haar wavelet of the previous exercise.

Exercise 12.5. Let $\pi = \oplus_{i \in I} \pi_i$ be a direct sum of discrete series representations of the locally compact group G, and let $\xi = (\xi_i)_{i \in I} \in V_\pi = \oplus_{i \in I} V_{\pi_i}$. For each pair $i, j \in I$ let $S_{ij} : V_{\pi_i} \to V_{\pi_j}$ be an intertwining operator (which must be 0 if π_i is not equivalent to π_j). Then ξ is admissible for π if and only if ξ_i is admissible for π_i for every $i \in I$ and

$$\langle C_{\pi_j} S_{ij} \xi_i, C_{\pi_j} \xi_j \rangle = 0$$

for all $i, j \in I$ with $i \neq j$.

Exercise 12.6 In this exercise we work out the two-dimensional analogue of the continuous wavelet transform of the real line. For this we consider the *similitude group* $G = \mathbb{R}^2 \rtimes (SO(2) \times \mathbb{R}^+)$ (with $\mathbb{R}^+$ the multiplicative group of positive reals) with respect to the action

$$(g, r) \cdot y = y + rgy$$

for $x \in \mathbb{R}^2$, $g \in SO(2)$, and $r \in \mathbb{R}^+$. Show that the representation $(\pi, L^2(\mathbb{R}^2))$ given by

$$\left(\pi(y, g, r)\phi\right)(x) = \frac{1}{r}\phi\left(\frac{1}{r}g^{-1}(x - y)\right)$$

is an irreducible discrete series representation of G with

$$D_\pi = \{\phi \in L^2(\mathbb{R}^2) : \left(x \mapsto \frac{\hat{\phi}(x)}{|x|}\right) \in L^2(\mathbb{R}^2)\}$$

and Duflo-Moore operator given by

$$\widehat{C_\pi \phi}(x) = \frac{\hat{\phi}(x)}{|x|},$$

where $|x|$ denotes the Euclidean norm of x.

Notes

The examples of continuous wavelet transforms we considered so far are all constructed with the help of irreducible representations. But there are interesting
examples of transforms that arise from more general representations. For instance, a
frequently used wavelet transform is the *dyadic wavelet transform*, which is given by
the restriction of the representation $(\pi, L^2(\mathbb{R}))$ of the continuous wavelet transform
of the real line via the $ax + b$-group G to the subgroup

$$\{(2^k, b) : k \in \mathbb{Z}, b \in \mathbb{R}\} \subset G.$$

The resulting representation is not irreducible and the results of the previous section
are not applicable for the analysis of the corresponding transforms. We refer to Führ's
book [Füh05] for more details on this more general setting. We should point out that
the content of this chapter bases to a good extend on the first two chapters of that
book. Other sources for the material of this chapter are the original articles [DM76],
[GMP85] and the more basic introduction into wavelet theory given in Blattner's
book [Bla98].

Chapter 13
p-Adic Numbers and Adeles

The majority of the examples of topological groups in this book given so far, are locally euclidean, meaning that the groups are locally homeomorphic to $\mathbb{R}^n$. In this chapter the reader will see some examples which are not of this type. These examples, the p-adic numbers and the adeles, resp. ideles, are not only interesting as examples of this theory, but they also carry great importance for other areas of mathematics, in particular number theory.

13.1 p-Adic Numbers

The set $\mathbb{R}$ of real numbers is the completion of $\mathbb{Q}$ with respect to the usual absolute value

$$|x|_\infty = \begin{cases} x & \text{if } x \geq 0, \\ -x & \text{if } x < 0. \end{cases}$$

We shall see, that there are more "absolute values" defined on $\mathbb{Q}$. But first we have to give this notion a precise meaning.

Absolute Values

By an *absolute value* on a field K we mean a map $|\cdot| : K \to [0,\infty)$, such that for all $a, b \in K$ one has

- $|a| = 0 \Leftrightarrow a = 0$, (definiteness)
- $|ab| = |a||b|$, (multiplicativity)
- $|a + b| \leq |a| + |b|$. (triangle inequality)

Remark Every absolute value maps ± 1 to 1, i.e., one has $|1| = |-1| = 1$. For a proof consider $|1| = |1 \cdot 1| = |1|^2$ so $|1| = 1$ and $|-1|^2 = |(-1)^2| = |1| = 1$ so that $|-1| = 1$.

A. Deitmar, S. Echterhoff, *Principles of Harmonic Analysis*, Universitext,
DOI 10.1007/978-3-319-05792-7_13, © Springer International Publishing Switzerland 2014

Lemma 13.1.1 *If $|\cdot|$ is an absolute value on the field K, then $d(x, y) = |x - y|$ is a metric on K.*

Proof The map d is positive definite. It is symmetric, too, since

$$d(y, x) = |y - x| = |(-1)(x - y)| = |-1||x - y| = |x - y| = d(x, y).$$

Finally. it satisfies the triangle inequality, since for $x, y, z \in K$ one has

$$d(x, y) = |x - z + z - y| \leq |x - z| + |z - y| = d(x, z) + d(z, y). \qquad \square$$

Examples 13.1.2

- For $K = \mathbb{Q}$ the usual absolute value $|\cdot|_\infty$ is an example.
- The *discrete absolute value* exists for every field and is given by

$$|x|_{\text{triv}} = \begin{cases} 0 & \text{if } x = 0, \\ 1 & \text{if } x \neq 0. \end{cases}$$

The metric generated by this absolute value is the *discrete metric*

$$d(x, y) = \begin{cases} 0 & \text{if } x = y \\ 1, & \text{if } x \neq y. \end{cases}$$

The discrete metric induces the discrete topology, as for every $x \in K$ the open ball $B_{1/2}(x)$ of radius $1/2$ equals the set $\{x\}$, which therefore is open.

Definition Consider the field $K = \mathbb{Q}$ of rational numbers and fix a prime number p. Every rational can be written in the form

$$r = p^k \frac{m}{n}, \qquad n \neq 0,$$

where $m, n \in \mathbb{Z}$ are coprime to p. The exponent $k \in \mathbb{Z}$ is uniquely determined by r, if $r \neq 0$. We define the *p-adic absolute value* by

$$|r|_p = \left| p^k \frac{m}{n} \right|_p := \begin{cases} p^{-k} & \text{if } r \neq 0, \\ 0 & \text{if } r = 0. \end{cases}$$

Lemma 13.1.3 *Let p be a prime number. Then $|\cdot|_p$ is an absolute value on $\mathbb{Q}$, which satisfies the strong triangle inequality*

$$|x + y|_p \leq \max(|x|_p, |y|_p).$$

Here we have equality, if $|x|_p \neq |y|_p$.

Proof Definiteness follows from the definition. To show multiplicativity, write $x = p^k \frac{m}{n}$ and $y = p^{k'} \frac{m'}{n'}$, where m, n, m', n' are coprime to p. Then $xy = p^{k+k'} \frac{mm'}{nn'}$, and this yields $|xy|_p = |x|_p |y|_p$ in the case $xy \neq 0$. The case $xy = 0$ is trivial. For a proof of the strong triangle inequality, we can assume $xy \neq 0$ and $k \leq k'$. Then we have

$$x + y = p^k \left(\frac{m}{n} + p^{k'-k} \frac{m'}{n'} \right) = p^k \frac{mn' + p^{k'-k} nm'}{nn'}.$$

If $|x|_p \neq |y|_p$, i.e., $k' - k > 0$, then the number $mn' + p^{k'-k} nm'$ is coprime to p and we have $|x + y| = p^{-k} = \max(|x|_p, |y|_p)$. If on the other hand $|x|_p = |y|_p$, then the enumerator $mn' + p^{k'-k} nm' = mn' + nm'$ is of the form $p^l N$, where $l \geq 0$ and N is coprime to p. This means that $|x+y|_p = |p^{k+l} \frac{N}{nn'}|_p = p^{-k-l} \leq \max(|x|_p, |y|_p)$. □

In what follows we denote by $R^\times$ the group of units in a ring R. Of course, if K is a field, we have $K^\times = K \setminus \{0\}$.

Proposition 13.1.4 *For every $x \in \mathbb{Q}^\times$ we have the product formula*

$$\prod_{p \leq \infty} |x|_p = 1.$$

The product is extended over all prime numbers and $p = \infty$. For a given number $x \in \mathbb{Q}^\times$, almost all factors in the product are equal to 1.

Remark When we say *almost all*, we mean *all, up to finitely many exceptions.*

Proof Write x as a fraction of coprime integers and write these integers as product of primes. Then one has $x = \pm p_1^{k_1} \cdots p_n^{k_n}$ for pairwise different primes $p_1, \ldots, p_n$ and $k_1, \ldots, k_n \in \mathbb{Z}$. The p-adic absolute value $|x|_p$ equals 1, if p is a prime not occurring among the above. So the product indeed has only finitely many factors $\neq 1$. Further one has $|x|_{p_j} = p_j^{-k_j}$ and $|x|_\infty = p_1^{k_1} \cdots p_n^{k_n}$. Hence

$$\prod_{p \leq \infty} |x|_p = \left(\prod_{j=1}^n p_j^{-k_j} \right) \cdot p_1^{k_1} \cdots p_n^{k_n} = 1.   \qquad\square$$

Remark One can show that every non-trivial absolute value $|\cdot|$ on $\mathbb{Q}$ is of the form $|x| = |x|_p^a$ for a uniquely determined $p \leq \infty$ and a uniquely determined real number $a > 0$.

$\mathbb{Q}_p$ *as Completion of* $\mathbb{Q}$

We now give the first construction of the set $\mathbb{Q}_p$ of p-adic numbers. This set is the completion of $\mathbb{Q}$ in the p-adic metric

$$d_p(x, y) = |x - y|_p.$$

Proposition 13.1.5 *Let $p \leq \infty$. Then $\mathbb{Q}$ is not complete in the metric d_p. We denote the completion by $\mathbb{Q}_p$. Addition and Multiplication of $\mathbb{Q}$ can be extended in a unique way to continuous maps $\mathbb{Q}_p \times \mathbb{Q}_p \to \mathbb{Q}_p$. With these operations, $\mathbb{Q}_p$ is a field, called the field of p-adic numbers. The absolute value $|\cdot|_p$ can be extended in a unique way to a continuous map on $\mathbb{Q}_p$, which is an absolute value, again.*

Proof We consider this proposition known for $p = \infty$. In this case one has $\mathbb{Q}_p = \mathbb{Q}_\infty = \mathbb{R}$. We now let $p < \infty$. We write $|\cdot| = |\cdot|_p$. The non-completeness of $\mathbb{Q}$ follows from another description of $\mathbb{Q}_p$, which will be shown in the next section. We now extend the operations. Let $x, y \in \mathbb{Q}_p$. As $\mathbb{Q}$ is dense in the metric space $\mathbb{Q}_p$, there are sequences (x_n) and (y_n) in $\mathbb{Q}$, converging to x, resp. y in $\mathbb{Q}_p$. These sequences are Cauchy sequences in $\mathbb{Q}$. The estimate

$$|(x_n + y_n) - (x_m + y_m)| \leq |x_n - x_m| + |y_n - y_m|$$

implies that $(x_n + y_n)$ is a Cauchy sequence as well. So it converges in $\mathbb{Q}_p$ to an element z. This element does not depend on the choice of the sequences, since if (x_n') and (y_n') is another choice, then the sequence $(x_n' + y_n')$ also is a Cauchy sequence, which differs from $(x_n + y_n)$ only by a sequence converging to zero, hence gives the same element in the completion. We set $x + y = z$ and have thus extended the addition to $\mathbb{Q}_p$. It is easy to see that addition is a continuous map from $\mathbb{Q}_p \times \mathbb{Q}_p \to \mathbb{Q}_p$. The multiplication is extended analogously and it is not difficult to show that $\mathbb{Q}_p$ is a field with these operations and that the absolute value extends as well. We leave the details as an exercise. $\square$

The strong triangle inequality $|x + y| \leq \max(|x|, |y|)$ still holds on $\mathbb{Q}_p$. It has astonishing consequences, for example, the set

$$\mathbb{Z}_p = \{x \in \mathbb{Q}_p : |x|_p \leq 1\},$$

which contains $\mathbb{Z}$, is a subring of the field $\mathbb{Q}_p$. This ring is called the *ring of p-adic integers*.

Power Series

Let p be a prime. We now give a second construction of p-adic numbers. Every integer $n \geq 0$ can be written in the *p-adic expansion*,

$$n = \sum_{j=0}^{N} a_j p^j,$$

with uniquely determined coefficients $a_j \in \{0, 1, \dots, p - 1\}$. The sum of n and a second number $m = \sum_{i=0}^{M} b_i p^i$ is

$$n + m = \sum_{j=0}^{\max(M,N)+1} c_j p^j,$$

where each c_j only depend on $a_0, \ldots, a_j$ and $b_0, \ldots, b_j$. More precisely, these co-efficients are computed as follows: First one sets $c'_j = a_j + b_j$. Then one has $0 \le c'_j \le 2p - 2$ and it may happen, then $c'_j \ge p$. Let j be the smallest index, for which this happens. One replaces c'_j by the remainder modulo p and increases c'_{j+1} by one. Then one repeats this step until all coefficients are $\le p - 1$.

For the multiplication one has

$$nm = \sum_{j=0}^{M+N+1} d_j p^j,$$

where again the coefficient d_j only depends on $a_0, \ldots, a_j$ and $b_0, \ldots, b_j$.

These properties of multiplication and addition make it possible, to extend them to the set Z of formal power series

$$\sum_{j=0}^{\infty} a_j p^j,$$

with $0 \le a_j < p$. A *formal power series* may be considered simply as the sequence of its coefficients $(a_0, a_1, \ldots)$. The multiplicative unit 1 is represented by the sequence $(1, 0, 0, \ldots)$. One only uses the notation of a series for convenience.

Lemma 13.1.6 *With these operations, the set Z is a ring. An element $x = \sum_{j=0}^{\infty} a_j p^j$ is invertible in Z if and only if $a_0 \neq 0$.*

Proof Associativity and distributivity are inherited from $\mathbb{Z}$, as it suffices to check them on finite parts of the series. To show that Z is a ring, we are left with showing that an additive inverse exists. So let $x = \sum_{j=0}^{\infty} a_j p^j$ in Z. We have to show the existence of some $y = \sum_{j=0}^{\infty} b_j p^j$ in Z, such that $x + y = 0$. We construct the coefficients b_j inductively. In the case $a_0 = 0$ we set $b_0 = 0$ and $b_0 = p - a_0$ otherwise. Assume $b_0, \ldots, b_n$ already constructed with the property, that the element $y_n = \sum_{j=0}^{n} b_j p^j$ satisfies

$$x + y_n = \sum_{j=n+1}^{\infty} c_j p^j, \qquad 0 \le c_j < p.$$

If $c_{n+1} = 0$, then one sets $b_{n+1} = 0$. Otherwise one sets $b_{n+1} = p - c_{n+1}$. In this way one gets an element $y = \sum_{j=0}^{\infty} b_j p^j$, which satisfies $x + y = 0$.

We show the second assertion. If $x = \sum_{j=0}^{\infty} a_j p^j$ is invertible, then $a_0 \neq 0$, since otherwise the series xy would have vanishing zeroth coefficient for every $y \in Z$. For the converse direction, let $x = \sum_{j=0}^{\infty} a_j p^j$ with $a_0 \neq 0$. We construct a multiplicative inverse $y = \sum_{j=0}^{\infty} b_j p^j$ by giving the coefficients b_j successively. Since $\mathbb{F}_p = \mathbb{Z}/p\mathbb{Z}$ is a field, there exists exactly one $1 \le b_0 < p$ such that $a_0 b_0 \equiv 1 \bmod p$. Assume next that $0 \leq b_0, \ldots, b_n < p$ are already constructed with the property that

$$\underbrace{\left(\sum_{0 \leq j} a_j p^j \right)}_{=A} \underbrace{\left(\sum_{0 \leq j \leq n} b_j p^j \right)}_{=B} \equiv 1 \bmod p^{n+1}.$$

Then one has $\frac{AB-1}{p^{n+1}} \in \mathbb{Z}$, so there exists exactly one $0 \leq b_{n+1} < p$, such that $\frac{AB-1}{p^{n+1}} + a_0 b_{n+1} = pc$, $c \in \mathbb{Z}$, or $AB - 1 + a_0 b_{n+1} p^{n+1} = p^{n+2} c$. In other words, one has

$$\left(\sum_{0 \leq j} a_j p^j \right) \left(\sum_{0 \leq j \leq n+1} b_j p^j \right) \equiv 1 \bmod p^{n+2}.$$

The element $y = \sum_{j=0}^{\infty} b_j p^j$ constructed in this way satisfies the equation $xy = 1$.   □

Lemma 13.1.7 *Let (a_j) be a sequence in $\{0, 1, \ldots, p-1\}$. Then the series $\sum_{j=0}^{\infty} a_j p^j$ converges in $\mathbb{Q}_p$. We map the formal series to this limit and get a map $\psi : Z \to \mathbb{Q}_p$. This map is an isomorphism of rings*

$$Z \xrightarrow{\cong} \mathbb{Z}_p.$$

Proof Let $x_n = \sum_{j=0}^{n} a_j p^j$. We have to show that (x_n) is a Cauchy sequence in $\mathbb{Q}_p$. For $m \geq n \geq n_0$ one has

$$|x_m - x_n| = \left| \sum_{j=n+1}^{m} a_j p^j \right| \leq \max_{n < j \leq m} |a_j|_p |p^j|_p \leq p^{-n_0}.$$

Therefore the sequence is Cauchy, so the map ψ is well-defined. It is easy to show that ψ is a ring homomorphism. It remains to show bijectivity of $\phi : Z \to \mathbb{Z}_p$.

Injectivity: Let $x = \sum_{j=0}^{\infty} a_j p^j \neq 0$. Then there is a minimal j_0 such that $a_{j_0} \neq 0$. We have

$$|\psi(x)| = \left| a_{j_0} p^{j_0} + \sum_{j=j_0+1}^{\infty} a_j p^j \right| = p^{-j_0},$$

since $\left| \sum_{j=j_0+1}^{\infty} a_j p^j \right| \leq \max_{j > j_0} |a_j| p^{-j} < p^{-j_0}$ (use continuity of $| \cdot |_p$). So it follows $\psi(x) \neq 0$ and therefore ψ has trivial kernel, thus is injective.

Surjectivity: We define an absolute value on Z by

$$|z| = |\psi(z)|_p.$$

We claim that Z is complete in this absolute value. Let (z_j) be a Cauchy-sequence in Z. For each $k \in \mathbb{N}$ there exists a $j_0(k) \in \mathbb{N}$, such that for all $i, j \geq j_0(k)$ one has

$|z_i - z_j| \leq p^{-k}$, which means, that $\psi(z_i) - \psi(z_j) \in p^k \mathbb{Z}_p$, so $z_i - z_j \in p^k Z$. We conclude, that the coefficients of the power series z_i and z_j coincide up to the index $k-1$. Therefore there are coefficients a_ν for $\nu = 0, 1, 2, \ldots$, such that for every $k \in \mathbb{N}$ and every $j \geq j_0(k)$ one has $z_j \equiv \sum_{\nu=0}^{k-1} a_\nu p^\nu \mod p^k Z$. Set $z = \sum_{\nu=0}^{\infty} a_\nu p^\nu \in Z$. The sequence (z_j) converges to z, so Z is complete. To finish the proof, it suffices to show that $\psi(Z)$ contains a dense subset of $\mathbb{Z}_p$. Such a set is given by the set of all rational numbers in $\mathbb{Z}_p$, i.e., the set of all $q = \pm p^k \frac{m}{n}$ where $k \geq 0$ and m, n coprime to p. As Z is a ring, it suffices to show that $\frac{1}{n} \in Z$, if $n \in \mathbb{N}$ is coprime to p. But for n coprime to p the zeroth coefficient of the p-adic expansion is non-zero and therefore n is invertible in Z. $\square$

We now can identify $\mathbb{Z}_p$ with the set of all power series in p. As $\mathbb{Z}_p$ equals the set of all $z \in \mathbb{Q}_p$ with $|z| \leq 1$, the set $p^{-j}\mathbb{Z}_p$ is the set of all $z \in \mathbb{Q}_p$ with $|z| \leq p^j$. Therefore,

$$\mathbb{Q}_p = \bigcup_{j=0}^{\infty} p^{-j}\mathbb{Z}_p.$$

So we can write $\mathbb{Q}_p$ as the set of all Laurent-series in p with only finitely many negative entries, i.e.,

$$\mathbb{Q}_p = \left\{ \sum_{j=-N}^{\infty} a_j p^j : N \in \mathbb{N}, \, 0 \leq a_j < p \right\}.$$

This also implies that $\mathbb{Q}_p$ is uncountable. In particular, $\mathbb{Q} \neq \mathbb{Q}_p$ and therefore $\mathbb{Q}$ is not complete in the p-adic metric.

Proposition 13.1.8. (a) *The topological spaces $\mathbb{Q}_p$ and $\mathbb{Q}_p^\times$ are locally compact and totally disconnected. So together with Proposition 13.1.5 this implies that these are totally disconnected LCA-groups.*

(b) *The open compact subgroups $p^n\mathbb{Z}_p$, $n \in \mathbb{N}$ form a basis of the unit-neighbourhoods of the additive group $(\mathbb{Q}_p, +)$.*

(c) *The compact open subgroups $1 + p^n\mathbb{Z}_p$, $n \in \mathbb{N}$ form a basis of the unit-neighbourhoods of the multiplictive group $(\mathbb{Q}_p^\times, \times)$.*

Proof It suffices to show (b) and (c), for these imply (a) as well. The subgroup $p^n\mathbb{Z}_p$ coincides with the open ball $B_r(0)$ for any $r > 0$ with $p^{-n} < r < p^{-n+1}$, so these sets clearly form a neighborhood basis of zero. Likewise, the set $1 + p^n\mathbb{Z}_p$ equals the open Ball $B_r(1)$ around 1 of radius $r > 0$, if $p^{-n} < r < p^{-n+1}$. Hence the claim follows as soon as we have shown that $\mathbb{Z}_p$, and hence $p^n\mathbb{Z}_p$, is compact. But if (x_n) is a sequence in $\mathbb{Z}_p$ and if we write each x_n as a power series $\sum_{j=0}^{\infty} a_j^n p^j$ with $a_j^n \in \{0, \ldots, p-1\}$ we may pass inductively to subsequences $(x_{n_k}^l)$ of (x_n) such that the first l coefficients $a_1, \ldots, a_l$ of all elements in the l-th subsequence agree. It is then easy to check that the diagonal subsequence $(x_{n_l}^l)$ of (x_n) converges in $\mathbb{Z}_p$. $\square$

p-Adic Numbers as Limits

Fix a prime p and let $m, n \in \mathbb{N}$ with $m \geq n$. Then the natural projection

$$\pi_n^m : \mathbb{Z}/p^m\mathbb{Z} \to \mathbb{Z}/p^n\mathbb{Z}$$

is a ring homomorphism. The family $\left(\pi_m^n\right)_{m,n}$ satisfies the axioms of a projective system of rings as in Sect. 1.8.

Proposition 13.1.9 *The ring $\mathbb{Z}_p$ is canonically isomorphic with the projective limit of the $\mathbb{Z}/p^n\mathbb{Z}$.*

Proof If we view $\mathbb{Z}_p$ as ring of power series, we get natural projections $\mathbb{Z}_p \to \mathbb{Z}/p^n\mathbb{Z}$ by cutting off a power series beyond its n-th entry. These projections are compatible with the projections $\pi_m^n : \mathbb{Z}/p^m\mathbb{Z} \to \mathbb{Z}/p^n\mathbb{Z}$, so these maps fit together to give a ring homomorphism

$$\mathbb{Z}_p \to \varprojlim_n \mathbb{Z}/p^n\mathbb{Z}.$$

Interpreting the right hand side as set of compatible elements in the product $\prod_n \mathbb{Z}/p^n\mathbb{Z}$ easily shows that this map is a bijection. $\qquad\square$

13.2   Haar Measures on p-adic Numbers

The absolute value $|\cdot|_p$ defines a metric, which yields a topology on $\mathbb{Q}_p$. We showed above that with this topology the groups $(\mathbb{Q}_p, +)$ and $(\mathbb{Q}_p^\times, \cdot)$ are locally compact abelian groups. We now determine their Haar measures.

Note that the group of units $\mathbb{Z}_p^\times$ in $\mathbb{Z}_p$ is exactly the set of all $x \in \mathbb{Q}_p$, which satisfy $|x|_p = 1$.

Let μ be the Haar measure on the group $(\mathbb{Q}_p, +)$, which gives the compact open subgroup $\mathbb{Z}_p$ the volume 1, so $\mu(\mathbb{Z}_p) = 1$. Invariance of μ means $\mu(x + A) = \mu(A)$ for every measurable $A \subset \mathbb{Q}_p$ and every $x \in \mathbb{Q}_p$.

Lemma 13.2.1 *For every measurable subset $A \subset \mathbb{Q}_p$ and every $x \in \mathbb{Q}_p$ one has $\mu(xA) = |x|_p \mu(A)$. In particular, for every integrable function f and $x \neq 0$ one has*

$$\int_{\mathbb{Q}_p} f\left(x^{-1}y\right) \, d\mu(y) = |x|_p \int_{\mathbb{Q}_p} f(y) \, d\mu(y).$$

Proof Let $x \in \mathbb{Q}_p \setminus \{0\}$. The measure μ_x, defined by $\mu_x(A) = \mu(xA)$, is a Haar measure again, as is easily seen. By uniqueness of Haar measures, there exists some $M(x) > 0$ such that $\mu_x = M(x)\mu$. We show that $M(x) = |x|_p$. It suffices to show

$\mu(x\mathbb{Z}_p) = |x|_p$. Assume that $|x|_p = p^{-k}$. Then $x = p^k y$ for some $y \in \mathbb{Z}_p^\times$, and so $x\mathbb{Z}_p = p^k \mathbb{Z}_p$. Therefore it suffices to show $\mu\left(p^k\mathbb{Z}_p\right) = p^{-k}$. We start with the case $k \geq 0$. Then $[\mathbb{Z}_p : p^k\mathbb{Z}_p] = p^k$, so there is a disjoint decomposition of $\mathbb{Z}_p$, $\mathbb{Z}_p = \bigcup_{j=1}^{p^k} (x_j + p^k\mathbb{Z}_p)$. By invariance of Haar measure we have

$$1 = \mu(\mathbb{Z}_p) = \sum_{j=1}^{p^k} \mu\left(x_j + p^k\mathbb{Z}_p\right) = p^k \mu\left(p^k\mathbb{Z}_p\right),$$

which implies the claim. If $k < 0$, then one uses $\left[p^k\mathbb{Z}_p : \mathbb{Z}_p\right] = p^{-k}$ in an analogous way. $\qquad\square$

For simplification, we write integration according to a Haar measure as dx, so

$$\int_{\mathbb{Q}_p} f(x)\, d\mu(x) = \int_{\mathbb{Q}_p} f(x)\, dx.$$

Proposition 13.2.2 *The measure $\frac{dx}{|x|_p}$ is a Haar measure of the multiplicative group $\mathbb{Q}_p^\times$.*

Proof Let $f \in C_c(\mathbb{Q}_p^\times)$ and $y \in \mathbb{Q}_p^\times$. Then one has

$$\int_{\mathbb{Q}_p^\times} f\left(y^{-1}x\right) \frac{dx}{|x|_p} = |y|_p^{-1} \int_{\mathbb{Q}_p^\times} f\left(y^{-1}x\right) \frac{1}{|y^{-1}x|_p}\, dx = \int_{\mathbb{Q}_p^\times} f(x) \frac{dx}{|x|_p}$$

by Lemma 13.2.1. $\qquad\square$

The subgroup $\mathbb{Z}_p^\times$ of $\mathbb{Q}_p^\times$ is the kernel of the group homomorphism $\mathbb{Q}_p^\times \to \mathbb{Z}$; $x \mapsto \frac{\log(|x|_p)}{\log p}$. Hence $\mathbb{Q}_p^\times$ can be written as disjoint union: $\mathbb{Q}_p^\times = \bigcup_{k\in\mathbb{Z}} p^k\mathbb{Z}_p^\times$. One has $\mathrm{vol}_{\frac{dx}{|x|_p}}\left(p^k\mathbb{Z}_p^\times\right) = \mathrm{vol}_{\frac{dx}{|x|_p}}\left(\mathbb{Z}_p^\times\right)$. It is therefore of interest, to compute the measure $\mathrm{vol}_{\frac{dx}{|x|_p}}\left(\mathbb{Z}_p^\times\right)$. One has $\mathrm{vol}_{\frac{dx}{|x|_p}}\left(\mathbb{Z}_p^\times\right) = \int_{\mathbb{Z}_p^\times} \frac{dx}{|x|_p} = \int_{\mathbb{Z}_p^\times} dx = \mathrm{vol}_{dx}\left(\mathbb{Z}_p^\times\right)$. Consider the power series representation of $\mathbb{Z}_p$ and order the elements of $\mathbb{Z}_p^\times$ by their first coefficient. We get a disjoint decomposition

$$\mathbb{Z}_p^\times = \bigcup_{\substack{a \bmod p \\ a\neq 0 \bmod p}} (a + p\mathbb{Z}_p).$$

This means that the subgroup $1 + p\mathbb{Z}_p$ of $\mathbb{Z}_p^\times$ has index $p - 1$, so

$$\mathrm{vol}_{dx}\left(\mathbb{Z}_p^\times\right) = (p-1)\mathrm{vol}_{dx}\left(p\mathbb{Z}_p\right) = \frac{p-1}{p}.$$

We define the *normalized multiplicative Haar measure* on $\mathbb{Q}_p$ by

$$d^\times x = \frac{p}{p-1}\frac{dx}{|x|_p}.$$

This Haar measure is determined by the property that the volume of the compact open subgroup $\mathbb{Z}_p^\times$ is one.

Self-duality

The group $\mathbb{R}$ is self-dual in the way that there is a character $\chi_0(x) = e^{2\pi i x}$ such that every character χ can be written as $\chi(x) = \chi_0(ax)$ for a unique $a \in \mathbb{R}$. Actually, the choice of χ_0 was arbitrary, so, given any non-trivial character ω, any character can uniquely be written as $x \mapsto \omega(ax)$. We will now find that $\mathbb{Q}_p$ is self-dual as well.

Theorem 13.2.3 (Self-duality of $\mathbb{Q}_p$). *Fix any non-trivial character $\omega \in \widehat{\mathbb{Q}_p}$. Then the map $\Phi : \mathbb{Q}_p \rightarrow \widehat{\mathbb{Q}_p}$, given by $\Phi(a) = \omega_a$, where $\omega_a(x) = \omega(ax)$, is an isomorphism of LCA groups.*

Proof The computation

$$\omega_{a+b}(x) = \omega(ax + bx) = \omega(ax)\omega(bx) = \omega_a(x)\omega_b(x)$$

shows that Φ is a group homomorphism. For injectivity, assume $\Phi(a) = 1$, then $\omega(ax) = 1$ for every $x \in \mathbb{Q}_p$ and as ω is non-trivial, this implies $a = 0$.

For surjectivity, we construct a standard character $\chi_0 : \mathbb{Q}_p \rightarrow \mathbb{T}$ and show that $\chi(x) = \chi_0(ax)$ for some $a \in \mathbb{Q}_p$. Since the same holds for ω we get $\omega(x) = \chi_0(bx)$ and as ω is non-trivial, it follows $b \neq 0$. Then we infer $\chi(x) = \chi_0(ax) = \chi_0(bb^{-1}ax) = \omega(b^{-1}ax)$. We use the power series representation of elements of $\mathbb{Q}_p$ to define χ_0 as follows

$$\chi_0\left(\sum_{k=-N}^{\infty} a_k p^k\right) = e^{2\pi i \sum_{k=-N}^{-1} a_k p^k}.$$

This is easily seen to be a character with $\chi_0(\mathbb{Z}_p) = 1$ and $\chi_0\left(p^{-N}\right) = e^{2\pi i p^{-N}}$. Let now χ be any character. By continuity, there exists $k \in \mathbb{Z}$ such that the open subgroup $p^k\mathbb{Z}_p$ is mapped into the open unit-neighborhood $\{\mathrm{Re}(z) > 0\}$ in $\mathbb{T}$. The latter set contains only one subgroup of $\mathbb{T}$, the trivial group. So we get $\chi\left(p^k\mathbb{Z}_p\right) = 1$. Replacing $\chi(x)$ with $\chi(p^k x)$, we can assume $\chi(\mathbb{Z}_p) = 1$. Let $N \in \mathbb{N}$. Then we have $\chi\left(p^{-N}\right)^{p^N} = 1$, so there are uniquely determined coefficients $a_k \in \{0, \dots, p-1\}$, such that

$$\chi\left(p^{-N}\right) = e^{2\pi i \left(\sum_{k=0}^{N-1} a_k p^k\right) p^{-N}}.$$

Since $\chi(p^{-N}) = \chi\left(p^{-(N+1)}p\right) = \chi\left(p^{-(N+1)}\right)^p$, these coefficients do not depend on N, so there is a number $a = \sum_{k=0}^{\infty} a_k p^k$ in $\mathbb{Z}_p$ with $\chi(p^{-N}) = \chi_0(ap^{-N})$ for every $N \in \mathbb{N}$. We apply this to varying N to conclude

$$\chi\left(\sum_{k=-N}^{\infty} a_k p^k\right) = \chi\left(\sum_{k=-N}^{-1} a_k p^k\right) = \prod_{k=-N}^{-1} \chi(a_k p^k) = \prod_{k=-N}^{-1} \chi(p^k)^{a_k}$$

$$= \prod_{k=-N}^{-1} \chi_0(ap^k)^{a_k} = \prod_{k=-N}^{-1} \chi_0(aa_k p^k) = \chi_0\left(a \sum_{k=-N}^{\infty} a_k p^k\right).$$

To establish continuity of ϕ, recall that the topology of $\widehat{\mathbb{Q}_p}$ is the topology of the structure space of $L^1(\mathbb{Q}_p)$. So it suffices to show that the map $a \mapsto \hat{f}(\omega_a)$ is continuous for every $f \in L^1(\mathbb{Q}_p)$. This map, however, is $\hat{f}(\omega_a) = \int_{\mathbb{Q}_p} f(x)\overline{\omega(ax)}\,dx$, and is seen to be continuous by means of the Theorem on Dominated Convergence, as for a sequence $a_j \to a$ the sequence $\hat{f}(\omega_{a_j})$ converges dominatedly to $\hat{f}(\omega_a)$.

The continuity of the inverse map follows from the Open Mapping Theorem 4.2.10.

$\square$

13.3 Adeles and Ideles

In this section, we compose all the completions $\mathbb{Q}_p$ to a big ring, called the adele ring, which contains number theoretical information on all primes. The naive idea would be to simply take the product of all $\mathbb{Q}_p$. This, however, will not give a locally compact space, as we show in the first section. The construction has to be refined to the so-called restricted product.

Restricted Products

By the Theorem of Tychonov, direct products of compact spaces are compact. For "locally compact", this does not hold in general, as Lemma 1.8.10 shows. Let $(X_i)_{i \in I}$ be a family of locally compact spaces and for each $i \in I$ let there be given a compact open subset $K_i \subset X_i$. Define the *restricted product* as

$$X = \widehat{\prod_{i \in I}}^{K_i} X_i := \left\{ x \in \prod_{i \in I} X_i : x_i \in K_i \text{ for almost all } i \in I \right\}$$

$$= \bigcup_{\substack{E \subset I \\ \text{finite}}} \left\{ \prod_{i \in E} X_i \times \prod_{i \notin E} K_i. \right\}$$

If it is clear, which sets K_i to take, one leaves them out of the notation and simply writes $X = \widehat{\prod}_{i \in I} X_i$.

On the restricted product we introduce the *restricted product topology* as follows. A *restricted open rectangle* is a subset of the restricted product of the form $\prod_{i \in E} U_i \times \prod_{i \notin E} K_i$, where $E \subset I$ is a finite subset and for each $i \in E$ the set $U_i \subset X_i$ is an arbitrary open subset of X_i. A subset $A \subset \widehat{\prod}_{i \in I} X_i$ is called open, if it can be written as a union of restricted open rectangles. Note that the intersection of two restricted open rectangles is again a restricted open rectangle, since the sets K_i have been assumed to be open.

Lemma 13.3.1 (a) *If I is finite, then* $\prod_i X_i = \widehat{\prod}_i X_i$ *and the restricted product topology is the usual product topology.*

(b) *For every disjoint decomposition of the index set* $I = A \cup B$ *one has a homeomorphism*

$$\widehat{\prod}_{i \in I} X_i \cong \left(\widehat{\prod}_{i \in A} X_i \right) \times \left(\widehat{\prod}_{i \in B} X_i \right).$$

(c) *The inclusion map* $\widehat{\prod}_i X_i \hookrightarrow \prod_i X_i$ *is continuous, but the restricted product topology only equals the subspace topology, if* $X_i = K_i$ *for almost all* $i \in I$.

(d) *If all the spaces* X_i *are locally compact, then so is* $X = \widehat{\prod}_i X_i$.

Proof (a) is trivial. For (b) note that both sides of the equation describe the same set. The definition of the restricted product topology implies that the left hand side carries the product topology of the two factors on the right.

(c) For continuity we have to show that the pre-image of a set of the form $\prod_{i \in E} U_i \times \prod_{i \notin E} X_i$ is open in $\widehat{\prod}_i X_i$, where $E \subset I$ is a finite subset and every $U_i \subset X_i$ is open. This follows from (a) and (b). The second assertion is clear.

We finally show (d). Let $x \in X$. Then there exists a finite set $E \subset I$ such that $x_i \in K_i$, if $i \notin E$. For every $i \in E$ choose a compact neighborhood U_i of x_i. Then $\prod_{i \in E} U_i \times \prod_{i \notin E} K_i$ is a compact neighborhood of x, so X is locally compact. $\square$

Adeles

By a *place* of $\mathbb{Q}$ we either mean a prime number or ∞, the latter we call the *infinite place*. Write $p < \infty$, if p is a prime and $p \leq \infty$ if p is an arbitrary place. This manner of speaking comes from algebraic geometry, as these "places" behave in many ways like points on a curve. We write $\mathbb{Q}_\infty = \mathbb{R}$.

The set of *finite adeles* is the restricted product

$$\mathbb{A}_{\text{fin}} = \widehat{\prod}_{p < \infty}^{\mathbb{Z}_p} \mathbb{Q}_p.$$

The set of *adeles* is the set $\mathbb{A} = \mathbb{A}_{\text{fin}} \times \mathbb{R}$. We also write $\mathbb{A} = \widehat{\prod}_{p \leq \infty} \mathbb{Q}_p$, although this is not a restricted product, as there is no restriction at the infinite place. For an arbitrary set of places S we write $\mathbb{A}_S = \widehat{\prod}_{p \in S} \mathbb{Q}_p$ and $\mathbb{A}^S = \widehat{\prod}_{p \notin S} \mathbb{Q}_p$. Note that $\mathbb{A} = \mathbb{A}_S \times \mathbb{A}^S$.

Theorem 13.3.2

(a) *For every set of places S the ring* $\mathbb{A}_S$ *is a locally compact topological ring.*

(b) *The set $\mathbb{Q}$, embedded diagonally into $\mathbb{A}$, is a discrete subgroup and the quotient of abelian groups $\mathbb{A}/\mathbb{Q}$ is compact.*

(c) $\mathbb{Q}$ *is dense in* $\mathbb{A}_{\mathrm{fin}}$.

Proof The space $\mathbb{A}_S$ is locally compact by Lemma 13.3.1. For (a) we have to show that addition and multiplication are continuous maps from $\mathbb{A}_S \times \mathbb{A}_S$ to $\mathbb{A}_S$. We only show this for addition, as the proof for multiplication is analogous. Let $a, b \in \mathbb{A}_S$ and let U be an open neighborhood of $a + b$. We have to show that there are open neighborhoods V, W of a and b such that $V + W \subset U$. Choosing U smaller, we can assume that $U = \prod_{p \in E} U_p \times \prod_{p \in S \setminus E} \mathbb{Z}_p$ for a finite set $E \subset S$. For given $p \in E$ the addition is continuous on $\mathbb{Q}_p$, so there are open neighborhoods $V_p, W_p \subset \mathbb{Q}_p$ of a_p and b_p, such that $V_p + W_p \subset U_p$. Set $V = \prod_{p \in E} V_p \times \prod_{p \in S \setminus E} \mathbb{Z}_p$ and $W = \prod_{p \in E} W_p \times \prod_{p \in S \setminus E} \mathbb{Z}_p$. Then V and W are open neighborhoods of a and b, and one has $V + W \subset U$ as claimed.

For part (b) let $U = \left(-\frac{1}{2}, \frac{1}{2} \right) \times \prod_{p < \infty} \mathbb{Z}_p$. The set U is an open neighborhood of zero in $\mathbb{A}$. For $r \in \mathbb{Q} \cap U$ one has $|r|_p \le 1$ for every $p < \infty$ and therefore $r \in \mathbb{Z}$. Further one has $|r|_\infty < \frac{1}{2}$ and so $r = 0$. We have thus found an open neighborhood of zero with $U \cap \mathbb{Q} = \{0\}$. As $\mathbb{Q}$ is a subgroup of the additive group $\mathbb{A}$, it is discrete in $\mathbb{A}$. For compactness it suffices to show, that the compact set $K = [0, 1] \times \prod_{p < \infty} \mathbb{Z}_p$ contains a set of representatives of $\mathbb{A}/\mathbb{Q}$, because then the projection $P : K \to \mathbb{A}/\mathbb{Q}$ is surjective, so $\mathbb{A}/\mathbb{Q}$ is the continuous image of a compact set, hence compact.

So let $x \in \mathbb{A}$. There is a finite set E of places with $\infty \in E$, such that $p \notin E \Rightarrow x_p \in \mathbb{Z}_p$. For $p \in E$ with $p < \infty$ we write $x_p = \sum_{j=-N}^{\infty} a_j p^j$. Then

$$x_p - \underbrace{\sum_{j=-N}^{-1} a_j p^j}_{=r \in \mathbb{Q}} \in \mathbb{Z}_p.$$

For a prime $q \ne p$ one has $|r|_q = \left| \sum_{j=-N}^{-1} a_j p^j \right|_q \le \max\{|a_j p^j|_q\} \le 1$. We replace x by $x - r$ and thus reduce E to $E \setminus \{p\}$. Repeating this argument, we end up with $E = \{\infty\}$, so $x_p \in \mathbb{Z}_p$ for every prime number p. This means that $x \in \mathbb{R} \times \prod_{p < \infty} \mathbb{Z}_p$. Modulo $\mathbb{Z}$ one can move x to $[0, 1] \times \prod_{p < \infty} \mathbb{Z}_p = K$.

Note that the above argument implies in particular that $\mathbb{A}_{\mathrm{fin}} = \mathbb{Q} + \widehat{\mathbb{Z}}$ for $\widehat{\mathbb{Z}} = \prod_{p < \infty} \mathbb{Z}_p$. Hence, for (c) it suffices to show that $\mathbb{Z}$ is dense in $\widehat{\mathbb{Z}}$. We have to show that $\mathbb{Z}$ meets every open subset of $\widehat{\mathbb{Z}}$. Every such set is a union of sets of the form $U = \prod_{p \in E} B_p \times \prod_{p \notin E} \mathbb{Z}_p$, where E is a finite set of places and every B_p is an open ball in $\mathbb{Z}_p$. This means that B_p is of the form $B_p = n_p + p^{k_p} \mathbb{Z}_p$ for some $n_p \in \mathbb{Z}$ and some $k_p \in \mathbb{N}_0$. We have to show that there is $l \in \mathbb{Z}$, such that for every $p \in E$ one has $l \in n_p + p^{k_p} \mathbb{Z}_p$, or $l \equiv n_p \bmod p^{k_p}$. The existence of such l is a consequence of the Chinese Remainder Theorem. $\qquad\square$

The ring $\mathbb{A}$ is locally compact, so in particular a locally compact group with respect to addition, so there is an additive Haar measure dx on $\mathbb{A}$. To describe it, we need the following definition.

Definition A *simple function* f on $\mathbb{A}$ is a function of the form $f = \prod_{p \leq \infty} f_p$ with $f_p = \mathbf{1}_{\mathbb{Z}_p}$ for almost all p. Likewise, a *simple function* on $\mathbb{A}_{\mathrm{fin}}$ is a function of the form $f = \prod_{p < \infty} f_p$ with $f_p = \mathbf{1}_{\mathbb{Z}_p}$ for almost all p.

Theorem 13.3.3 *The Haar measure dx on $(\mathbb{A}, +)$ can be chosen such that for every integrable simple function $f = \prod_p f_p$ one has the product formula*

$$\int_{\mathbb{A}} f(x)\, dx = \prod_p \int_{\mathbb{Q}_p} f_p(x_p)\, dx_p.$$

The Haar measure dx_p on $\mathbb{Q}_p$ is normalized such that $\mathrm{vol}(\mathbb{Z}_p) = 1$ for $p < \infty$ and dx_∞ equals the Lebesgue measure. The product is alway finite, i.e., almost all factors are equal to 1,

This theorem also holds for $\mathbb{A}_S$ for an arbitrary set of places S. In the sequel, we will always use the normalization of the theorem.

Proof Since $\mathbb{A} = \mathbb{A}_{\mathrm{fin}} \times \mathbb{R}$, any Haar measure on $\mathbb{A}$ is a product of the Lebesgue measure and some Haar measure on $\mathbb{A}_{\mathrm{fin}}$. It therefore suffices to show that the Haar measure on $\mathbb{A}_{\mathrm{fin}}$ can be normalized in a way that for every simple function f on $\mathbb{A}_{\mathrm{fin}}$ one has

$$\int_{\mathbb{A}_{\mathrm{fin}}} f(x)\, dx = \prod_p \int_{\mathbb{Q}_p} f_p(x_p)\, dx_p.$$

Lemma 13.3.4 *Let $f \in C_c(\mathbb{A}_{\mathrm{fin}})$ be a continuous function with compact support. Then there is a compact subset $K \subseteq \mathbb{A}_{\mathrm{fin}}$ and a sequence of simple functions (f_n) on $\mathbb{A}_{\mathrm{fin}}$ with supports in K which converges uniformly to f.*

Proof Let L be the support of f. Since f is uniformly continuous, we find a neighborhood U_n of zero of the form $\prod_{p < \infty} B_p$ with $B_p = p^{k_p} \mathbb{Z}_p$ for all $p < \infty$ with $k_p \in \mathbb{Z}$ for all p and $k_p = 0$ for almost all p such that $|f(x + y) - f(x)| < \frac{1}{n}$ for all $x \in \mathbb{A}_{\mathrm{fin}}$ and $y \in U_n$. Then U_n is a compact open subgroup of $\mathbb{A}_{\mathrm{fin}}$, so L can be covered by a disjoint union of a finite number of translates of U_n, so $L \subset \bigsqcup_{i=1}^{l} (x_i + U_n)$ for suitable $x_i \in K$. Define $g_n(x) = f(x_i)$ if $x \in x_i + U_n$. Then $\mathrm{supp}\, g_n \subseteq \mathrm{supp} f + U_n$ and $\| f - g_n \|_{\mathbb{A}_{\mathrm{fin}}} \leq \frac{1}{n}$. Doing this construction for all n and taking care that $U_{n+1} \subseteq U_n$ for all n, we obtain the desired sequence (g_n).   □

Proof of Theorem 13.3.3 If $f \in C_c(A_{\mathrm{fin}})$ choose a sequence (f_n) of simple functions as in the lemma. Then it is easy to check that $\left(\int_{\mathbb{A}_{\mathrm{fin}}} f_n(x)\, dx \right)$ is a Cauchy sequence in $\mathbb{C}$, and that the limit

$$\int_{\mathbb{A}} f(x)\,dx := \lim_n \int_{\mathbb{A}} f_n(x)\,dx$$

does not depend on the chosen sequence. Then $\int_{\mathbb{A}} : C_c(\mathbb{A}) \to \mathbb{C}$ is a positive Radon integral which is left invariant, since it is left invariant on the set of simple functions. □

We will finally show that $\mathbb{A}$ is self-dual as well. For this let χ be a character of the LCA-group $\mathbb{A}$. For $p \le \infty$, we define the character χ_p of $\mathbb{Q}_p$ as the composition $\mathbb{Q}_p \hookrightarrow \mathbb{A} \xrightarrow{\chi} \mathbb{T}$. If p is a prime, we say that χ_p is *unramified*, if $\chi_p(\mathbb{Z}_p) = 1$.

Lemma 13.3.5 *For almost all p, the character χ_p is unramified. For an adele a one has $\chi(a) = \prod_{p \le \infty} \chi_p(a_p)$, where the product is finite, i.e., almost all factors are equal to one.*

Proof As χ is continuous, there exists a unit-neighborhood U in $\mathbb{A}$ such that $\chi(U) \subset \{\mathrm{Re}(z) > 0\}$. Then U contains a restricted open rectangle, therefore U contains $\mathbb{Z}_p$ for almost all p. The image $\chi(\mathbb{Z}_p) = \chi_p(\mathbb{Z}_p)$ is a subgroup of $\mathbb{T}$ contained in $\{\mathrm{Re}(z) > 0\}$, hence trivial. So almost all χ_p are unramified. Finally, let $a \in \mathbb{A}$ and let S be a finite set of places outside which $a_p \in \mathbb{Z}_p$ and χ_p is unramified. This implies that outside S one has $\chi_p(a_p) = 1$. Let a_S be the product of all a_p with $p \in S$ and a^S the product of all a_p with $p \notin S$. Then $a = a_S a^S$ and we have $\chi(a^S) = 1$ as well as $\chi(a_S) = \prod_{p \in S} \chi_p(a_p)$ as the product is finite. □

Definition We say that a character ω is *nowhere trivial*, if $\omega_p \ne 1$ for every $p \le \infty$.

Theorem 13.3.6 (Self-duality of adeles). *There are characters ω of $\mathbb{A}$ which are nowhere trivial. For any such, the map $\Phi : \mathbb{A} \to \widehat{\mathbb{A}}$ given by $\Phi(a) = \omega_a$ with $\omega_a(x) = \omega(ax)$ is an isomorphism of locally compact groups.*

Proof At each $p \le \infty$, fix a non-trivial character ω_p in a way that ω_p is unramified for almost all p. One can, for instance, choose ω_p to be the standard character used in the proof of Theorem 13.2.3, which was called χ_0 there. It is easy to verify that the prescription

$$\omega(a) = \prod_{p \le \infty} \omega_p(a_p)$$

defines a nowhere trivial character.

As in the case of $\mathbb{Q}_p$ in Theorem 13.2.3, we observe that Φ is a group homomorphism. For injectivity, let $a \in \mathbb{A}$ with $\Phi(a) = 1$. Then $\omega(ax) = 1$ for every $x \in \mathbb{A}$, which implies $\omega_p(a_p x_p) = 1$ for every $x_p \in \mathbb{Q}_p$, hence $a_p = 0$ and so $a = 0$.

To show surjectivity, let χ be a character. By the corresponding local result, for each $p \le \infty$, there exists a unique $a_p \in \mathbb{Q}_p$ with $\chi_p(x_p) = \omega_p(a_p x_p)$ for every $x_p \in \mathbb{Q}_p$. At places p, where χ and ω are both unramified, we get $a_p \in \mathbb{Z}_p$. Hence the a_p are the coordinates of an adele a and we have $\chi(x) = \omega(ax)$ for all $x \in \mathbb{A}$.

Continuity and openness follows exactly as in the proof of the corresponding result for $\mathbb{Q}_p$ in Theorem 13.2.3.   □

The ring of adeles $\mathbb{A}$ can be used to describe the dual group $\widehat{\mathbb{Q}}$ of the discrete additive group $\mathbb{Q}$. For this recall that $\mathbb{Q}$ imbeds diagonally into $\mathbb{A}$ as a discrete subgroup. Thus we obtain a short exact sequence

$$0 \rightarrow \mathbb{Q} \overset{\iota}{\rightarrow} \mathbb{A} \overset{q}{\rightarrow} \mathbb{A}/\mathbb{Q} \rightarrow 0$$

which dualizes to the short exact sequence

$$0 \rightarrow \mathbb{Q}^\perp \rightarrow \widehat{\mathbb{A}} \overset{\text{res}}{\rightarrow} \widehat{\mathbb{Q}} \rightarrow 0$$

For each $p < \infty$ let $e_p : \mathbb{Q}_p \rightarrow \mathbb{T}$ denote the standard character given by

$$e_p \left(\sum_{k=-N}^{\infty} a_k p^k \right) = e^{2\pi i \sum_{k=-N}^{-1} a_k p^k}$$

and let $e_\infty : \mathbb{R} \rightarrow \mathbb{T}$ be the character $e_\infty(x) = e^{-2\pi i x}$. Then the character $\omega = \prod_{p \leq \infty} e_p : \mathbb{A} \rightarrow \mathbb{T}$ is nowhere trivial and by Theorem 13.3.6 we have an isomorphism $\mathbb{A} \cong \widehat{\mathbb{A}}$ given by $a \mapsto \omega_a$ with $\omega_a(x) = \omega(xa)$. If we compose this with the restriction map res $: \widehat{\mathbb{A}} \rightarrow \widehat{\mathbb{Q}}$ we obtain a surjective homomorphism $\phi : \mathbb{A} \rightarrow \widehat{\mathbb{Q}}$ by $a \mapsto \omega_a|_{\mathbb{Q}}$.

Theorem 13.3.7 *The kernel* $\ker \phi \subseteq \mathbb{A}$ *is precisely the image of* $\mathbb{Q}$ *under the diagonal embedding. Therefore* ϕ *factors through an isomorphism of topological groups* $\mathbb{A}/\mathbb{Q} \cong \widehat{\mathbb{Q}}$ *given by* $a + \mathbb{Q} \mapsto \omega_a|_{\mathbb{Q}}$.

Proof An element $a \in \mathbb{A}$ lies in the kernel of ϕ if and only if $a\mathbb{Q} \subseteq \ker \omega$ with $\omega = \prod_{p \leq \infty} e_p$ as above. We first show that $\mathbb{Q} \subseteq \ker \omega$, which then implies that $\mathbb{Q} \subseteq \ker \phi$. For this let $x \in \mathbb{Q}$. Let E be the finite set of primes p such that $|x|_p > 1$ and for $p \in E$ let $x_p = \sum_{k=-N}^{\infty} a_k p^k$ denote the p-adic expansion of x and let $r_p := \sum_{k=-N}^{-1} a_k p^k \in \mathbb{Q}$. Then $|r_p|_q \leq \max_{k \leq -1} |a_k p^k|_q \leq 1$ for all $q \neq p$ and therefore $r_p \in \ker e_q$ for all primes $q \neq p$. It follows that

$$\omega(x - r_p) = \omega(x) e_\infty (-r_p) e_p (-r_p) = \omega(x) e^{-2\pi i r_p} e^{2\pi i r_p} = \omega(x).$$

Thus, replacing x by $x + \sum_{p \in E} r_p$, we may assume without loss of generality that $|x_p|_p \leq 1$ for all $p < \infty$. But this implies that $x \in \mathbb{Z}$, hence $x_p \in \ker e_p$ for all $p \leq \infty$ and $\omega(x) = 1$.

Assume now that $a \in \mathbb{A}$ such that $\omega(ax) = 1$ for all $x \in \mathbb{Q}$. We need to show that $a \in \mathbb{Q}$. Let E be the finite set of primes p with $|a_p|_p = p^{k_p} > 1$. By passing from a to $a' = a \cdot \prod_{p \in E} p^{k_p}$ if necessary we may assume that $E = \emptyset$, hence $|a_p|_p \leq 1$ for all $p < \infty$. It follows that $a_p \in \ker e_p$ for all $p \leq \infty$ and $1 = \omega(a) = e_\infty(a_\infty)$, hence $a_\infty \in \mathbb{Z}$. Writing $a_\infty = \pm p_1^{k_1} \cdots p_l^{k_l}$ with $k_1, \ldots, k_l \geq 0$ and after passing to

$a'' = a \cdot (\pm p_1^{-k_1} \cdots p_l^{-k_l})$ we may even assume that $a_\infty = 1$. (Note that multiplication of a with $p_i^{-k_i}$ only alters the norm of a_{p_i}. But since $1 = e_\infty(a_\infty \cdot p_i^{-k_i})$ and $1 = \omega\left(a \cdot p_i^{-k_i}\right) = e_\infty\left(a_\infty \cdot p_i^{-k_i}\right) \cdot e_{p_i}\left(a_p \cdot p_i^{-k_i}\right)$ we still have $|a_{p_i} \cdot p_i^{-k_i}|_{p_i} \le 1$, hence $|a_p''|_p \le 1$ for all $p < \infty$.)

After these reductions we need to show that $a_p = 1$ for all $p < \infty$. To see this let $a_p = \sum_{k=0}^\infty b_k p^k$. We need to show that $b_0 = 1$ and $b_k = 0$ for all $k > 0$. To see this we consider for all $l \in \mathbb{N}$

$$1 = \omega\left(a \cdot p^{-l}\right) = e_\infty\left(p^{-k}\right) e_p\left(\sum_{k=0}^l b_k p^{k-l}\right)$$

$$= \exp\left(\frac{2\pi i}{p^l}\left((b_0 - 1) + \sum_{k=1}^{l-1} b_k p^k\right)\right)$$

which is only possible if $b_0 = 1$ and $b_k = 0$ for all $1 \le k < l$. Since l is arbitrary, the result follows. $\qquad\square$

Ideles

The group $\mathbb{A}^\times$ of invertible elements of the adele ring $\mathbb{A}$ can be described as follows

$$\mathbb{A}^\times = \left\{a \in \mathbb{A} : \begin{array}{l} a_p \ne 0 \, \forall p \le \infty \\ |a_p|_p = 1 \text{ for almost all } p \end{array}\right\}.$$

Equipping $\mathbb{A}^\times$ with the subspace topology makes the multiplication continuous, but not the map $x \mapsto x^{-1}$. In order to make $\mathbb{A}^\times$ a topological group, we need more open sets. We have to insist, that with each open set U, the set $U^{-1} = \{u^{-1} : u \in U\}$ is open as well. The topology of $\mathbb{A}$ is generated by all sets of the form $\prod_{p \in E} U_p \times \prod_{p \notin E} \mathbb{Z}_p$, where E is a finite set of places and U_p is open in $\mathbb{Q}_p$ for every $p \in E$. The subspace topology of $\mathbb{A}^\times$ therefore is generated by all sets of the form

$$U = \left\{a \in \mathbb{A}^\times : \begin{array}{l} a_p \in U_p, \, p \in E \\ |a_p| \le 1, p \notin E \end{array}\right\},$$

where we can insist, that every U_p lies in $\mathbb{Q}_p^\times$. So we have to ask that sets of the form

$$U^{-1} = \left\{a \in \mathbb{A}^\times : \begin{array}{l} a_p^{-1} \in U_p, \, p \in E \\ |a_p| \ge 1, p \notin E \end{array}\right\}$$

be open as well. This implies that the intersection of sets of the form U and another of the form $(U')^{-1}$ be open. Such intersections are of the form

$$W = \left\{a \in \mathbb{A}^\times : \begin{array}{l} a_p \in W_p, \, p \in E \\ |a_p| = 1, p \notin E \end{array}\right\},$$

where W_p is any open subset of $\mathbb{Q}_p^\times$. On the other hand, sets of the form U or U^{-1} above can be written as unions of sets of the form W.

Lemma 13.3.8 *The coarsest topology on $\mathbb{A}^\times$, which contains the subspace topology of $\mathbb{A}$ and makes $\mathbb{A}^\times$ a topological group is the topology generated by all sets of the form W above with W_p any open set in $\mathbb{Q}_p^\times$. This topology is a restricted product topology, i.e., one can write $\mathbb{A}^\times$ as restricted product*

$$\mathbb{A}^\times = \left(\widehat{\prod_{p<\infty}}^{\mathbb{Z}_p^\times} \mathbb{Q}_p^\times\right) \times \mathbb{R}^\times.$$

With this topology, $\mathbb{A}^\times$ is a locally compact group, called the idele group of $\mathbb{Q}$. The elements are referred to as ideles.

Proof This is clear by what we have said above. $\square$

Definition The *absolute value of an idele* $a \in \mathbb{A}^\times$ is defined as $|a| = \prod_p |a_p|_p$. This product is well defined, since almost all factors are equal to 1. We extend the definition to all of $\mathbb{A}$ by setting $|a| = 0$, if $a \in \mathbb{A} \setminus \mathbb{A}^\times$. Note that the identity $|a| = \prod_p |a_p|_p$ also holds in this case, if one interprets the product as $(|a_\infty|_\infty \lim_{N\to\infty} \prod_{p\leq N} |a_p|_p)$. Let

$$\mathbb{A}^1 = \{a \in \mathbb{A}^\times : |a| = 1\}.$$

Proposition 13.1.4 says that $\mathbb{Q}^\times \subset \mathbb{A}^1$. Recall that we write

$$\widehat{\mathbb{Z}} = \prod_{p<\infty} \mathbb{Z}_p.$$

Then $\widehat{\mathbb{Z}}$ is a compact subring of $\mathbb{A}_{\mathrm{fin}}$. Its unit group is $\widehat{\mathbb{Z}}^\times = \prod_{p<\infty} \mathbb{Z}_p^\times$.

Theorem 13.3.9 *The subgroup $\mathbb{Q}^\times$ is discrete in $\mathbb{A}^\times$, it lies in the closed subgroup $\mathbb{A}^1$, and the quotient $\mathbb{A}^1/\mathbb{Q}^\times$ is compact. More precisely, there is a canonical isomorphism*

$$\mathbb{A}^1/\mathbb{Q}^\times \cong \widehat{\mathbb{Z}}^\times.$$

The absolute value induces an isomorphism of topological groups: $\mathbb{A}^\times \cong \mathbb{A}^1 \times (0, \infty)$ given by $x \mapsto (\tilde{x}, |x|_\infty)$, where $\tilde{x} \in \mathbb{A}^1$ is defined by

$$\tilde{x}_p = \begin{cases} x_p & \text{if } p < \infty, \\ \frac{x_\infty}{|x|} & \text{if } p = \infty. \end{cases}$$

Further one has $\mathbb{A}^1 \cong \mathbb{A}_{\mathrm{fin}}^\times \times \{\pm 1\}$.

Proof Choose $0 < \varepsilon < 1$ and set $U = (1 - \varepsilon, 1 + \varepsilon) \times \prod_{p<\infty} \mathbb{Z}_p^\times$. Then U is an open unit neighborhood in $\mathbb{A}^\times$. With $r \in \mathbb{Q} \cap U$ we get $|r|_p = 1$ for every prime number p, so $r \in \mathbb{Z}$ and $r^{-1} \in \mathbb{Z}$. We have $r \in (1 - \varepsilon, 1 + \varepsilon)$, so $r = 1$.

Consider the map $\eta : \prod_p \mathbb{Z}_p^\times \to \mathbb{A}^1/\mathbb{Q}^\times$ given by $x \mapsto (x, 1)\mathbb{Q}^\times$. We claim that η is an isomorphism of topological groups. The map η is a group homomorphism, and since the map $\prod_p \mathbb{Z}_p^\times \hookrightarrow \mathbb{A}^\times$ is continuous, η is continuous. The inverse map is given by $x = (x_{\text{fin}}, x_\infty) \mapsto \frac{1}{x_\infty} x_{\text{fin}}$, where we note, that for $x \in \mathbb{A}^1$ one has $x_\infty \in \mathbb{Q}^\times$.

We leave it as an exercise to check that the map $x \mapsto (\tilde{x}, |x|_\infty)$ gives an isomorphism $\mathbb{A}^\times \cong \mathbb{A}^1 \times (0, \infty)$. Finally, the map $\phi : \mathbb{A}_{\text{fin}}^\times \times \{\pm 1\} \to \mathbb{A}^1$ given by $\phi(a_{\text{fin}}, \varepsilon) = (a_{\text{fin}}, \varepsilon |a_{\text{fin}}|^{-1})$ is easily seen to be an isomorphism. $\square$

Proposition 13.3.10 (a) *The set $\mathbb{A}_{\text{fin}}^\times$ is the disjoint union*

$$\mathbb{A}_{\text{fin}}^\times = \coprod_{q \in \mathbb{Q}_{>0}^\times} q\widehat{\mathbb{Z}}^\times.$$

The set $\widehat{\mathbb{Z}} \cap \mathbb{A}_{\text{fin}}^\times$ is the disjoint union

$$\widehat{\mathbb{Z}} \cap \mathbb{A}_{\text{fin}}^\times = \coprod_{k \in \mathbb{N}} k\widehat{\mathbb{Z}}^\times.$$

(b) *For every $s \in \mathbb{C}$ with $\mathrm{Re}(s) > 1$ the integral $\int_{\widehat{\mathbb{Z}}} |x|^s \, d^\times x$ converges absolutely and equals the Riemann zeta function $\zeta(s)$. Here $d^\times x$ is the uniquely determined Haar measure on $\mathbb{A}_{\text{fin}}^\times$, which gives the compact open subgroup $\widehat{\mathbb{Z}}^\times$ the measure 1. We consider this measure also as a measure on $\mathbb{A}_{\text{fin}}$, which is zero outside $\mathbb{A}_{\text{fin}}^\times$.*

Proof For given $x \in \mathbb{A}_{\text{fin}}^\times$ the absolute value $|x|$ lies in $\mathbb{Q}_{>0}^\times$. Consider the element $|x|x \in \mathbb{A}_{\text{fin}}^\times$. Let p be a prime number. One has $x_p = p^k u$ for some $k \in \mathbb{Z}$ and some $u \in \mathbb{Z}_p^\times$. So one has $|x| = p^{-k} r$, where $r \in \mathbb{Q}$ is coprime to p. We infer that $||x|x_p|_p = 1$, so $|x|x \in \widehat{\mathbb{Z}}^\times$. With $q = |x|^{-1}$ we have $x \in q\widehat{\mathbb{Z}}^\times$. If $x \in \widehat{\mathbb{Z}}$ we have $k_p \geq 0$ for all p, which implies that $q \in \mathbb{Z}$. This concludes the proof of (a).

We use (a) to show (b) as follows

$$\int_{\widehat{\mathbb{Z}}} |x|^s \, d^\times x = \sum_{k \in \mathbb{N}} \int_{k\widehat{\mathbb{Z}}^\times} |x|^s \, d^\times x$$

$$= \sum_{k \in \mathbb{N}} \int_{\widehat{\mathbb{Z}}^\times} |kx|^s \, d^\times x = \sum_{k \in \mathbb{N}} k^{-s} \underbrace{\int_{\widehat{\mathbb{Z}}^\times} |x|^s \, d^\times x}_{=1}.$$

The convergence follows from the convergence of the Dirichlet series $\zeta(s)$. $\square$

13.4 Exercises

Exercise 13.1 For $a \in \mathbb{Q}_p$ and $r > 0$ let $B_r(a)$ be the open ball $B_r(a) = \{x \in \mathbb{Q}_p : |a - x|_p < r\}$. Show:

(a) If $b \in B_r(a)$, then $B_r(a) = B_r(b)$.

(b) Two open balls are either disjoint or one is contained in the other.

Exercise 13.2 Show that there is a canonical ring isomorphism $\mathbb{Q}_p \cong \mathbb{Q} \otimes_\mathbb{Z} \mathbb{Z}_p$.

Exercise 13.3 Show that $\sum_{j=-N}^\infty a_j p^j \mapsto \sum_{j=-N}^\infty a_j p^{-j}$, $0 \le a_j < p$, defines a continuous map $\mathbb{Q}_p \to \mathbb{R}$. Is this a ring homomorphism? Describe its image.

Exercise 13.4 Let $\mathbb{T} = \{z \in \mathbb{C} : |z| = 1\}$ denote the circle group and let $\chi : \mathbb{Z}_p \to \mathbb{T}$ be a continuous group homomorphism, i.e., $\chi(a + b) = \chi(a)\chi(b)$.

Show that there exists $k \in \mathbb{N}$ with $\chi(p^k\mathbb{Z}_p) = 1$. It follows that χ factors through the finite group $\mathbb{Z}_p/p^k\mathbb{Z}_p \cong \mathbb{Z}/p^k\mathbb{Z}$, so the image of χ is finite.
(Hint: Let $U = \{z \in \mathbb{T} : \mathrm{Re}(z) > 0\}$. Then U is an open neighborhood of the unit, so $\chi^{-1}(U)$ is an open neighborhood of zero.)

Exercise 13.5 Let $e_p : \mathbb{Q}_p \to \mathbb{T}$ be defined by

$$e_p\left(\sum_{j=-N}^\infty a_j p^j\right) = \exp\left(2\pi i \sum_{j=-N}^{-1} a_j p^j\right),$$

where $a_j \in \mathbb{Z}$ with $0 \le a_j < p$. Show that e_p is a continuous group homomorphism.

Exercise 13.6 (For this exercise it helps to have some familiarity with number theory.) Let p be a prime number and let $\mathcal{O}$ be the polynomial ring $\mathbb{F}_p[t]$. As one can perform division with remainder, the ring $\mathcal{O}$ is a factorial principal domain. The prime ideals of $\mathcal{O}$ are the principal ideals of the form 0 or (η), where $\eta \ne 0$ is an irreducible polynomial in $\mathcal{O}$.

(a) For such η let $v_\eta : \mathcal{O} \to \mathbb{N}_0 \cup \{\infty\}$ be defined by $v_\eta(f) = \sup\{k : f \in (\eta^k)\}$. Show that $|f|_\eta = p^{-\deg(\eta)v_\eta(f)}$ defines an absolute value on the ring $\mathcal{O}$.

(b) Let $v_\infty(f) = -\deg(f)$. Show that $|f|_\infty = p^{-v_\infty(f)}$ is an absolute value.

(c) Prove the product formula $\prod_{\eta \le \infty} |f|_\eta = 1$.

Exercise 13.7 (a) Show that the family $(N\widehat{\mathbb{Z}})_{N \in \mathbb{N}}$ is a neighborhood basis of zero in $\mathbb{A}_{\mathrm{fin}}$. That is, show that every $N\widehat{\mathbb{Z}}$ is a neighborhood of zero and that every zero neighborhood contains a set of the form $N\widehat{\mathbb{Z}}$ for some N.

(b) Show that the sets of the form $(1 + N\widehat{\mathbb{Z}}) \cap \widehat{\mathbb{Z}}^\times$, $N \in \mathbb{N}$ are a neighborhood basis of the unit 1 in $\mathbb{A}_{\mathrm{fin}}^\times$.

Exercise 13.8 Let p be a prime number, $n \in \mathbb{N}$ and let dx be the additive Haar measure on $M_n(\mathbb{Q}_p)$, so

$$\int_{M_n(\mathbb{Q}_p)} f(x)\,dx = \int_{\mathbb{Q}_p} \cdots \int_{\mathbb{Q}_p} f(x_{i,j})\,dx_{1,1} \cdots dx_{n,n}.$$

(a) Show that $\frac{dx}{|\det x|^n}$ is a left- and right-invariant Haar measure on the group $GL_n(\mathbb{Q}_p)$. Conclude that the group $GL_n(\mathbb{Q}_p)$ is unimodular.

(b) Show that the group $GL_n(\mathbb{A})$ is unimodular.

Exercise 13.9 Let $n, N \in \mathbb{N}$ and let K_N be the set of all invertible $n \times n$ matrices g with entries in $\widehat{\mathbb{Z}}$ such that $g \equiv I \bmod N$. Show

- K_N is a compact open subgroup of $GL_n(\widehat{\mathbb{Z}})$,

- $K_N \subset K_d$ if $d|N$,

- the K_N form a neighborhood basis of the unit in $GL_n(\widehat{\mathbb{Z}})$.

Exercise 13.10 Let U be a compact open subgroup of the locally compact group G. Show that for every $g \in G$ the set UgU/U is finite.

Exercise 13.11 Let G be a totally disconnected locally compact group. For a compact open subgroup U and a compact set K let $L(U, K)$ be the set of all functions $f : G \to \mathbb{C}$ with

- $\operatorname{supp} f \subset K$ and

- $f(ux) = f(x)$ for every $x \in G$ and every $u \in U$.

Further let $R(U, K)$ be the set of all functions $f : G \to \mathbb{C}$ with

- $\operatorname{supp} f \subset K$ and

- $f(xu) = f(x)$ for every $x \in G$ and every $u \in U$.

Show that in general one has $L(U, K) \neq R(U, K)$, but

$$\bigcup_{U,K} L(U, K) = \bigcup_{U,K} R(U, K).$$

Appendix A
Topology

In the appendix, we have collected notions and results from topology, measure theory, and functional analysis, which are used in the body of the book. With a few exceptions, we have included proofs, so that the appendix can be read as an independent introduction into the named topics.

We will use standard notations of set theoretic topology. For the convenience of the reader, we will recall a few here. To name a reference, we give [Gaa64].

First recall the definition of a topology and a topological space. A *topology* on a set X is a system of subsets $\mathcal{O} \subset \mathcal{P}(X)$, which contains $\emptyset$ and X, and is closed under finite intersections and arbitrary unions. A *topological space* is a pair $(X, \mathcal{O})$ consisting of a set X and a topology $\mathcal{O}$ on X. The sets in $\mathcal{O}$ are called *open sets*; their complements are *closed sets*. For every set $A \subset X$ there is a smallest closed subset $\overline{A}$ that contains A, it is called the *closure* of A, and its existence is secured by the fact that the intersection of all closed sets that contain A is a closed set.

There are two immediate examples, firstly the set $\mathcal{P}(X)$ of all subsets of X is a topology on X for every set X, called the *discrete topology* on X. Secondly, every set X carries the *trivial topology* $\{\emptyset, X\}$. Our standard example of a topological space is the real line $\mathbb{R}$ where open sets are the unions of open intervals, or more generally a *metric space* (X, d) where the open sets are arbitrary unions of open balls $B_r(x) = \{y \in X : d(x, y) < r\}, r \geq 0$.

Let $x \in X$ be a point. Any open set $U \subset X$, which contains x, is called an *open neighborhood* of x. A *neighborhood* V of x is a subset of X, which contains an open neighborhood of x.

Lemma A.0.1 *Let A be a subset of the topological space X. A point $x \in X$ belongs to the closure of A if and only if $A \cap U \neq \emptyset$ for every neighborhood U of x.*

Proof The claim is equivalent to saying that $x \notin \overline{A}$ if and only if there exists a neighborhood U of x with $A \cap U = \emptyset$. We may assume U to be open.

A. Deitmar, S. Echterhoff, *Principles of Harmonic Analysis*, Universitext, DOI 10.1007/978-3-319-05792-7, © Springer International Publishing Switzerland 2014

If U is an open neighborhood of x with $A \cap U = \emptyset$, then A is a subset of the closed set $X \setminus U$, so $x \notin \overline{A}$. Conversely, assume $x \notin \overline{A}$. Then $U = X \setminus \overline{A}$ is an open neighborhood of x with $A \cap U = \emptyset$.   □

A.1 Generators and Countability

For a given system of subsets $\mathcal{E} \subset \mathcal{P}(X)$ there exists a smallest topology containing $\mathcal{E}$ given by

$$\mathcal{O}_{\mathcal{E}} = \bigcap_{\substack{\mathcal{O} \supset \mathcal{E} \\ \mathcal{O} \text{ topology}}} \mathcal{O}.$$

One calls $\mathcal{O}_{\mathcal{E}}$ the *topology generated by* $\mathcal{E}$.

Lemma A.1.1 *Let $\mathcal{E} \subset \mathcal{P}(X)$, and let $\mathcal{S} \subset \mathcal{P}(X)$ be the system of all sets*

$$A_1 \cap \cdots \cap A_n,$$

where $A_1, \ldots, A_n \in \mathcal{E}$. Next let $\mathcal{T}'$ be the system of all sets of the form

$$\bigcup_{i \in I} S_i,$$

where $S_i \in \mathcal{S}$ for every $i \in I$. Finally, let $\mathcal{T} = \mathcal{T}' \cup \{\emptyset, X\}$. Then $\mathcal{O}_{\mathcal{E}} = \mathcal{T}$.

Proof Every topology containing $\mathcal{E}$ must contain $\mathcal{S}$ and $\mathcal{T}$, so $\mathcal{T} \subset \mathcal{O}_{\mathcal{E}}$. On the other hand one verifies that $\mathcal{T}$ is a topology containing $\mathcal{E}$, so $\mathcal{T} \supset \mathcal{O}_{\mathcal{E}}$.   □

Definition For a point $x \in X$ a *neighborhood base* is a family $(U_i)_{i \in I}$ of neighborhoods of x such that every neighborhood of x contains one of them, i.e., for every neighborhood U there exists an index $i \in I$ with $U_i \subset U$. If moreover every U_i is open, then the family (U_i) is called an *open neighborhood base*. For example, an open neighborhood base of zero in $\mathbb{R}$ is given by the family of open intervals $(-1/n, 1/n)$, where the index n runs in $\mathbb{N}$.

A topological space X is called *first countable* if every point $x \in X$ possesses a countable neighborhood base.

Examples A.1.2

- Let (X, d) be a metric space. Then for any $x \in X$ the family of balls $(B_{1/n}(x))_{n \in \mathbb{N}}$ forms a neighborhood base for x. It follows that every metric space is first countable.

- A product $\prod_{i \in I} X_i$ is not first countable, provided all X_i have non-trivial topologies, and the index set I is uncountable.

Definition A *topology base* is a family $(U_i)_{i \in I}$ of open sets such that every open set $U \subset X$ can be written as a union of members U_i of the family. For example, the family of open intervals (a, b), where a and b are rational numbers with $a < b$ forms a topology base for $\mathbb{R}$.

A topological space X is called *second countable* if it possesses a countable topology base. Every second countable space is first countable. It is a direct consequence of Lemma A.1.1 above that a space is second countable if and only if its topology possesses a countable generating set.

A.2   Continuity

A map $f : X \to Y$ between topological spaces X and Y is called *continuous* if $f^{-1}(U)$ is open in X for every open set $U \subset Y$. This is equivalent to the condition that $f^{-1}(C)$ is closed in X for every closed $C \subset Y$.

From the definition it is clear that a composition $f \circ g$ is continuous if f and g are.

In this book we shall always equip the complex numbers $\mathbb{C}$ with the topology given by the metric $d(z, w) = |z - w|$. So continuity for a function $f : X \to \mathbb{C}$ is always understood with respect to this topology on $\mathbb{C}$. Similar holds for real functions $f : X \to \mathbb{R}$ on X.

A map $f : X \to Y$ between topological spaces X and Y is called *open* if $f(U) \subset Y$ is open in Y for all open sets U in X and f is called *closed* if $f(C)$ is closed in Y for all closed $C \subset X$. If f is bijective, then both conditions coincide, but in general an open map doesn't have to be closed and vice versa.

A bijective map $f : X \to Y$ is called a *homeomorphism* if it is continuous and open. It is clear that openness of f is equivalent to the continuity of the inverse map $f^{-1} : Y \to X$. In particular, $f^{-1} : Y \to X$ is also a homeomorphism. We say that X and Y are *homeomorphic* if a homeomorphism $f : X \to Y$ exists.

In topology one often identifies two topological spaces X and Y if they are homeomorphic, since they both carry the same topological information. But in Analysis it is usually necessary to keep track of the given homeomorphism if one wants identify two spaces, since other structures, like differentiability are not carried over in general by homeomorphisms.

Examples A.2.1

- Every nonempty open interval $(a, b) \subset \mathbb{R}$ is homeomorphic to the real line $\mathbb{R}$ when both are equipped with the usual topology.

- Every solid rectangle $[a, b] \times [c, d] \subset \mathbb{R}^2$ with $a < b, c < d$ is homeomorphic to the solid ball $\overline{B}_1(0) = \{(x, y) \in \mathbb{R}^2 : x^2 + y^2 \leq 1\}$. We leave it as an interesting exercise to construct explicit homeomorphisms.

A.3 Compact Spaces

Recall that a topological space X is called *compact* if every open covering contains a finite sub-cover. In other words, X is compact if for any family $(U_i)_{i \in I}$ of open sets in X, which satisfy $X = \bigcup_{i \in I} U_i$, there exists a finite subset $F \subset I$ with $X = \bigcup_{i \in F} U_i$. Switching to the complements, this can be reformulated to

Lemma A.3.1 *A topological space X is compact if and only if for every family $(A_i)_{i \in I}$ of closed sets in X with $\bigcap_{i \in F} A_i \neq \emptyset$ for every finite subset $F \subset I$, one has*

$$\bigcap_{i \in I} A_i \neq \emptyset.$$

This property of compact spaces is called the *finite intersection property*.

Proof This is a straightforward reformulation of the compactness property obtained by switching to complements. $\square$

We now state some important facts for compact spaces, which are used throughout this book without further reference.

Lemma A.3.2 *Let X be a topological space. Then*

(a) *If X is compact and $C \subset X$ is closed, then C is compact.*

(b) *If X is Hausdorff and $C \subset X$ is compact, then C is closed.*

(c) *If $f : X \to Y$ is continuous and $C \subset X$ is compact, then so is $f(C) \subset Y$.*

Proof For (a) let $(U_i)_{i \in I}$ be an open cover of C. Then $(U_i)_{i \in I} \cup \{X \setminus C\}$ is an open cover of X. Thus there exist indices $i_1, \ldots, i_l$ such that $X \subset X \setminus C \cup \bigcup_{j=1}^{l} U_{i_j}$, which implies $C \subset \bigcup_{j=1}^{l} U_{i_j}$.

For (b) we have to show that for any $x \in X \setminus C$ there exists an open neighborhood U of x with $U \cap C = \emptyset$. Since X is Hausdorff we can find for every $y \in C$ open neighborhoods V_y of y and U_y of x with $V_y \cap U_y = \emptyset$. Since $(V_y)_{y \in C}$ is an open covering of C we find $y_1, \ldots, y_l \in C$ with $C \subseteq \bigcup_{j=1}^{l} V_{y_j}$. Then $U = \bigcap_{j=1}^{l} U_{y_j}$ is an open neighborhood of x with $U \cap C = \emptyset$.

Finally, for (c) let $(U_i)_{i \in I}$ be an open cover of the set $f(C)$. Then $(f^{-1}(U_i))_{i \in I}$ is an open cover of C and there exist $i_1, \ldots, i_l$ with $C \subset \bigcup_{j=1}^{l} f^{-1}(U_{i_j})$, which implies that $f(C) \subset \bigcup_{j=1}^{l} U_{i_j}$. □

Definition In this book, a space X is called *locally compact* if every point $x \in X$ possesses a compact neighborhood. For instance $\mathbb{R}$ is locally compact with its usual topology.

A.4 Hausdorff Spaces

A topological space X is called a *Hausdorff space*, if any two different points can be separated by disjoint neighborhoods, i.e., if for any two $x \neq y$ in X there are open sets $U, V \subset X$ with $x \in U$ and $y \in V$ and $U \cap V = \emptyset$.

Examples A.4.1

- Any metric space is Hausdorff.

- The discrete topology $\mathcal{P}(X)$ is Hausdorff for every X but the trivial topology $\{\emptyset, X\}$ is not Hausdorff if X has more than one element.

- Let X be an infinite set. A subset $A \subset X$ is called *cofinite* if its complement $X \setminus A$ is finite. On X we install the *cofinite topology*, which consists of the empty set and all cofinite sets. With this topology, X is not Hausdorff.

Definition For any two topological spaces X, Y one defines the *product space* $X \times Y$ to be the cartesian product of the sets X and Y, equipped with the *product topology*, where every open set is a union of sets of the form $U \times W$ for open sets $U \subset X$ and $W \subset Y$. If $X = Y$ one has the diagonal $\Delta \subset X \times X$, which is the set of all elements of the form (x, x) for $x \in X$.

Lemma A.4.2 *A topological space X is a Hausdorff space if and only if the diagonal Δ is closed in $X \times X$.*

Proof Suppose X is Hausdorff, and let (x, y) be in $(X \times X) \setminus \Delta$, that means $x \neq y$. Then there exist open neighborhoods $U \ni x$ and $V \ni y$ with $U \cap V = \emptyset$. The latter condition means that $U \times V \cap \Delta = \emptyset$, i.e., $U \times V$ is an open neighborhood of (x, y) contained in $(X \times X) \setminus \Delta$, the latter set therefore is open, so Δ is closed. The converse direction is similar. □

Definition A Hausdorff space is also called a T_2-*space*, or a *separated space*. A topological space X is called a T_1-*space* if for any two elements $x, y \in X$ there are open sets U, V with $x \in U$ and $y \in V$ such that $x \notin V$ and $y \notin U$. A space is T_1 if and only if all singletons $\{x\}$, $x \in X$, are closed subsets. Every Hausdorff space is also T_1.

A.5 Initial- and Final-Topologies

Let X be a set, and let $f_i : X \to Y_i$ be a family of maps into topological spaces Y_i. The *initial topology* on X given by the family f_i is the smallest topology such that all f_i are continuous, so it is the topology generated by the inverse images $f_i^{-1}(U)$ of open sets $U \subset Y_i$.

Let X be a set, and let $g_i : W_i \to X$ be a family of maps from topological spaces W_i. The *final-topology* on X given by the g_i is the biggest topology on X that makes all g_i continuous. A set $U \subset X$ is open in this topology if and only if its inverse image $g_i^{-1}(U) \subset W_i$ is open for every $i \in I$. A special case of a final-topology is the *quotient topology* on the set $E/\sim$ of classes of an equivalence relation $\sim$ on a topological space E. It is the final-topology given by the single map $E \to E/\sim$.

Examples A.5.1

- Let $A \subset X$ be a subset of the topological space X. The topology on A induced by the inclusion map $i : A \hookrightarrow X$ is called the *subspace topology* on A. The open sets in A are precisely the sets of the form $A \cap U$, where U is open in X.

- Let G be a topological group and H a subgroup. Then one equips the coset space G/H with the quotient topology, i.e., the final topology of the projection map $G \to G/H$.

- Let $(X_i)_{i \in I}$ be a family of topological spaces, and let $X = \prod_{i \in I} X_i$ be the cartesian product of the X_i. Then the *product topology* on X is the initial topology induced by all projections $p_i : X \to X_i$.

- Let X be a locally compact Hausdorff space, and let $C_c(X)$ denote the set of all continuous complex valued functions on X with compact supports. The space $C_c(X)$ can be equipped with the *inductive limit topology* defined as follows. For a compact subset K let $C_K(X)$ be the set of all continuous functions on X with support in K, then $C_c(X)$ is the union of all $C_K(X)$ as K varies. We equip $C_K(X)$ with the topology given by the supremum-norm

$$\|f\|_K = \sup_{x \in K} |f(x)|$$

and give $C_c(X)$ the final topology induced by all inclusions $i_K : C_K(X) \hookrightarrow C_c(X)$.

Proposition A.5.2

(a) *Let X be a set equipped with the initial topology given by the family $f_i : X \to Y_i$. Then a map $\alpha : W \to X$ from a topological space W is continuous if and only if $f_i \circ \alpha$ is continuous for every index i.*

(b) *Likewise, let X be equipped with the final topology given by $g_i : W_i \to X$, then a map $\beta : X \to Y$ is continuous if and only if $\beta \circ g_i$ is continuous for every i.*

Proof For (a), let α be continuous then $f_i \circ \alpha$ is continuous as a composition of continuous maps. Conversely, assume that $f_i \circ \alpha$ is continuous for every i. Let $\mathcal{E}$ be the system of subsets of X of the form $f_i^{-1}(U)$ where U is an open subset of Y_i. Then $\mathcal{E}$ generates the topology $\mathcal{O}$ on X. Let $\mathcal{O}_\alpha$ be the biggest topology on X that makes α continuous, then, as $f_i \circ \alpha$ is continuous, it follows $\mathcal{E} \subset \mathcal{O}_\alpha$; therefore $\mathcal{O} \subset \mathcal{O}_\alpha$, so α is continuous. Part (b) is proved along the same lines.   □

Examples A.5.3

- Let $X = \prod_{i \in I} X_i$ be the direct product of the topological spaces X_i equipped with the product topology. Let $p_i : X \to X_i$ be the projections. Then a map $f : W \to X$ is continuous if and only if all maps $p_i \circ f : W \to X_i$ are continuous. This implies, for instance, that for two topological spaces X, Y and $y_0 \in Y$ the map $X \to X \times Y$ that maps x to (x, y_0) is continuous.

- Let H be a closed subgroup of the topological group G. Then a map $f : G/H \to \mathbb{C}$ is continuous if and only if $q \circ f : G \to \mathbb{C}$ is continuous, where $g : G \to G/H$ denotes the quotient map.

- In the case of the inductive limit topology on the space $C_c(X)$ for a locally compact Hausdorff space this means, for instance, that the restriction $T_K :$ $C(K) \cap C_c(X) \to \mathbb{C}$ is continuous for every compact set $K \subseteq X$. In terms of sequences, this means that a map $T : C_c(X) \to \mathbb{C}$ is continuous if and only if for every compact set $K \subset X$ and every sequence $(f_i)_{i \in \mathbb{N}}$ of functions in $C_c(X)$ with support in K, which converges uniformly to $f \in C_c(X)$, one has $\lim_{i \to \infty} T(f_i) = T(f)$.

A.6   Nets

Every student of mathematics is taught convergence of sequences in $\mathbb{R}$ or more generally, in metric spaces. When dealing with more general topological spaces, which naturally arise in analysis, sequences are insufficient, because a point can have "too many" open neighborhoods. In this book, we use the concept of nets, which generalizes the concept of sequences in a way sufficient for all needs. In the beginning, it takes a little practice to get used to the notion of nets. They then reveal their beauty and utility in the applications. Once having understood the differences to sequences, which mainly lie in the possible uncountability of a net, one can basically use nets just like sequences. Before starting seriously on the definition of nets, a word of caution is required here, as just the possible uncountability of nets renders them useless for measure-theoretic conclusions. For example, the Theorem of monotonic convergence is false for nets.

Let I be a set. Recall a *partial order* on I is a relation $\leq$ which is

- reflexive: $x \leq x$,

- anti-symmetric: $x \leq y$ and $y \leq x \Rightarrow x = y$,
- transitive: $x \leq y$ and $y \leq z \Rightarrow x \leq z$.

Examples A.6.1

- The natural order $\leq$ on $\mathbb{R}$.
- Let X be a set. On the set $\mathcal{P}(X)$ of all subsets of X one has a natural order by inclusion, so for $A, B \subset X$,

$$A \leq B \iff A \subset B.$$

A partially ordered set $(I, \leq)$ is called a *directed set* if for any two elements there is an upper bound, i.e., for any two elements $x, y \in I$ there exists $z \in I$ such that $x \leq z$ and $y \leq z$. By induction it follows that every finite set has an upper bound if I is directed.

A *net* in a topological space X is a map

$$\alpha : I \to X,$$

where I is a directed set. It is a convention to write the images as α_i, $i \in I$, instead of $\alpha(i)$.

We say that a net *converges* to a point $x \in X$ if for every neighborhood U of x there exists an index $i_0 \in I$ such that

$$i \geq i_0 \Rightarrow \alpha_i \in U.$$

In the special case of a sequence, $I = \mathbb{N}$, this notion coincides with the notion of convergence of a sequence.

Note that a net can converge to more than one point. The extreme case is the trivial topology, in which every net converges to every point. Indeed, the uniqueness of limits is equivalent to the Hausdorff property, as the following lemma shows.

Proposition A.6.2 *A topological space X is a Hausdorff space if and only if limits are unique, i.e., if any net has at most one limit.*

Proof Suppose X is Hausdorff and (x_i) is a convergent net. Assume it converges to x and y with $x \neq y$. By the Hausdorff property, there are open sets $U \ni x$ and $V \ni y$ such that $U \cap V = \emptyset$. As (x_i) converges to x and y there exists an index i such that $x_i \in U$ and $x_i \in V$, a contradiction. This means that the limit of a convergent net indeed is unique.

For the converse assume uniqueness of limits and let $x \neq y$ be in X. Let S be the system of pairs (U, V) such that U, V are open subsets of X with $U \ni x$ and $V \ni y$. The set S is partially ordered by inverse inclusion, i.e., $(U, V) \leq (U', V')$ if and only if $U \supset U'$ and $V \supset V'$. The set is directed, as one can take intersections. We show the Hausdorff property by contradiction, so assume that $U \cap V \neq \emptyset$ for every $(U, V) \in S$. Choose an element z_{UV} in $U \cap V$ for each $(U, V) \in S$. Then the z_{UV} form a net with index set S. As z_{UV} lies in U and V, it follows that this net converges to x and y. The uniqueness of limits then implies $x = y$, a contradiction. $\qquad\square$

To get used to the concept of nets, we will prove the following proposition.

Proposition A.6.3 *Let X be a topological space, and let $A \subset X$, then the closure $\overline{A}$ coincides with the set of all limits of nets in A. In other words, a point $x \in X$ lies in $\overline{A}$ if and only if there exists a net $(\alpha_i)_{i \in I}$ in A which converges to x.*

Proof The closure $\overline{A}$ is the set of all $x \in X$ such that A intersects every neighborhood of x. So let $x \in \overline{A}$ and U a neighborhood of x. Then the intersection $A \cap U$ is non-empty. Choose an element α_U in $A \cap U$. Let I be the set of all neighborhoods of x with the partial order

$$U \leq U' \Leftrightarrow U \supset U'.$$

The set I is directed as the intersection of two neighborhoods of x is a neighborhood of x. The net $(\alpha_U)_{U \in I}$ converges by construction to x. This proves the 'only if' part.

For the converse direction let $x \in X$ and $\alpha_i \in A$, $i \in I$ a net converging to x. For any given neighborhood U of x there exists $i \in I$ with $\alpha_i \in U$. Since α_i also lies in A, it follows that $A \cap U$ is non-empty. As U was arbitrary, it follows that $x \in \overline{A}$. $\qquad\square$

Proposition A.6.4 *Let $f : X \rightarrow Y$ be a map between topological spaces. The map f is continuous if and only if for every net (x_j) in X, which converges to, say, $x \in X$, the net $f(x_j)$ converges to $f(x)$.*

Proof Observe that the following proof is almost verbatim the same as in the case of maps on $\mathbb{R}$, where sequences can be used. Let f be continuous and let $(x_i)_{i \in I}$ be a net in X convergent to $x \in X$. We have to show $f(x_i) \rightarrow f(x)$. For this let U be an open neighborhood of $f(x)$ in Y, then $V = f^{-1}(U)$ is an open neighborhood of x. As $x_i \rightarrow x$, there exists i_0 such that $x_i \in V$ for every $i \geq i_0$ and so $f(x_i) \in U$ for every $i \geq i_0$.

For the converse direction assume that f satisfies the convergence condition. Let $A \subset Y$ be closed and let $B \subset X$ be the inverse image of A. We have to show that B is closed in X. For this let b_i be a net in B, convergent to $x \in X$. As $f(b_i)$ converges to $f(x)$, which therefore lies in A, it follows that x lies in $B = f^{-1}(A)$, so B indeed is closed. $\qquad\square$

Example A.6.5 Let G be a group with a topology. Then G is a topological group if and only if for any two nets x_i and y_i, which converge to x resp. $y \in G$, the net $x_i y_i^{-1}$ converges to xy^{-1}. This is a consequence of the last proposition.

Definition A map $\phi : J \to I$ between two directed sets is called *strictly cofinal* if for every $i_0 \in I$ there exists a $j_0 \in J$ such that $j \geq j_0$ implies $\phi(j) \geq i_0$. This means that the map ϕ is allowed to jump back and forth, but "in the limit" it is monotonically increasing and "exhausts" the set I.

Definition Let $\alpha : I \to X$ be a net. A *subnet* is a net $\beta : J \to X$ together with a factorization

such that the map ϕ is strictly cofinal. In other words, subnets of α can be identified with strictly cofinal maps to I. If α converges to $x \in X$, then every subnet converges to $x \in X$.

Proposition A.6.6 *A topological space X is compact if and only if every net in X has a convergent subnet.*

Proof Let X be compact and let $(x_i)_{i \in I}$ be a net in X. For each $i \in I$ let A_i be the closure of the set $\{x_j : j \geq i\}$. Any finite intersection of the A_i is nonempty, so by the finite intersection property, there exists $x \in X$ which lies in every A_i. That means, that to every neighborhood U of x and every $i \in I$, one can choose an index $i' = \phi(U, i) \in I$ such that $i' \geq i$ and $x_{i'} = x_{\phi(U,i)} \in U$. Let J be the directed set of all pairs (U, i), where U is a neighborhood of x and $i \in I$. We order J as follows: $(U, i) \leq (U', i')$ if $U \supset U'$ and $i \leq i'$. We have constructed a map $\phi : J \to I$ of which we claim that it is strictly cofinal. For this let $i \in I$ and in J choose any element of the form $j = (U, i)$ with i as the second component. Then by construction, $\phi(j') \geq i$ for every $j' \geq j$, so ϕ is indeed strictly cofinal. We claim that the induced subnet $\phi : J \to X$ converges. For this let U be a neighborhood of x and choose an element $j_0 = (U, i)$. Then for every $j \geq j_0$ one has $\phi(j) \in U$, so indeed (x_i) has a convergent subnet.

For the converse assume that every net in X has a convergent subnet. Let $\mathcal{A}$ be a collection of closed subsets such that every finite intersection of members of $\mathcal{A}$ is non-empty. We need to show that the intersection of the elements in $\mathcal{A}$ is non-empty, too. For this let $\mathcal{B}$ denote the set of all finite intersections of elements in $\mathcal{A}$. Then $\mathcal{B}$ is a directed set via $B_1 \geq B_2 \Leftrightarrow B_1 \subset B_2$. Choose $x_B \in B$ for all $B \in \mathcal{B}$. Then $(x_B)_{B \in \mathcal{B}}$ is a net in X and by assumption there exists a subnet $(x_{B_j})_{j \in J}$ which converges to some $x \in X$. But then $x \in B$ for all $B \in \mathcal{B}$, for if B is fixed we can choose j_0 such that $B_j \subset B$ for all $j \geq j_0$. But this implies that $x_{B_j} \in B$ for all $j \geq j_0$. Since B is closed, the limit x of (x_{B_j}) also lies in B. $\square$

A.7 Tychonov's Theorem

Let I be an index set, and for each $i \in I$ let X_i be a topological space. As mentioned above, one equips the product $X = \prod_{i \in I} X_i$ with the initial topology of the projections $p_i : X \to X_i$. This topology is generated by all sets of the form

$$U_i \times \prod_{j \neq i} X_j,$$

where U_i is an open subset of X_i. Let's call such a set a *simple set*. Recall that this means that every open set is a union of finite intersections of simple open sets. Dually, every closed set is an intersection of finite unions of simple closed sets.

Theorem A.7.1 (Tychonov). *Assume the space $X = \prod_{i \in I} X_i$ to be non-empty. Then X is compact if and only if each factor X_i is a compact space.*

Proof The projections $p_i : X \to X_i$ being continuous, the X_i are all compact, if X is. The difficult part is the converse. So assume that every X_i is compact. Let $\mathcal{F} = (F_v)_{v \in N}$ be a family of closed subsets in X satisfying the finite intersection property. Then there exists a maximal family $\mathcal{F}^* = (F_v)_{v \in N^*}$ of subsets with $\mathcal{F}^* \supset \mathcal{F}$ having the finite intersection property. This is easily seen by means of Zorn's lemma.

(A) If $F_1, \ldots, F_n \in \mathcal{F}^*$, then $F_1 \cap \cdots \cap F_n$ is in $\mathcal{F}^*$, as is clear by maximality of $\mathcal{F}^*$.

(B) If $S \subset X$ is any subset with the property that $S \cap F_v \neq \emptyset$ for every $F_v \in \mathcal{F}^*$, then $S \in \mathcal{F}^*$, as follows from maximality again.

Let $i \in I$. The family of closed sets $(\overline{p_i(F_v)})_{v \in N^*}$ has the finite intersection property, so there exists an element z_i in their intersection. Let

$$U = U_{i_1} \times \cdots \times U_{i_n} \times \prod_{i \neq i_1, \ldots, i_n} X_i$$

be an open neighborhood of the element $z = (z_i)_{i \in I}$ in X. Let $k \in \{1, \ldots, n\}$. For every $F_v \in \mathcal{F}^*$ there is an $f \in F_v$ with $p_{i_k}(f) \in U_{i_k}$, so with $S_k = p_{i_k}^{-1}(U_{i_k})$ we have $S_k \cap F_v \neq \emptyset$. By (B) the set S_k is in $\mathcal{F}^*$. By (A) we have $U = S_1 \cap \cdots \cap S_n \in \mathcal{F}^*$. In particular, U has non-empty intersection with every $F \in \mathcal{F}^*$, so with every $F \in \mathcal{F}$. Since these sets form a neighborhood base, the point z lies in the closure of F_v so in F_v for every $v \in N$. We infer that $\bigcap_{n \in N} F_n$ is non-empty and X is compact. $\square$

A.8 The Lemma of Urysohn

Recall that a Hausdorff space is called locally compact if every point has a compact neighborhood, and a subset $S \subset X$ is called relatively compact if its closure $\overline{S} \subset X$ is compact.

Lemma A.8.1 (Lemma of Urysohn). *Let X be a locally compact Hausdorff space. Let $K \subset X$ be compact and $A \subset X$ closed with $K \cap A = \emptyset$.*

(i)   *There exists a relatively compact open neighborhood U of K such that $K \subset U \subset \overline{U} \subset X \setminus A$.*

(ii)   *There exists a continuous function of compact support $f : X \to [0, 1]$ with $f \equiv 1$ on K and $f \equiv 0$ on A.*

(iii) *Let $B \subset X$ be closed, and let $h : B \to [0, \infty)$ be in $C_0(B)$ and suppose $h(x) \geq 1$ for every $x \in K \cap B$. Then there exists a continuous function f as in (ii) with the additional property that $f(b) \leq h(b)$ for every $b \in B$.*

This lemma is of fundamental importance, we will give the proof.

Proof To prove the first assertion, let $a \in A$. Then for every $k \in K$ there is an open, relatively compact neighborhood U_k of k that is disjoint from some neighborhood $U_{k,a}$ of a. The family (U_k) forms an open covering of K. As K is compact, finitely many suffice. Let V be their union and W be the intersection of the corresponding $U_{k,a}$. Then V and W are open disjoint neighborhoods of K and a, and V is relatively compact. Repeating this argument with K in the role of a and $\overline{V} \cap A$ in the role of K one gets open disjoint neighborhoods U' of K and W' of $\overline{V} \cap A$, respectively. Then the set $U = U' \cap V$ does the trick.

For (ii), choose U as in the first part and replace A with $A \cup (X \setminus U)$. One sees that it suffices to prove the claim without f having compact support. So again let U be as in the first part and rename this open set to $U_{\frac{1}{2}}$. Next there is a relatively compact neighborhood $U_{\frac{1}{4}}$ of K such that $U_{\frac{1}{2}} \subset \overline{U}_{\frac{1}{2}} \subset U_{\frac{1}{4}}$. Let R be the set of all numbers of the form $\frac{k}{2^n}$ in the interval $[0, 1)$. Formally set $U_0 = X \setminus A$. Iterating the above construction we get open sets $U_r, r \in R$, with $K \subset U_r \subset \overline{U}_r \subset U_s \subset X \setminus A$ for all $r > s$ in R. We now define f. For $x \in A$ set $f(x) = 0$ and otherwise set $f(x) = \sup\{r \in R : x \in \overline{U}_r\}$. Then $f \equiv 1$ on K. For $r > s$ in R one gets

$$f^{-1}(s, r) = \bigcup_{s < s' < s'' < r} U_{s'} \setminus \overline{U}_{s''},$$

which is open. Similarly, $f^{-1}([0, s))$ and $f^{-1}((r, 1])$ are open. Since the intervals of the form (r, s), $[0, s)$, and $(r, 1]$ generate the topology on $[0, 1]$ it follows that f is continuous.

The proof of (iii) is an obvious variant of the last proof, where at each step of the construction of the U_r one applies part (i) with A replaced by $A \cup \{b \in B : h(b) \leq r\}$. $\square$

Corollary A.8.2 *In a locally compact Hausdorff space, every point possesses a neighborhood base consisting of compact sets.*

Proof Let X be a locally compact Hausdorff space, and let $x \in X$. We have to show that every open neighborhood V of x contains a compact neighborhood. Let $K = \{x\}$, then K is a compact subset of X. Let A be the closed set $X \setminus V$. Then $K \cap A = \emptyset$ and by Lemma A.8.1 (i) there exists a relatively compact open neighborhood U of x with $\overline{U} \subset V$. $\qquad\square$

Theorem A.8.3 (Tietze's Extension Theorem). *Let X be a locally compact Hausdorff space, and let $B \subset X$ be a closed subset. Then the restriction $C_0(X) \to C_0(B)$ is surjective.*

Proof Let $h \in C_0(B)$. Decomposing h into real and imaginary parts we see that it suffices to assume that h is real valued. Writing $h = h^+ - h^-$ for positive functions $h^\pm$, we restrict to the case $h \geq 0$. Scaling h, we can assume $0 \leq h \leq 1$.

Let $K = \{b \in B : h(b) \geq \frac{1}{2}\}$. Then K is compact in X and the Lemma of Urysohn gives a function $0 \leq f \leq 1$ in $C_c(U)$ with $f \equiv 1$ on K and $f(b) \leq 2\,h(b)$ for every $b \in B$. So the function $g_1 = \frac{1}{2}f$ is equal to $\frac{1}{2}$ on K and satisfies $0 \leq g_1 \leq \frac{1}{2}$ as well as $g_1 \leq h_1 = h$. Set $h_2 = h_1 - g_1$. Then $0 \leq h_2 \leq \frac{1}{2}$. By repeating the argument we get $g_2 \in C_c(X)$ with $0 \leq g_2 \leq \frac{1}{4}$ and $g_2 \leq h_2$. Set $h_3 = h_2 - g_2$. Iteration gives a sequence $g_n \in C_c(X)$ such that $0 \leq g_n \leq 2^{-n}$. The function

$$g(x) = \sum_{n=1}^{\infty} g_n(x)$$

is continuous, vanishes at infinity, and satisfies $g(b) = h(b)$ for every $b \in B$. $\qquad\square$

A.9 Baire Spaces

A subset D of a topological space X is *dense* in X if $\overline{D} = X$. This is equivalent to saying that $U \cap D \neq \emptyset$ for every open $U \subset X$. In this section we show that every locally compact topological space X is a *Baire space*, i.e., whenever we have a countable family $(U_n)_{n \in \mathbb{N}}$ of open and dense subsets of X, then the intersection $D = \bigcap_{n \in \mathbb{N}} U_n$ is also dense in X. Equivalently, X is a Baire space if, whenever X can be written as a countable union $X = \bigcup_{n \in \mathbb{N}} A_n$ with A_n closed in X, then there exists at least one $n_0 \in \mathbb{N}$ such that A_{n_0} has nonempty interior.

Proposition A.9.1 *Every locally compact Hausdorff space and every complete metric space is a Baire space.*

Proof Let X be either a locally compact Hausdorff space or a complete metric space. Suppose $V_1, V_2, \ldots$ are dense open subsets of X. We define a sequence $B_0, B_1, \ldots$ of open sets as follows. Let B_0 be an arbitrary nonempty open set in X. Inductively, assume $n \geq 1$ and an open set $B_{n-1} \neq \emptyset$ has been chosen, then, as V_n is dense, there is an open set $B_n \neq \emptyset$ with

$$\overline{B_n} \subset V_n \cap B_{n-1}.$$

If X is a locally compact Hausdorff space, one can choose $\overline{B_n}$ to be compact. If X is a complete metric space, B_n may be taken to be a ball of radius $< 1/n$. Set $K = \bigcap_{n=1}^{\infty} \overline{B_n}$. If X is a locally compact Hausdorff space, then $K \neq \emptyset$ by the finite intersection property. If X is a complete metric space, then the centers of the balls B_n form a Cauchy sequence that converges to some point of K, so $K \neq \emptyset$ in either case. One has $K \subset B_0$ and $K \subset V_n$ for each n, so $B_0 \cap \bigcap_n V_n \neq \emptyset$. $\square$

A.10   The Stone-Weierstraß Theorem

For a locally compact Hausdorff space X it is very often necessary to decide whether a given subalgebra of the algebra $C_0(X)$ is dense with respect to sup-norm $\|\cdot\|_X$. The Stone-Weierstraß Theorem gives an easy criterion that applies in many situation.

Theorem A.10.1 (Stone-Weierstraß Theorem). *Let X be a locally compact Hausdorff space and suppose that $A \subset C_0(X)$ is a subalgebra of $C_0(X)$ such that*

(a) *A separates the points, i.e., for $x \neq y$ in X there exists $f \in A$ with $f(x) \neq f(y)$,*

(b) *for every $x \in X$ there is $f \in A$ such that $f(x) \neq 0$, and*

(c) *A is closed under complex conjugation.*

Then A is dense in $C_0(X)$.

The above complex version is a consequence of the following real version, in which $C_0^{\mathbb{R}}(X)$ denotes the real Banach algebra consisting of all real continuous functions on X which vanish at infinity equipped with $\|\cdot\|_X$.

Theorem A.10.2 (Stone-Weierstraß Theorem over $\mathbb{R}$). *Let X be a locally compact Hausdorff space and suppose that the set $A \subset C_0^{\mathbb{R}}(X)$ is a (real) subalgebra of $C_0^{\mathbb{R}}(X)$ such that*

(a) *A separates the points, and*

(b) *for every $x \in X$ there is $f \in A$ such that $f(x) \neq 0$.*

Then A is dense in $C_0^{\mathbb{R}}(X)$.

We first show how the complex version follows from the real version: For this suppose that $A \subset C_0(X)$ is as in Theorem A.10.1. Then $A = A^{\mathbb{R}} + iA^{\mathbb{R}}$, where $A^{\mathbb{R}} =$

$A \cap C_0^{\mathbb{R}}(X)$. This follows from the decomposition $f = \mathrm{Re}(f) + i\mathrm{Im}(f)$ with $\mathrm{Re}(f) = \frac{1}{2}(f + \bar{f})$ and $\mathrm{Im}(f) = \frac{1}{2i}(f - \bar{f})$ in $A^{\mathbb{R}}$. If A satisfies the conditions of Theorem A.10.1 one easily checks that $A^{\mathbb{R}}$ satisfies the conditions of Theorem A.10.2. But then $\overline{A^{\mathbb{R}}} = C_0^{\mathbb{R}}(X)$ and $\overline{A} = \overline{A^{\mathbb{R}}} + i\overline{A^{\mathbb{R}}} = C_0(X)$.

So we only have to show the real version. We need two lemmas.

Lemma A.10.3 (Dini's Theorem). *Let X be a compact topological space, and let $(f_n)_{n \in \mathbb{N}}$ be a monotonically increasing sequence of continuous functions $f_n : X \to \mathbb{R}$ that converges point-wise to the continuous function $f : X \to \mathbb{R}$. Then (f_n) converges uniformly to f.*

Proof Let $\varepsilon > 0$ be given. For every $x \in X$ there exists $n_x \in \mathbb{N}$ with $f(x) - \varepsilon < f_n(x) \leq f(x)$ for every $n \geq n_x$. Let $U_x := \{y \in K : f(y) - \varepsilon < f_{n_x}(y)\}$. Then $\{U_x : x \in X\}$ is an open cover of X. Since X is compact, we find $x_1, \ldots, x_l \in X$ with $X = \bigcup_{j=1}^l U_{x_j}$. Then $\|f - f_n\|_X < \varepsilon$ for ever $n \geq N = \max\{n_{t_1}, \ldots, n_{t_l}\}$. □

Lemma A.10.4 *Let A be a subalgebra of $C_0^{\mathbb{R}}(X)$. If $f \in \overline{A}$, then $|f| \in \overline{A}$. Moreover, if $f, g \in \overline{A}$, then also $\max(f, g), \min(f, g) \in \overline{A}$.*

Proof Let $0 \neq f \in A$. By passing to $\frac{1}{\|f\|_X} f$ we may assume that $f(X) \subset [-1, 1]$, and therefore $f(x)^2 \in [0, 1]$ for every $x \in X$. We define inductively a sequence (p_n) of polynomials on $[0, 1]$ such that $p_1 \equiv 0$ and

$$p_{n+1}(t) = p_n(t) - \frac{1}{2}(p_n(t)^2 - t), \quad t \in [0, 1].$$

We claim that $(p_n(t))$ is a monotonically increasing sequence that converges to $\sqrt{t}$ for every $t \in [0, 1]$. Indeed, we show first by induction that $0 \leq p_n(t) \leq \sqrt{t}$ and $p_n(0) = 0$ for every $n \in \mathbb{N}$. This is clear for $n = 1$ and if it is shown for n the result for $n + 1$ follows from

$$p_{n+1}(t) - \sqrt{t} = \left(p_n(t) - \sqrt{t}\right) - \frac{1}{2}\left(p_n(t) - \sqrt{t}\right)\left(p_n(t) + \sqrt{t}\right)$$
$$= \left(p_n(t) - \sqrt{t}\right)\left(1 - \frac{1}{2}\left(p_n(t) + \sqrt{t}\right)\right) \leq 0,$$

since $p_n(t) - \sqrt{t} \leq 0$ and $p_n(t) + \sqrt{t} \leq 2\sqrt{t} \leq 2$. Thus, since $p_{n+1}(t) - p_n(t) = \frac{1}{2}(t - p_n(t)^2) \geq 0$, the sequence $(p_n(t))$ is monotonically increasing and bounded by $\sqrt{t}$. It therefore converges to some $0 \leq g(t) \leq \sqrt{t}$. We then get

$$0 = g(t) - g(t) = \lim_n (p_{n+1}(t) - p_n(t)) = \lim_n \frac{1}{2}\left(t - p_n(t)^2\right) = \frac{1}{2}\left(t - g(t)^2\right),$$

from which it follows that $g(t) = \sqrt{t}$.

Since g is continuous, it follows from Dini's Theorem that (p_n) converges uniformly to g on $[0, 1]$. Define $f_n(x) = p_n(f(x)^2)$ for every $x \in X$. Then (f_n) converges uniformly to $\sqrt{f^2} = |f|$ on X. Since f_n is a linear combination of powers of f it lies in $\overline{A}$ for every $n \in \mathbb{N}$. Thus $|f| \in \overline{A}$.

The final assertion now follows from the fact that $\overline{A}$ is a real algebra and $\max f, g = \frac{1}{2}(f + g + |f - g|)$ and $\min(f, g) = \frac{1}{2}(f + g - |f - g|)$. $\square$

Proof of Theorem A.10.2 We first show that for all pairs $x, y \in X$ with $x \neq y$ there exists a $g \in A$ with $g(x) \neq g(y)$ and $g(x), g(y) \neq 0$. Assume that this is not the case for the given pair x, y. Choose $g_1 \in A$ with $g_1(x) \neq g_1(y)$ and assume that $g_1(y) \neq 0$. Then $g_1(x) = 0$. Choose $g_2 \in A$ with $g_2(x) \neq 0$. Then $g_2(x) = g_2(y)$ or $g_2(y) = 0$. If $g_2(x) = g_2(y)$ define $g = g_1 + g_2$ and if $g_2(y) = 0$ put $g = g_1 + \mu g_2$ with $\mu \in \mathbb{R}$ such that $g_1(y) \neq \mu g_2(x) \neq 0$. In both cases one checks that $0 \neq g(x) \neq g(y) \neq 0$, a contradiction.

In the next step we show that for all pairs $x, y \in X$ with $x \neq y$ and any given $\alpha, \beta \in \mathbb{R}$ there exists a function $f \in A$ with $f(x) = \alpha$ and $f(y) = \beta$. For this choose g as above. We make the Ansatz $f = \lambda g + \mu g^2$ for $\lambda, \mu \in \mathbb{R}$. Then $f(x) = \alpha$, $f(y) = \beta$ is equivalent to

$$\begin{pmatrix} g(x) & g(x)^2 \\ g(y) & g(y)^2 \end{pmatrix} \begin{pmatrix} \lambda \\ \mu \end{pmatrix} = \begin{pmatrix} \alpha \\ \beta \end{pmatrix}.$$

But since $0 \neq g(x) \neq g(y) \neq 0$ one computes $\det \begin{pmatrix} g(x) & g(x)^2 \\ g(y) & g(y)^2 \end{pmatrix} = g(x)g(y)$ $(g(y) - g(x)) \neq 0$, and the above linear equation has a unique solution.

Now let $h \in C_0^{\mathbb{R}}(X)$ be given, and let $\varepsilon > 0$. We need to show that there exists some $f \in \overline{A}$ with $\|h - f\|_X < \varepsilon$. For each pair $x, y \in X$ with $x \neq y$ we choose $g_{x,y} \in A$ with $h(x) = g_{x,y}(x)$ and $h(y) = g_{x,y}(y)$. For fixed y we define

$$U_x := \{z \in X : h(z) < g_{x,y}(z) + \varepsilon\}.$$

Then U_x is an open neighborhood of x and $x \setminus U_x = \{z \in X : (h - g_{x,y})(z) \geq \varepsilon\}$ is compact, since $h - g_{x,y}$ vanishes at infinity. Thus, if we fix $x_1 \in X$, then there exist $x_2, \dots, x_l \in X \setminus U_{x_1}$ with $X \setminus U_{x_1} \subset \bigcup_{j=2}^{l} U_{x_j}$, from which it follows that $X \subset \bigcup_{j=1}^{l} U_{x_j}$. Set

$$f_y = \max\left(g_{x_1,y}, \dots, g_{x_l,y}\right).$$

It lies in $\overline{A}$ by Lemma A.10.4 and by the construction it follows that $h(z) - f_y(z) < \varepsilon$ for every $z \in X$, since if $z \in U_{x_j}$, then $h(z) < g_{x_j,y}(z) + \varepsilon \leq f_y(z) + \varepsilon$.

Now, for $y \in X$ define

$$V_y = \{z \in X : f_y(z) < h(z) + \varepsilon\}.$$

Since $f_y(y) = h(y)$, this is an open neighborhood of y, and similarly as above we can show that there exist $y_1, \ldots, y_k \in X$ such that $X \subset \bigcup_{j=1}^{k} V_{y_j}$. Put

$$f = \min(f_{y_1}, \ldots, f_{y_k}).$$

Then $f \in \overline{A}$ and one easily checks that $f(z) - \varepsilon < h(z) < f(z) + \varepsilon$ for every $z \in X$. This finishes the proof.

Appendix B
Measure and Integration

In this Appendix we want to review some basic facts on integration theory, following mainly the approach given in Rudin's beautiful book [Rud87]. In the first section we recall the basic definitions of measure spaces, measurable functions and integration of functions with respect to positive measures. In the second section we review the Riesz Representation Theorem, which, for any locally compact Hausdorff space X, gives a one-to-one correspondence between Borel measure spaces with certain regularity properties and positive linear functionals $I : C_c(X) \to \mathbb{C}$, called Radon integrals on X. In further sections we review some basic facts on L^p-spaces and give a proof of the Radon Nikodym Theorem.

B.1 Measurable Functions and Integration

Let X be a set. A *σ-algebra* in X is a set $\mathcal{A}$ of subsets of X, which satisfies the following axioms:

- $\emptyset \in \mathcal{A}$, and if $A \in \mathcal{A}$ then so is its complement $X \smallsetminus A$,

- $\mathcal{A}$ is closed under countable unions.

It follows immediately from the above axioms that a σ-algebra is closed under countable intersections and that $A \smallsetminus B \in \mathcal{A}$ for all $A, B \in \mathcal{A}$.

The set $\mathcal{P}(X)$ of all subsets of X is a σ-algebra, and since the intersection of two (or infinitely many) σ-algebras is again a σ-algebra, we see that for every given set $\mathcal{S}$ of subsets of X there exists a smallest σ-algebra $\mathcal{A}$ that contains $\mathcal{S}$. We then say that $\mathcal{S}$ generates $\mathcal{A}$. If X is a topological space, then $\mathcal{B} = \mathcal{B}(X)$ denotes the σ-algebra generated by the topology on X. It is called the *Borel σ-algebra*, its elements are called the *Borel sets* in X. If $\mathcal{A}$ is a σ-algebra in X, then we call the pair $(X, \mathcal{A})$ a *measurable space*. In particular, for each topological space we obtain the measurable space $(X, \mathcal{B})$.

A. Deitmar, S. Echterhoff, *Principles of Harmonic Analysis,* Universitext,
DOI 10.1007/978-3-319-05792-7, © Springer International Publishing Switzerland 2014

A map $f : X \to Y$ between two measurable spaces $(X, \mathcal{A}_X)$ and $(Y, \mathcal{A}_Y)$ is called a *measurable map* if $f^{-1}(A) \in \mathcal{A}_X$, whenever $A \in \mathcal{A}_Y$. The composition of two measurable maps is measurable.

We consider the real line $\mathbb{R}$ furnished with its Borel σ-algebra. We now list some basic facts on measurable functions, which are easy to prove and will be used without further mention.

Lemma B.1.1 *Let* $(X, \mathcal{A})$ *be a measurable space.*

(a) *A Function* $f : X \to \mathbb{R}$ *is measurable if and only if* $f^{-1}((a, \infty)) \in \mathcal{A}$ *for every* $a \in \mathbb{R}$.

(b) *A function* $f : X \to \mathbb{C}$ *is measurable if and only if* $\mathrm{Re}\, f$ *and* $\mathrm{Im}\, f$ *are measurable.*

(c) *If* $f, g : X \to \mathbb{C}$ *are measurable, then so are* $f + g$, $f \cdot g$, *and* $|f|^p$ *for every* $p > 0$.

(d) *If* $f, g : X \to \mathbb{R}$ *are measurable, then so are* $\max(f, g)$ *and* $\min(f, g)$.

(e) *If* $f_n : X \to \mathbb{C}$ *is measurable for every* $n \in \mathbb{N}$ *and* $(f_n)_{n \in \mathbb{N}}$ *converges point-wise to* $f : X \to \mathbb{C}$, *then* f *is measurable*

Proof For the proof of the first assertion, we only have to show that $\mathcal{S} = \{(a, \infty) : a \in \mathbb{R}\}$ generates $\mathcal{B}(\mathbb{R})$. But since $[b, \infty) = \bigcap_{n \in \mathbb{N}} (b - \frac{1}{n}, \infty)$ and $(a, b) = (a, \infty) \setminus [b, \infty)$ we see that the σ-algebra generated by $\mathcal{S}$ must contain all open intervals in $\mathbb{R}$. Thus $\mathcal{S}$ generates $\mathcal{B}(\mathbb{R})$.

The other assertions are easy, except for the last one. For this, we first consider the case where $(f_n)_{n \in \mathbb{N}}$ is monotonically increasing. Then $f^{-1}((a, \infty]) = \bigcup_{n=1}^{\infty} f_n^{-1}((a, \infty]) \in \mathcal{A}$, hence f is measurable. A similar argument applies if $(f_n)_{n \in \mathbb{N}}$ is decreasing. For the general case we can first use (b) to restrict to the case of real functions, and then we can write f as the limit of the decreasing sequence $(g_k)_{k \in \mathbb{N}}$ with $g_k = \sup\{f_n : n \geq k\}$, which itself is a limit of the increasing sequence of measurable functions $h_n = \max\{f_k, \ldots, f_{k+n}\}$.   □

In what follows, it is sometimes useful to consider functions $f : X \to [0, \infty]$, where we equip $[0, \infty]$ with the obvious topology. As in the lemma, such a function is measurable if and only if $f^{-1}((a, \infty]) \in \mathcal{A}$ for every $a \in \mathbb{R}$. Assertions (c), (d) and (e) of the lemma remain valid for functions $f : X \to [0, \infty]$.

Definition. A *measure* μ on a measurable space $(X, \mathcal{A})$ is a function $\mu : \mathcal{A} \to [0, \infty]$ that satisfies $\mu(\emptyset) = 0$ and

- $\mu\left(\bigcup_{n=1}^{\infty} A_n\right) = \sum_{n=1}^{\infty} \mu(A_n)$ for any sequence $(A_n)_{n \in \mathbb{N}}$ of pairwise disjoint elements $A_n \in \mathcal{A}$.

From this condition one can derive the following facts

- $\mu(A \cup B) = \mu(A) + \mu(B) - \mu(A \cap B)$ for all $A, B \in \mathcal{A}$.

- If $(A_n)_{n \in \mathbb{N}}$ is a sequence in $\mathcal{A}$ such that $A_n \subseteq A_{n+1}$ for every $n \in \mathbb{N}$, then $\mu(A_n) \to \mu(A)$ for $A = \bigcup_{n=1}^{\infty} A_n$.

- If $(A_n)_{n \in \mathbb{N}}$ is a sequence in $\mathcal{A}$ such that $A_n \supseteq A_{n+1}$ for every $n \in \mathbb{N}$ and $\mu(A_1) < \infty$, then $\mu(A_n) \to \mu(A)$ for $A = \bigcap_{n=1}^{\infty} A_n$.

If $\mu : \mathcal{A} \to [0, \infty]$ is a measure on $(X, \mathcal{A})$ we call $(X, \mathcal{A}, \mu)$ a *measure space*.

If $(X, \mathcal{A}, \mu)$ is a measure space, then a *step function* $g : X \to \mathbb{C}$ is a function of the form $g = \sum_{i=1}^{m} a_i 1_{A_i}$ with $a_i \in \mathbb{C}$ and $A_i \in \mathcal{A}$ with $\mu(A_i) < \infty$ for every $1 \le i \le m$. For such step function g we define

$$\int_X g \, d\mu \overset{\text{def}}{=} \sum_{i=1}^{m} a_i \mu(A_i) \in \mathbb{C}.$$

For a positive measurable function $f : X \to [0, \infty]$ we define

$$\int_X f \, d\mu = \sup \left\{ \int_X g \, d\mu : g \le f, g \text{ a positive step function} \right\}.$$

We say that f is *integrable* if $\int_X f \, d\mu < \infty$. A measurable function $f : X \to \mathbb{R}$ is integrable if and only if $f^+ = \max(f, 0)$ and $f^- = -\min(f, 0)$ are integrable, and then we put $\int_X f \, d\mu = \int_X f^+ \, d\mu - \int_X f^- \, d\mu$. Similarly, we define the integral of complex measurable functions via its real and imaginary part in the obvious way.

The set $\mathcal{L}^1(X)$ of integrable complex valued functions on X is a complex vector space such that the integral $f \mapsto \int_X f \, d\mu$ is a *positive functional* on $\mathcal{L}^1(X)$ in the sense that $f \ge 0$ implies $\int_X f \, d\mu \ge 0$. We shall frequently use the following proposition.

Proposition B.1.2 *Let $(X, \mathcal{A}, \mu)$ be a measure space. Then a measurable function $f : X \to \mathbb{C}$ is integrable if and only if $|f|$ is integrable and then we have*

$$\left| \int_X f \, d\mu \right| \le \|f\|_1 \overset{\text{def}}{=} \int_X |f| \, d\mu.$$

As a consequence, if $f, g : X \to \mathbb{C}$ are measurable with $|f| \le |g|$ and g is integrable, then so is f.

Proof It follows from the definition of the integral that the positive measurable function dominated by an integrable function must be integrable. If f is real, then $|f| = f^+ + f^-$ is integrable if f is, and if f is complex, we can use the equation $|f| \le |\mathrm{Re}\, f| + |\mathrm{Im}(f)|$ to see that $|f|$ is integrable if f is. The equation $\left| \int_X f \, d\mu \right| \le \int_X |f| \, d\mu$ follows by replacing f by zf if $z \in \mathbb{C}$ with $|z| = 1$ such that $z \int_X f \, d\mu \ge 0$, which then implies that $\int_X f \, d\mu = \int_X \mathrm{Re}\, f \, d\mu \le \int_X |f| \, d\mu$. $\square$

Remark B.1.3 If $f : X \to [0, \infty]$ is an integrable function, the value ∞ can only occur on a null-set, i.e., a set N with $\mu(N) = 0$. Indeed, if $N = \{x \in X : f(x) = \infty\}$ has measure $a \geq 0$, then $n \cdot a = \int_X n 1_A \, d\mu \leq \int_X f \, d\mu < \infty$ for every $n \in \mathbb{N}$, which implies $a = 0$.

Similarly, if f is integrable, then $A = \{x \in X : f(x) \neq 0\}$ is always σ-finite, since $A = \bigcup_{n=1}^{\infty} A_n$ with $A_n = \{x \in X : f(x) \geq \frac{1}{n}\}$ and $\mu(A_n) \leq n \int_X f \, d\mu < \infty$. (A set $A \in \mathcal{A}$ is called σ-*finite* if it can be covered by a countable collection of sets with finite measure).

The following two theorems are used frequently.

Theorem B.1.4 (Monotone Convergence Theorem). *Let* $(f_n)_{n \in \mathbb{N}}$ *be a point-wise monotonically increasing sequence of positive integrable functions. For* $x \in X$ *let* $f(x) = \lim_n f_n(x)$ *(with possible value* ∞*). Then f is measurable and one has*

$$\int_X f \, d\mu = \lim_n \int_X f_n \, d\mu.$$

Proof See [Rud87, Theorem 1.26]. □

Theorem B.1.5 (Dominated Convergence Theorem). *If the sequence of measurable functions* $(f_n)_{n \in \mathbb{N}}$ *converges point-wise to some function f, and there exists an integrable function g such that* $|f_n| \leq |g|$ *for every* $n \in \mathbb{N}$*, then f is integrable and*

$$\int_X f \, d\mu = \lim_n \int_X f_n \, d\mu.$$

Proof (See [Rud87], Theorem 1.34). □

A measure space $(X, \mathcal{A}, \mu)$ is called *complete* if $\mathcal{A}$ is closed under taking subsets of sets of measure zero, i.e., if $A \in \mathcal{A}$ with $\mu(A) = 0$ and $B \subseteq A$, then $B \in \mathcal{A}$ (and then it follows automatically that $\mu(B) = 0$).

Given an arbitrary measure space $(X, \mathcal{A}, \mu)$, one can always pass to its completion $(X, \widehat{\mathcal{A}}, \hat{\mu})$, which is obtained by adding all subsets of null-sets in $\mathcal{A}$ with the obvious extension $\hat{\mu}$ of μ. A function is integrable with respect to $\hat{\mu}$ if and only if it differs only on a null-set from a μ-integrable function, and then the two integrals coincide. So passing to the completion does not change the theory substantially, but it gives more freedom in dealing with measurable and integrable functions.

Thus we will always consider measurable functions and their integrals with respect to the completion $(X, \widehat{\mathcal{A}}, \hat{\mu})$, without further mention.

B.2   The Riesz Representation Theorem

Let X be a locally compact Hausdorff space. By a *Radon integral* on X we understand a positive linear functional $I : C_c(X) \to \mathbb{C}$, i.e., I is linear, and $I(\phi) \geq 0$ if $\phi \geq 0$. In this section we want to associate to each Radon integral I a certain Borel measure $\mu : \mathcal{B} \to [0, \infty]$ such that the functional I can be recovered via integration of functions with respect to μ. Recall the notion of an outer Radon measure from Sect. 1.3.

Lemma B.2.1 *For an outer Radon measure μ on X and every measurable set $A \subset X$ with $\mu(A) < \infty$ one has*

$$\mu(A) = \sup_{\substack{K \subset A \\ K \text{ compact}}} \mu(K).$$

Proof First let A be a subset of a compact set L and set $T = L \smallsetminus A$. For any given $\delta > 0$ there exists an open set $W \supset T$ with $\mu(W \smallsetminus T) < \delta$. The set $K = L \smallsetminus W$ is closed in L, hence compact, and we have

$$K = L \smallsetminus W \subset L \smallsetminus T = L \smallsetminus (L \smallsetminus A) = A$$

and

$$\mu(A \smallsetminus K) = \mu(A \smallsetminus (L \smallsetminus W)) \leq \mu(W \smallsetminus (L \smallsetminus A)) = \mu(W \smallsetminus T) < \delta.$$

Let now A be arbitrary with $\mu(A) < \infty$ and $\varepsilon > 0$. By outer regularity there exists an open set $U \supset A$ of finite measure. As U is inner regular, there is a compact set $L \subset U$ such that $\mu(U \smallsetminus L) < \varepsilon/2$. Then $\mu(A) = \mu(A \cap L) + \mu(A \smallsetminus L)$ and $\mu(A \smallsetminus L) \leq \mu(U \smallsetminus L) < \varepsilon/2$. Set $B = A \cap L$. By the first part there is a compact set $K \subset B$ with $\mu(B \smallsetminus K) < \varepsilon/2$ and we conclude

$$\mu(A \smallsetminus K) = \mu(B \smallsetminus K) + \mu(A \smallsetminus L) < \varepsilon/2 + \varepsilon/2 = \varepsilon. \qquad \square$$

Theorem B.2.2 (Riesz Representation Theorem). *Let $I : C_c(X) \to \mathbb{C}$ be a positive linear functional. Then there exists a unique outer Radon measure $\mu : \mathcal{B} \to [0, \infty]$ such that*

$$I(\phi) = \int_X \phi \, d\mu$$

for every $\phi \in C_c(X)$.

Proof Using Lemma B.2.1 the conditions of outer regularity imply the conditions given in [Rud87, Theorem 2.14], where a proof of the theorem is given. $\square$

In the literature one will find the notion *Radon measure* sometimes attached to what we call an *inner Radon measure*. A locally finite Borel measure μ on $\mathcal{B}$ is called an *inner Radon measure* if

- $\mu(A) = \sup_{K \subset A} \mu(K)$ holds for every $A \in \mathcal{B}$. Here the supremum is extended over all subsets K of A that are compact in the space X.

there is a representation theorem of the above type for inner Radon measures as well, so for every positive linear functional I on $C_c(X)$ there exists a unique inner Radon measure μ_{inn} such that

$$I(f) = \int_X f \, d\mu_{inn}$$

holds for every $f \in C_c(X)$. Thus the two representation theorems give a bijection $\mu \mapsto \mu_{inn}$ from the set of all outer Radon measures to the set of all inner Radon measures on X. This map can be made explicit, as for a given outer Radon measure μ one can describe the corresponding inner Radon measure μ_{inn} by

$$\mu_{inn}(A) = \sup_{\substack{K \subset A \\ K \text{ compact}}} \mu(K).$$

Remark B.2.3 A topological space X is called σ-*compact* if X is a countable union of compact sets. If μ is an outer Radon measure on the σ-compact space X, then every measurable set $A \subset X$ is σ-finite, from which it follows that the equation $\mu(A) = \sup_{K \subset A} \mu(K)$ holds for every measurable set $A \subset X$. Thus for σ-compact spaces the two notions of Radon measures coincide.

This is not true in general. As an example one can look at the space $X = \mathbb{R} \times \mathbb{R}_d$, where the second factor is equipped with the discrete topology. On compact sets $K \subset X$ we define $\mu(K) = \sum_{y \in \mathbb{R}_d} \lambda(K_y)$ with $K_y = \{x \in \mathbb{R} : (x, y) \in K\}$ and λ denotes Lebesgue measure on $\mathbb{R}$. With the obvious extension formulas we obtain outer and inner Radon measures μ and μ_{inn} on X, both given on compact sets as above. Let $\Delta = \{(x, x) : x \in \mathbb{R}\} \subset X$. Then every open set containing Δ has infinite measure with respect to μ, since it is a union of uncountably many disjoint open sets with strictly positive measure. On the other hand, every compact subset of Δ is finite and has measure zero. It follows that $\mu(\Delta) = \infty$ while $\mu_{inn}(\Delta) = 0$.

B.3   Fubini's Theorem

If $(X, \mathcal{A}, \mu)$ and $(Y, \mathcal{C}, \nu)$ are σ-finite measure spaces, then there is a unique product measure $\mu \cdot \nu : \mathcal{A} \otimes \mathcal{C} \to [0, \infty]$ on the σ-algebra $\mathcal{A} \otimes \mathcal{C}$ generated by $\{A \times C : A \in \mathcal{A}, C \in \mathcal{C}\}$ such that the following theorem holds:

Theorem B.3.1 (Fubini's Theorem). *Let (X, μ) and (Y, ν) be two σ-finite measure spaces, and let f be a measurable function on $X \times Y$.*

(a) *If $f \geq 0$, then the two partial integrals $\int_X f(x, y)\, dx$ and $\int_Y f(x, y)\, dy$ define measurable functions and the Fubini formula*

$$\int_{X \times Y} f(x, y)\, d\mu \cdot \nu(x, y) = \int_X \int_Y f(x, y)\, dy\, dx$$

$$= \int_Y \int_X f(x, y)\, dx\, dy$$

holds.

(b) *If f is complex valued and one of the iterated integrals*

$$\int_X \int_Y |f(x, y)|\, dy\, dx \quad \text{or} \quad \int_Y \int_X |f(x, y)|\, dx\, dy$$

is finite, then f is integrable and the Fubini formula holds.

Proof ([Rud87], Theorem 7.8). □

Unfortunately, the Haar measures we use in this book fail to be σ-finite in general, so the above theorem cannot be applied in all situations. In order to fix this, we shall prove a version of Fubini's Theorem that works for all outer Radon measures on locally compact spaces. As preparation we need.

Lemma B.3.2 *Let μ be an outer Radon measure on X, and let $\mathcal{F}$ be any subset of $C_c(X)$ consisting of functions $\phi \geq 0$ such that for $\phi, \psi \in \mathcal{F}$ there exists a function $\eta \in \mathcal{F}$ with $\eta \geq \phi, \psi$. Then*

$$\sup_{\phi \in \mathcal{F}} \int_X \phi(x)\, dx = \int_X \sup_{\phi \in \mathcal{F}} \phi(x)\, d\mu(x).$$

It is part of the assertion that the integrand on the right is a measurable function.

Proof We first show that the function $g(x) = \sup_{\phi \in \mathcal{F}} \phi(x)$ is measurable. As the Borel σ-algebra on $\mathbb{R}$ is generated by the intervals (a, ∞) for $a \in \mathbb{R}$ it suffices to show that $g^{-1}(a, \infty)$ is measurable for every $a \in \mathbb{R}$. Then

$$g^{-1}(a, \infty) = \bigcup_{\phi \in \mathcal{F}} \phi^{-1}(a, \infty)$$

is a union of open sets, hence open and thus measurable.

For the integral formula, note that the estimate "$\leq$" is trivially satisfied. Let $s = \sum_{i=1}^m a_i 1_{A_i}$ be a step function with $s \leq g$. By the definition of step functions,

$\mu(A_i) < \infty$ for every i. So we can find compact sets $K_i \subset A_i$ such that for a given $\varepsilon > 0$ one has

$$\int_X s \, d\mu < \sum_{i=1}^m a_i \mu(K_i) + \varepsilon.$$

Let $K = \bigcup_{i=1}^m K_i$ and write $s_0 = \sum_{i=1}^m a_i 1_{K_i}$. For given $0 < \delta < 1$ one has $(1 - \delta)s_0(x) < g(x)$ for every $x \in K$. This implies that for every $x \in K$ there exists $\phi_x \in \mathcal{F}$ such that $(1 - \delta)s_0(x) < \phi_x(x)$. The open sets $U_x = \{y : (1 - \delta)s_0(y) < \phi_x(y)\}$ form an open covering of the compact set K, so there are $x_1, \ldots, x_n$ such that $K \subset U_{x_1} \cup \cdots \cup U_{x_n}$. By assumption, there exists $\phi \in \mathcal{F}$ with $\phi(x) \geq \phi_{x_1}(x), \ldots, \phi_{x_n}(x)$ for every $x \in X$. Then $\phi > (1 - \delta)s_0$, so that

$$\int_X s \, dx < \int_X s_0 \, dx + \varepsilon < \frac{1}{1 - \delta} \int_X \phi \, dx + \varepsilon.$$

Varying first ϕ, then s, this implies

$$\int_X g \, dx \leq \frac{1}{1 - \delta} \sup_{\phi \in \mathcal{F}} \int_X \phi \, dx + \varepsilon.$$

Letting ε and δ tend to zero, we get the claim. $\qquad\square$

Theorem B.3.3 (Fubini's Theorem for Radon measures). *Let μ and ν be outer Radon measures on the Borel sets of the locally compact spaces X and Y, respectively. Then there exists a unique outer Radon measure $\mu \cdot \nu$ on $X \times Y$ such that*

(a) *If $f : X \times Y \to \mathbb{C}$ is $\mu \cdot \nu$-integrable, then the partial integrals $\int_X f(x, y) \, dx$ and $\int_Y f(x, y) \, dy$ define integrable functions such that Fubini's formula holds:*

$$\int_{X \times Y} f(x, y) \, d(x, y) = \int_X \int_Y f(x, y) \, dy \, dx = \int_Y \int_X f(x, y) \, dx \, dy.$$

(b) *If f is measurable such that $A = \{(x, y) \in X \times Y : f(x, y) \neq 0\}$ is σ-finite, and if one of the iterated integrals*

$$\int_X \int_Y |f(x, y)| \, dy \, dx \quad \text{or} \quad \int_Y \int_X |f(x, y)| \, dx \, dy$$

is finite, then f is integrable and the Fubini formula holds.

Proof Uniqueness is an immediate consequence of the Riesz Representation Theorem, since Fubini's formula determines the values of the integral on $C_c(X \times Y)$.

To show existence of $\mu \cdot \nu$ we first observe that by the classical Fubini theorem for each compact set $K = K_1 \times K_2$ there is a unique product measure $(\mu \cdot \nu)_K$ on K such that integration with respect to $(\mu \cdot \nu)_K$ is given by integration in parts. Since $(\mu \cdot \nu)_L$ restricts to $(\mu \cdot \nu)_K$ whenever $K \subseteq L$ with $L = L_1 \times L_2$ compact, it follows that integrating a continuous function with compact support in K with respect to $(\mu \cdot \nu)_K$ determines a well defined positive functional on $C_c(X \times Y)$. Let $\mu \cdot \nu$ denote the corresponding outer Radon measure, as in Riesz's Representation Theorem.

To show that Fubini's formula holds for all integrable functions $f : X \times Y \to \mathbb{C}$, we may assume, by linearity, that $f \geq 0$. Then, approximating f point-wise by an increasing sequence of step functions, it follows from the Monotone Convergence Theorem that it suffices to prove the theorem for step functions. By linearity, we may assume that $f = 1_A$ for some measurable set $A \subset X \times Y$ with finite measure.

If $A = U$ is open, we use Lemma B.3.2 several times to get

$$\int_X \int_Y 1_U(x, y) \, dy \, dx = \int_X \int_Y \sup_{0 \leq \phi \leq 1_U} \phi(x, y) \, dy \, dx$$

$$= \sup_{0 \leq \phi \leq 1_U} \int_X \int_Y \phi(x, y) \, dy \, dx$$

$$= \sup_{0 \leq \phi \leq 1_U} \int_{X \times Y} \phi(x, y) \, d(x, y)$$

$$= \int_{X \times Y} \sup_{0 \leq \phi \leq 1_U} \phi(x, y) \, d(x, y)$$

$$= \int_{X \times Y} 1_U(x, y) \, d(x, y).$$

If $A = K$ is a compact set, then let V be a relatively compact open neighborhood of K. Then $1_K = 1_V - 1_{V \smallsetminus K}$. The claim follows for $A = K$. Using the Monotone Convergence Theorem, the claim follows for A being an arbitrary σ-compact set.

Every measurable set A with finite measure is a disjoint union of a σ-compact set A' and a null-set N. To see this, note first that by Lemma B.2.1 we have $\mu(A) = \sup\{\mu(K) : K \subset A, K \text{ compact}\}$. Choose an increasing sequence of compact sets K_n such that $\mu(K_n) \to \mu(A)$. Then $A' = \bigcup_{n \in \mathbb{N}} K_n$ is σ-compact with $N = A \smallsetminus A'$ a null-set. It therefore remains to show the result for a null-set $N \subset X \times Y$. Let $\varepsilon > 0$, and let $U \supset N$ be an open set of measure $< \varepsilon$. Then

$$\int_X \int_Y 1_N \, dy \, dx \leq \int_X \int_Y 1_U \, dy \, dx = \int_{X \times Y} 1_U < \varepsilon.$$

Letting ε tend to zero, the first assertion of the theorem follows.

For the second assertion let $f : X \times Y \to \mathbb{C}$ be given such that the partial integral $\int_Y \int_X |f(x, y)| \, dx \, dy$ exists. It suffices to show that $|f|$ is integrable. Choose a sequence $(A_n)_{n \in \mathbb{N}}$ of measurable sets in $X \times Y$ with finite measure such that $A = \bigcup_{n=1}^\infty A_n$ and define $f_n : X \times Y \to \mathbb{C}$ by $f_n = \min(|f| 1_{A_n}, n)$. Then $(f_n)_{n \in \mathbb{N}}$ is an increasing sequence of integrable functions, which converges point-wise to $|f|$. It follows from the first part of this theorem that

$$\int_{X \times Y} f_n(x, y) \, d(x, y) = \int_Y \int_X f_n(x, y) \, dx \, dy \le \int_Y \int_X |f(x, y)| \, dx \, dy$$

for every $n \in \mathbb{N}$. The result follows now from the Monotone Convergence Theorem.   □

Remark B.3.4 We should note that the analogue of the above theorem does not hold for the inner Radon measures. Indeed, if $\Delta \subset \mathbb{R} \times \mathbb{R}_d$ is as in Remark B.2.3, then Δ is a null-set with respect to the inner Radon measure on $\mathbb{R} \times \mathbb{R}^d$ corresponding to Lebesgue measure on $\mathbb{R}$ and counting measure on $\mathbb{R}_d$. But $\int_{\mathbb{R}} \left(\sum_{y \in \mathbb{R}_d} 1_\Delta(x, y) \right) dx = \infty$.

B.4 L^p-Spaces and the Riesz-Fischer Theorem

Let $(X, \mathcal{A}, \mu)$ be a measure space. For $1 \le p < \infty$ let $\mathcal{L}^p(X)$ be the set of all measurable functions $f : X \to \mathbb{C}$ such that

$$\|f\|_p \overset{\text{def}}{=} \left(\int_X |f|^p \, d\mu \right)^{\frac{1}{p}} < \infty.$$

A function in $\mathcal{L}^2(X)$ is also called a *square integrable* function. Further, let $\mathcal{L}^\infty(X)$ be the set of measurable functions $f : X \to \mathbb{C}$ such that there is a null-set N, on the complement of which f is bounded. For such f let

$$\|f\|_\infty \overset{\text{def}}{=} \inf\{c > 0 : \exists \text{ null} - \text{set } N \text{ with } |f(X \setminus N)| \le c\}.$$

Note that $\|\cdot\|_\infty$ might differ from the sup-norm $\|\cdot\|_X$ defined by $\|f\|_X = \sup\{|f(x)| : x \in X\}$. A function $f \in \mathcal{L}^\infty(X)$ might even be unbounded. It is easily seen that $\|\cdot\|_\infty$ is a semi-norm on $\mathcal{L}^\infty(X)$, but it takes a bit of work to see that this is also true for $\|\cdot\|_p$ for $1 \le p < \infty$. For this we first need

Proposition B.4.1 (Hölder's inequality). *Let* $1 \le p, q \le \infty$ *be such that* $\frac{1}{p} + \frac{1}{q} = 1$ *(we set* $q = \infty$ *if* $p = 1$*). Then* $f \cdot g \in \mathcal{L}^1(X)$ *for every* $f \in \mathcal{L}^p(X)$ *and* $g \in \mathcal{L}^q(X)$ *and*

$$\|f \cdot g\|_1 \le \|f\|_p \|g\|_q.$$

Proof In case $p = 1$ and $q = \infty$ we have $f \cdot g : X \to \mathbb{C}$ a measurable function with $|f \cdot g| < |f| \|g\|_\infty$ almost everywhere. Thus $\|f \cdot g\|_1 \le \|f\|_1 \|g\|_\infty$ by Proposition B.1.2.

So let $1 < p < \infty$. We may assume that $\|f\|_p \ne 0 \ne \|g\|_q$ since otherwise $f \cdot g = 0$ almost everywhere and hence $\|fg\|_1 = 0$.

We first show that $ab \le \frac{a^p}{p} + \frac{b^q}{q}$ for all $a, b \ge 0$. To see this, consider the function $\phi : (0, \infty) \to \mathbb{R}, \phi(t) = t^{1/p} - \frac{1}{p}t$. It has a global maximum at $t = 1$, since $\phi'(t) = 0$ if and only if $t = 1$ and $\phi''(t) < 0$ for every t. This implies that

$$\frac{1}{q} = 1 - \frac{1}{p} = \phi(1) \ge \phi\left(a^p b^{-q}\right) = ab^{-q/p} - \frac{1}{p}a^p b^{-q}.$$

Multiplying this with b^q gives

$$\frac{1}{q}b^q + \frac{1}{p}a^p \ge ab^{q-q/p} = ab.$$

If we apply this inequality to $a = \frac{|f|}{\|f\|_p}$ and $b = \frac{|g|}{\|g\|_q}$ we get

$$\frac{|f \cdot g|}{\|f\|_p \|g\|_q} \le \frac{|f|^p}{p\|f\|_p^p} + \frac{|g|^q}{p\|g\|_q^q}.$$

The integral of the function on the right hand side exists by assumption, which then implies by Proposition B.1.2 that the integral of the function on the left hand side exists, too. Integrating both sides we get

$$\frac{1}{\|f\|_p \|g\|_q} \int_X |f \cdot g| \, d\mu \le \frac{1}{p\|f\|_p^p} \int_X |f|^p \, dx + \frac{1}{p\|g\|_q^q} \int_X |g|^q \, d\mu$$

$$= \frac{1}{p} + \frac{1}{q} = 1.$$

This finishes the proof.   □

Proposition B.4.2 (Minkowski's inequality). *Let $p \in [1, \infty]$. Then for all $f, g \in \mathcal{L}^p(X)$ we have $f + g \in \mathcal{L}^p(X)$ with*

$$\|f + g\|_p \le \|f\|_p + \|g\|_p.$$

Thus $\|\cdot\|_p$ is a semi-norm on $\mathcal{L}^p(X)$ for every $1 \le p \le \infty$.

Proof The result is clear for $p = 1$ or $p = \infty$, so let $1 < p < \infty$. By Hölder's inequality for $\mathbb{C}^2$ we get

$$|a + b| \le |a| \cdot 1 + |b| \cdot 1 \le \left(|a|^p + |b|^p\right)^{1/p} 2^{1/q}$$

for $a, b \in \mathbb{C}$ and $q = \frac{p}{p-1}$. We therefore get

$$|f + g|^p \le 2^{p/q}\left(|f|^p + |g|^q\right).$$

Since the right hand side is integrable, the same is true for the left hand side, which shows that $f + g \in \mathcal{L}^p(X)$. Since $p = (p - 1)q$ we have $|f + g|^{p-1} \in \mathcal{L}^q(X)$. Applying Hölder we get

$$\|f + g\|_p^p = \int_X |f + g| \cdot |f + g|^{p-1} d\mu$$

$$\le \int_X |f| \cdot |f + g|^{p-1} d\mu + \int_X |g| \cdot |f + g|^{p-1} d\mu$$

$$\le (\|f\|_p + \|g\|_p) \left(\int_X |f + g|^{(p-1)q} d\mu \right)^{1/q}$$

$$= (\|f\|_p + \|g\|_p) \|f + g\|_p^{p/q},$$

where the last equation follows from $p = (p - 1)q$. Dividing both sides by $\|f + g\|_p^{p/q}$ and using $p - \frac{p}{q} = 1$ gives the result. □

Definition A measurable function f is a *null function* if there exists a null-set N with $f(X \setminus N) = \{0\}$. Since integrals don't see null-sets, it follows that for every $p \in [1, \infty]$ the set $\mathcal{N}$ of null functions is a sub vector space of $\mathcal{L}^p(X)$ such that

$$f \in \mathcal{N} \Leftrightarrow \|f\|_p = 0.$$

Set

$$L^p(X) \overset{\text{def}}{=} \mathcal{L}^p(X)/\mathcal{N}.$$

Then $\|\cdot\|_p$ factorizes to give a norm on $L^p(X)$ for every $1 \le p \le \infty$. We need to know that $(L^p(X), \|\cdot\|_p)$ is a Banach space for every $p \in [1, \infty]$. This is easy for $p = \infty$ since for any Cauchy sequence $(f_n)_{n \in \mathbb{N}}$ in $\mathcal{L}^\infty(X)$ we can choose a null-set $N \subseteq X$ such that $(f_n)_{n \in \mathbb{N}}$ is Cauchy with respect to $\|\cdot\|_{X \setminus N}$, and hence converges uniformly to a bounded function $f : X \setminus N \to \mathbb{C}$. If we trivially extend f to X we obtain a limit $f \in \mathcal{L}^\infty(X)$ of the sequence $(f_n)_{n \in \mathbb{N}}$. Note that f is measurable as a point-wise limit of measurable functions. The cases $1 \le p < \infty$ require a bit more work.

Theorem B.4.3 (Riesz-Fischer). *Let $1 \le p < \infty$ and suppose that (f_n) is a sequence in $\mathcal{L}^p(X)$ that is a Cauchy sequence with respect to $\|\cdot\|_p$. Then there exists a function $f \in \mathcal{L}^p(X)$ such that*

(a) $\|f_n - f\|_p \to 0$.

(b) *There exists a subsequence $(f_{n_k})_{k \in \mathbb{N}}$ of $(f_n)_{n \in \mathbb{N}}$ such that $f_{n_k}(x)$ converges to $f(x)$ for every x outside a set of measure zero.*

Proof Choose indices $n_1 < n_2 < n_3 < \cdots$ so that $\|f_{n_{k+1}} - f_{n_k}\|_p < \frac{1}{2^k}$ and put $g_k = f_{n_{k+1}} - f_{n_k}$. Let $g = \sum_{k=1}^\infty |g_k|$. By Minkowski's inequality we have $\left(\sum_{k=1}^m |g_k|\right)^p$ integrable with integrals bounded above by $\left(\sum_{k=1}^\infty \|g_k\|_p\right)^p \le 1$. It then follows from the Monotone Convergence Theorem that g^p is integrable, hence $g \in \mathcal{L}^p(X)$. But then it follows from Remark B.1.3 that $g(x) \neq \infty$ outside a null-set $N \subseteq X$ and the series $\sum_{k=1}^\infty g_k(x)$ converges absolutely to some function $h(x)$ for every $x \notin N$. We trivially extend h to all of X. Then h is measurable as a point-wise limit of measurable functions, and since $|h(x)| \le g(x)$ for every $x \in X$ we have $h \in \mathcal{L}^p(X)$. Put $f = h + f_{n_1}$. Then

$$f(x) = f_{n_1}(x) + \lim_{m \to \infty} \sum_{k=1}^{m} (f_{n_{k+1}} - f_{n_k})$$

$$= \lim_{m \to \infty} f_{n_m}(x)$$

for every $x \notin N$, which proves assertion (b) of the theorem. To see (a) observe that

$$\|f - f_{n_m}\|_p = \left\| \sum_{k=m}^{\infty} (f_{n_{k+1}} - f_{n_k}) \right\|_p \le \sum_{k=m}^{\infty} \frac{1}{2^k} \to 0,$$

which implies that $(f_{n_k})_{k \in \mathbb{N}}$ converges to f with respect to $\|\cdot\|_p$. Since $(f_n)_{n \in \mathbb{N}}$ is Cauchy we also get $\|f - f_n\|_p \to 0$. $\qquad\square$

Corollary B.4.4 *The spaces $(L^p(X), \|\cdot\|_p)$ are Banach spaces for every $1 \le p \le \infty$.*

An important special case is the case $p = 2$, in which it follows from Hölder and the Riesz-Fischer theorem that $L^2(X)$ is a Hilbert space with respect to the inner product

$$\langle f, g \rangle = \int_X f(x)\overline{g(x)} \, d\mu(x).$$

Another direct corollary of the Riesz-Fischer Theorem is

Corollary B.4.5 *Suppose that $f_n \to f$ in $L^p(X)$. Then there exists a subsequence $(f_{n_k})_{k \in \mathbb{N}}$ that converges point-wise almost everywhere to f.*

We should point out that in general one cannot expect point-wise almost everywhere convergence of the original sequence $(f_n)_{n \in \mathbb{N}}$, as can be seen from

Example B.4.6 Consider $X = [0, 1]$ with the Lebesgue integral. Define a sequence $f_n : [0, 1] \to \mathbb{R}$ of integrable functions as follows: Put $f_1 = 1$, $f_2 = 1_{[0, \frac{1}{2}]}$, $f_3 = 1_{[\frac{1}{2}, 1]}$, $f_4 = 1_{[0, \frac{1}{4}]}$, $f_5 = 1_{[\frac{1}{4}, \frac{1}{2}]}$, $f_6 = 1_{[\frac{1}{2}, \frac{3}{4}]}$, $f_7 = 1_{[\frac{3}{4}, 1]}$, $f_8 = 1_{[1, \frac{1}{8}]}$, and so on. It is then clear that $\|f_n\|_1 \to 0$ but $f_n(x) \not\to 0$ for every $x \in [0, 1]$.

Remark B.4.7 It follows from Hölder's inequality that for $1 \le p, q \le \infty$ with $\frac{1}{p} + \frac{1}{q} = 1$ every function $g \in L^q(X)$ determines a continuous linear functional $T_g : L^p(X) \to \mathbb{C}$ by

$$T_g(f) = \int_X f \cdot g \, d\mu$$

such that $\|T_g\| \le \|g\|_q$ for every $g \in L^q(X)$. If $1 < q < \infty$, it is an important result in Functional Analysis that $g \mapsto T_g$ actually gives an isometric isomorphism from

$L^q(X)$ onto the space $L^p(X)'$ of all bounded linear functionals on X, equipped with the operator norm. The same holds for $q = \infty$ and $p = 1$ if X is σ-finite. However, the duality $L^1(X) \cong L^\infty(X)'$ almost never holds. (See [Rud87, Theorem 6.16]). For Radon integrals on locally compact spaces the duality $L^1(X)' = L^\infty(X)$ holds, but with $L^\infty(X)$ constructed with respect to a slightly different norm (See [Ped89, 6.4.5 and 6.5.11]).

Definition A topological space is called *separable*, if it has a countable dense subset.

Examples B.4.8

- The space $\mathbb{R}^n$ is separable as $\mathbb{Q}^n$ is a countable dense subset.

- A discrete space is separable if and only if it is countable.

Recall that a topological space is called *second countable* if its topology admits a countable base. The following Lemma on separability will be needed in the text.

Lemma B.4.9 (a) *A Hilbert space is separable if and only if it has a countable orthonormal basis.*

(b) *If X is a second countable compact Hausdorff space with a Radon measure μ, then the Hilbert space $L^2(X, \mu)$ is separable.*

Proof (a) Let the Hilbert space H be separable. Then there exists a family $(r_j)_{j \in \mathbb{N}}$ which is dense in H. The procedure of orthonormalization produces a countable orthonormal basis of H consisting of linear combinations of the r_j, hence H has a countable orthonormal basis. The other way round, let $(e_j)_{j \in \mathbb{N}}$ be a countable orthonormal basis. Then the $\mathbb{C}$-span of the e_j is dense in H. As the countable field $\mathbb{Q}(i) = \mathbb{Q} \oplus \mathbb{Q}i$ is dense in $\mathbb{C}$ the countable set of all $\mathbb{Q}(i)$-linear combinations of the e_i is dense in H.

(b) As X is second countable, there exists a countable set C of open sets such that every open set in X is the union of some elements of C. As the set of all finite intersections of the elements of C is countable again, we can extend C and assume that all finite intersections of elements of C lie in C. For $U, V \in C$ we have $\mathbf{1}_{U \cup V} = \mathbf{1}_U + \mathbf{1}_V - \mathbf{1}_{U \cap V}$ and therefore the space H spanned by all linear combinations of the functions $\mathbf{1}_U$, $U \in C$ contains all functions $\mathbf{1}_{U \cup V}$ for $U, V \in C$. A straightforward generalization of this argument implies that H contains $\mathbf{1}_A$ for any finite union A of elements of C. Let now U be an arbitrary open subset of X. By the countability of C the set U can be written as the union of a family $(U_j)_{j \in \mathbb{N}}$ where $U_j \in C$ for every $j \in \mathbb{N}$. So the sequence of functions $\left(\mathbf{1}_{U_1 \cup \cdots \cup U_j} \right)$ converges monotonically to $\mathbf{1}_U$ and so it converges in $L^2(X, \mu)$. It follows that the closure of H contains all functions $\mathbf{1}_U$ for open sets U. By the Radon property the latter span a space which contains all simple functions, hence is dense in $L^2(X, \mu)$, which therefore is separable. $\qquad \square$

B.5 The Radon-Nikodym Theorem

Recall that a measure μ on a set X is called σ-*finite* if there exists a sequence $A_n \subset X$ of measurable sets such that $X = \bigcup_n A_n$ and $\mu(A_n) < \infty$ for every $n \in \mathbb{N}$.

Theorem B.5.1 (Radon-Nikodym Theorem). *Given two σ-finite measures λ, μ on a σ-algebra $\mathcal{A}$ over a set X. Suppose that every λ-null-set is a μ-null-set. Then there exist a measurable function $h \geq 0$ such that for every measurable function $\phi : X \to \mathbb{C}$ we have ϕ is μ-integrable if and only if ϕh is λ-integrable, and then $\int_X \phi \, d\lambda = \int_X \phi h \, d\mu$. In particular, $\mu(A) = \int_A h \, d\lambda$ for every measurable set $A \subset X$.*

The function h is called the Radon-Nikodym derivative *of μ with respect to λ. It is unique up to changes on λ-null-sets. One also writes this as $d\mu = h d\lambda$.*

Proof As both measures are σ-finite, there is a pairwise disjoint decomposition $X = \bigcup_{j=1}^{\infty} X_j$ into measurable sets X_j, on which both measures are finite. Suppose there exist Radon-Nikodym derivatives on each X_j. Then these can be patched together to give a Radon-Nikodym derivative on the whole of X. Therefore it suffices to prove the theorem under the condition that both measures be finite.

Assume μ and λ are finite and set $\tau = \lambda + \mu$. For every τ-integrable function ϕ we have by the Cauchy-Schwarz inequality,

$$\left| \int_X \phi \, d\mu \right| \leq \int_X |\phi| \, d\mu \leq \int_X |\phi| \, d\tau \leq \left(\int |\phi|^2 \, d\tau \right)^{\frac{1}{2}} (\tau(X))^{\frac{1}{2}}.$$

Therefore the map $\phi \mapsto \int_X \phi \, d\mu$ is a continuous linear map on $L^2(X, \tau)$. The same is true for the map $\phi \mapsto \int_X \phi \, d\lambda$. By the completeness of the Hilbert space $L^2(X, \tau)$ there are unique (up to addition of null functions) measurable, τ-square integrable functions f and g such that $\int_X \phi \, d\mu = \int_X f\phi \, d\tau$ and $\int_X \phi \, d\lambda = \int_X g\phi \, d\tau$. Approximating a positive measurable function ϕ by an increasing sequence $(\phi_n)_{n \in \mathbb{N}}$ of positive step functions and using the Monotone Convergence Theorem on both sides of the equation $\int_X \phi_n \, d\lambda = \int_X \phi_n g \, d\tau$ shows that the function ϕ is λ-integrable if and only if ϕg is τ-integrable. A similar result holds for μ and τ.

Let N be the set $\{x \in X : g(x) = 0\}$. Then $\lambda(N) = 0$ and so also $\mu(N) = 0$. Let $h(x) = f(x)/g(x)$ if $x \in X \setminus N$ and $h(x) = 0$ for $x \in N$. Since null-sets can be neglected for integration it follows for any measurable function $\phi : X \to \mathbb{C}$ that ϕ is μ-integrable if and only if $\phi f = \phi g h$ is τ-integrable if and only if ϕh is λ-integrable, and then

$$\int_X \phi \, d\mu = \int_X \phi g h \, d\tau = \int_X \phi h \, d\lambda. \qquad \square$$

B.6   Vector-Valued Integrals

In this section we introduce the vector-valued integral, also known as the *Bochner integral*. For a Banach space valued function $f : X \to V$ one wants to define an integral $\int_X f \, d\mu \in V$, such that for every continuous linear functional α on V the formula

$$\alpha \left(\int_X f \, d\mu \right) = \int_X \alpha(f) \, d\mu$$

holds.

To be precise, let $(V, \|\cdot\|)$ be a Banach space, and let $(X, \mathcal{A}, \mu)$ be a measure space. A *simple function* is a function $s : X \to V$, that can be written as

$$s = \sum_{j=1}^{n} \mathbf{1}_{A_j} b_j,$$

where $A_1, \ldots, A_n$ are pairwise disjoint measurable sets of finite measure $\mu(A_j) < \infty$, and $b_j \in V$. We define the integral of the simple function s as

$$\int_X s \, d\mu \stackrel{\text{def}}{=} \sum_{j=1}^{n} \mu(A_j) b_j \in V.$$

Note that $\left\| \int_X s \, d\mu \right\| \le \int_X \|s\| \, d\mu$ and that for every linear functional $\alpha : V \to \mathbb{C}$ one has $\alpha(\int_X s \, d\mu) = \int_X \alpha(s) \, d\mu$.

We equip V with the Borel σ-algebra. A measurable function $f : X \to V$ is called *integrable* if there exists a sequence s_n of simple functions such that

$$\lim_{n \to \infty} \int_X \|f - s_n\| \, d\mu = 0.$$

In this case we call (s_n) an *approximating sequence*. The following proposition will justify the terminology.

Proposition B.6.1 (a) *If f is integrable and (s_n) an approximating sequence, then the sequence of vectors $\int_X s_n \, d\mu$ converges. Its limit does not depend on the choice of the approximating sequence.*

One defines the integral of f to be this limit:

$$\int_X f \, d\mu \stackrel{\text{def}}{=} \lim_{n \to \infty} \int_X s_n \, d\mu.$$

(b) *For every integrable function f one has*

$$\left\| \int_X f \, d\mu \right\| \le \int_X \|f\| \, d\mu < \infty.$$

(c) *Let f be integrable. For every continuous linear operator $T : V \to W$ to a Banach space W one has*

$$T\left(\int_X f \, d\mu\right) = \int_X T(f) \, d\mu.$$

(d) If $V = \mathbb{C}$, then the Bochner integral coincides with the usual integral.

Proof   It suffices to show that for every approximating sequence (s_n) the sequence $\int_X s_n \, d\mu$ converges, for if (t_n) is another approximating sequence, then the sequence (r_n) with $r_{2n} = s_n$ and $r_{2n-1} = t_n$ also is an approximating sequence. As $\int_X r_n \, d\mu$ must converge, the limits of $\int_X s_n \, d\mu$ and $\int_X t_n \, d\mu$ coincide.

To show the convergence, it suffices to show that the sequence of vectors $\int_X s_n \, d\mu$ is a Cauchy-sequence. For $m, n \in \mathbb{N}$ consider

$$\left\| \int_X s_m \, d\mu - \int_X s_n \, d\mu \right\| = \left\| \int_X s_m - s_n \, d\mu \right\|$$

$$\leq \int_X \| s_m - s_n \| \, d\mu$$

$$\leq \int_X \| s_m - f \| \, d\mu + \int_X \| f - s_n \| \, d\mu$$

and the latter tends to zero as $m, n \to \infty$. Therefore, $\int_X s_n \, d\mu$ indeed is a Cauchy-sequence. This proves (a).

To prove (b) note that the l inequality $\big| \|f\| - \|s_n\| \big| \leq \| f - s_n \|$ implies that the $\mathbb{C}$-valued function $\|f\|$ is integrable and that $\|s_n\|$ converges to $\|f\|$ in $L^1(X)$. It follows that

$$\left\| \int_X f \, d\mu \right\| = \lim_n \left\| \int_X s_n \, d\mu \right\| \leq \lim_n \int_X \| s_n \| \, d\mu = \int_X \| f \| \, d\mu.$$

Finally for part (c). The continuity and linearity of T implies that

$$T\left(\int_X f \, d\mu\right) = \lim_n \int_X T(s_n) \, d\mu.$$

We want to show that the right hand side equals $\int_X T(f) \, d\mu$. Since T is continuous, there exists $C > 0$ such that $|T(v)| \leq C\|v\|$ holds for every $v \in V$. In particular one has $\|T(f)\| \leq C\|f\|$ and so $T(f)$ is an integrable function. We can estimate

$$\left\| \int_X T(f)\,d\mu - \int_X T(s_n)\,d\mu \right\| \le \int_X \|T(f) - T(s_n)\|\,d\mu$$

$$= \int_X \|T(f - s_n)\|\,d\mu$$

$$\le C \int_X \|f - s_n\|\,d\mu.$$

As the latter tends to zero, the claim follows. Finally, (d) follows from the last estimate applied to the identity operator.   □

Definition A function $f : X \to V$ is called a *separable function* if there exists a countable set $C \subset V$ such that the image $f(X)$ is contained in the closure $\overline{C}$ of C. Recall that the Banach space V is called separable if it contains a countable dense set. So if V is separable, then every function $f : X \to V$ is separable. In applications we will, however, encounter non-separable Banach spaces, which is why we need the notion of separable functions.

A function $f : X \to V$ is called *essentially separable* if there exists a measurable set $N \subset X$ of measure zero such that f is separable on $X \setminus N$. This means that f is separable up to a negligible alteration.

Lemma B.6.2 *Let X be a topological space and $f : X \to V$ a continuous function of σ-compact support. Then f is separable.*

Proof The image $f(X)$ is a σ-compact set. So we have to show that every σ-compact subset $K \subset V$ has a dense countable subset. It suffices to assume that K is compact. By compactness it follows that for every $n \in \mathbb{N}$ there are elements $k_1^n, \ldots, k_{r(n)}^n \in K$ such that

$$K \subset \bigcup_{\nu=1}^{r(n)} B_{1/n}(k_\nu),$$

where for $r > 0$ and $a \in V$ we denote by $B_r(a)$ the open ball around a of radius r, i.e., the set of all $x \in V$ with $\|x - a\| < r$. The set of all k_ν^n for $n \in \mathbb{N}$ and $1 \le \nu \le r(n)$ then is a countable dense subset of K.   □

Proposition B.6.3 *For a measurable function $f : X \to V$ the following are equivalent:*

- *f is integrable.*
- *f is essentially separable and $\int_X \|f\|\,d\mu < \infty$.*

Proof If f is integrable, then by Proposition B.6.1 (b) the function $\|f\|$ is integrable as well. We need to show that it is essentially separable. So let (s_n) be an approximating sequence. As each s_n has finite image, the Banach space E generated by all the images $s_n(X)$, $n = 1, 2, \dots$ is separable. The set $N = f^{-1}(V \setminus E)$ is a countable union $N = \bigcup_n N_n$, where $N_n = \{x \in X : \|f(x) - e\| \geq \frac{1}{n} \ \forall e \in E\}$. Since $\int_X \|f - s_n\| \, d\mu$ tends to zero, the set N_n is a set of measure zero for every $n \in \mathbb{N}$. hence N is a null-set and so f is essentially separable.

For the converse direction assume that f is essentially separable and satisfies $\int_X \|f\| \, d\mu < \infty$. After altering f on a set of measure zero, we assume f is separable. Let $C = \{c_n : n \in \mathbb{N}\}$ be a countable subset of V with $f(X) \subset \overline{C}$. For $n \in \mathbb{N}$ and $\delta > 0$ let A_n^δ be the set of all $x \in X$ such that $\|f(x)\| \geq \delta$ and $\|f(x) - c_n\| < \delta$. As f is measurable, this is a measurable set. To have a sequence of pairwise disjoint sets, we define

$$D_n^\delta \overset{\text{def}}{=} A_n^\delta \setminus \bigcup_{k < n} A_k^\delta.$$

Then the set $\bigcup_{n \in \mathbb{N}} A_n^\delta = \bigcup_{n \in \mathbb{N}} D_n^\delta$ equals $f^{-1}(f(X) \setminus B_\delta(0))$, as C is dense in $f(X)$. As f is integrable, the set $\bigcup_{n \in \mathbb{N}} D_n^\delta$ is of finite measure. Let $s_n = \sum_{j=1}^n \mathbf{1}_{D_j^{1/n}} c_j$. Then s_n is a simple function. We show that the sequence (s_n) converges to f point-wise. Let $x \in X$. If $f(x) = 0$, then $s_n(x) = 0$ for every n. So suppose $f(x) \neq 0$. Then $\|f(x)\| \geq \frac{1}{n}$ for some $n \in \mathbb{N}$. For every $m \geq n$ one has $x \in \bigcup_{\nu \in \mathbb{N}} D_\nu^{1/m}$, so that for each $m \geq n$ there exists a unique ν_0 with $x \in D_{\nu_0}^{1/m}$, hence $s_m(x) = c_{\nu_0}$ and $\|f(x) - c_{\nu_0}\| < \frac{1}{m}$, which implies $s_n \to f$ as claimed. We also see that $\|s_n\| \leq 2\|f\|$ by construction. So we get $\|f - s_n\| \to 0$ point-wise and $\|f - s_n\| \leq \|f\| + \|s_n\| \leq 3\|f\|$, so by dominated convergence,

$$\int_X \|f - s_n\| \, d\mu \to 0. \qquad \square$$

Corollary B.6.4 *Suppose that X is a locally compact topological space and μ is a Radon measure. Then every continuous function $f : X \to V$ with compact support is integrable.*

Proof As the $\mathbb{C}$-valued function $\|f\|$ is again continuous of compact support, it is integrable. Thus the corollary follows immediately from Lemma B.6.2 and Proposition B.6.3. $\qquad \square$

An important case of a Bochner integral is given by the convolution $f * g$ in $L^1(G)$ for a locally compact group G.

Lemma B.6.5 *Let G be a locally compact group. If $f \in C_c(G)$ and $g \in L^1(G)$, then the Bochner integral $\int_G f(x) L_x g \, dx$ exists in the Banach space $L^1(G)$ and equals the convolution product $f * g$.*

Proof Consider the function $\phi : G \to L^1(G)$ given by $\phi(x) = f(x) L_x g$. It is continuous by Lemma 1.4.2 and it has compact support as f has compact support. So it follows that ϕ is continuous of compact support, hence integrable by Corollary B.6.4.

It follows, that the Bochner integral $\int_G \phi(x)\,dx$ exists in the Banach space $L^1(G)$. We want to show that it coincides with $f * g$. If we can show $h * \int_G \phi(x)\,dx = h * f * g$ for every $h \in C_c(G)$, the claim follows, as h might run through a Dirac net (See Lemma 1.6.6).

Let $h \in C_c(G)$. For $\phi \in L^1(G)$ the convolution product $h * \phi$ is continuous, as follows from the Theorem of Dominated Convergence. So it makes sense to evaluate this function at some given $y \in G$. One has

$$
\begin{aligned}
|h * \phi(y)| &\leq \int_G |h(z)||\phi(z^{-1}y)|\,dz \\
&= \int_G \Delta(z^{-1})|h(yz^{-1})||\phi(z)|\,dz \\
&\leq C\|\phi\|_1,
\end{aligned}
$$

with $C \geq 0$ an upper bound for the function $z \mapsto \Delta(z^{-1})|h(yz^{-1})|$. This implies that the linear functional $\alpha : \phi \mapsto h * \phi(y)$ is continuous. For $\phi = \int_G f(x)L_x g\,dx$ it follows

$$
\begin{aligned}
h * \phi(y) = \alpha(\phi) &= \alpha\left(\int_G f(x)L_x g\,dx \right) \\
&= \int_G f(x)\alpha(L_x g)\,dx \\
&= \int_G \int_G f(x)h(z)g\left(x^{-1}z^{-1}y\right)\,dz\,dx \\
&= h * f * g(y).
\end{aligned}
$$

So it follows that $\int_G f(x)L_x g\,dx = f * g$ as claimed. $\square$

Definition As an application of the Bochner integral, we will now prove the Cauchy Integral Formula for Banach space valued functions. Let $D \subset \mathbb{C}$ be an open set, and let $f : D \to V$ be holomorphic (Sect. 2.2), and let $\gamma : [0, 1] \to D$ be continuously differentiable. The *path integral* is then defined as

$$
\int_\gamma f(z)\,dz \stackrel{\text{def}}{=} \int_{[0,1]} \gamma'(t)f(\gamma(t))\,dt.
$$

Let $a \in D$, and let $B = B_r(a)$ be a disk whose closure is contained in the set D. We write $\int_{\partial B} f(z)\,dz$ for the integral over the positive oriented boundary of B, i.e., $\int_{\partial B} f(z)\,dz = \int_\gamma f(z)\,dz$, where $\gamma : [0, 1] \to D$ is given by $\gamma(t) = a + re^{2\pi i t}$.

Theorem B.6.6 (Cauchy's Integral Formula). *Let $D \subset \mathbb{C}$ be an open set and $f : D \to V$ be holomorphic, where V is a Banach space. Let $B \subset \mathbb{C}$ be an open disk with $\overline{B} \subset D$. Then for every $z \in B$ we have*

$$f(z) = \frac{1}{2\pi i} \int_{\partial B} \frac{f(\xi)}{\xi - z} \, d\xi.$$

Proof Let $\alpha : V \to \mathbb{C}$ be a continuous linear functional, then

$$\alpha \left(\frac{1}{2\pi i} \int_{\partial B} \frac{f(\xi)}{\xi - z} \, d\xi \right) = \frac{1}{2\pi i} \int_{\partial B} \frac{\alpha(f(\xi))}{\xi - z} \, d\xi = \alpha(f(z))$$

according to Cauchy's integral formula for holomorphic functions. So the two sides of our claimed equality coincide after the application of any continuous linear functional. By Corollary C.1.4, the claim follows. $\qquad\square$

Corollary B.6.7 *Let the situation be as in the theorem, and let B be an open ball around zero such that $\overline{B} \subset D$. Then there are $v_n \in V$ such that*

$$f(z) = \sum_{n=0}^{\infty} z^n v_n$$

holds for every $z \in B$, where the sum converges uniformly on every closed subset of B.

Proof If $z \in B$ and $\xi \in \partial B$, then $|z/\xi| < 1$, which means that the geometric series

$$\sum_{n=0}^{\infty} (z/\xi)^n = \frac{1}{1 - z/\xi}$$

converges uniformly for (z, ξ) in a given closed subset of $B \times \partial B$. We apply Cauchy's formula to get

$$f(z) = \frac{1}{2\pi i} \int_{\partial B} \frac{f(\xi)}{\xi - z} \, d\xi = \frac{1}{2\pi i} \int_{\partial B} \frac{1}{\xi} \frac{f(\xi)}{1 - z/\xi} \, d\xi$$

$$= \frac{1}{2\pi i} \int_{\partial B} \frac{f(\xi)}{\xi} \sum_{n=0}^{\infty} (z/\xi)^n \, d\xi = \frac{1}{2\pi i} \sum_{n=0}^{\infty} z^n \int_{\partial B} \frac{f(\xi)}{\xi^{n+1}} \, d\xi.$$

The last step is justified by uniform convergence. The claim follows. $\qquad\square$

Appendix C
Functional Analysis

C.1 Basic Concepts

A *topological vector space* is a complex vector space V together with a topology such that the additive group $(V, +)$ is a topological group and such that the scalar multiplication map $\mathbb{C} \times V \to V$, which maps (λ, v) to λv, is continuous. A Banach space is a special case of a topological vector space, in which the topology is induced by a norm such that the corresponding metric space is complete. Most spaces considered here are Banach spaces, but an important example of a topological vector space, which is not a Banach space is given by the space $\mathcal{S}(\mathbb{R})$ of Schwartz functions on $\mathbb{R}$. The topology is generated by all sets of the form $f + B_{m,n,C}$, where $f \in \mathcal{S}(\mathbb{R})$ and $B_{m,n,C}$ is the set of all $g \in \mathcal{S}(\mathbb{R})$ with $|g^{(m)}(x)x^n| < C$ for every $x \in \mathbb{R}$. For a topological vector space V we denote by V' the *continuous dual space*, i.e., the space of all continuous linear functionals $\alpha : V \to \mathbb{C}$. In most cases this is a proper subspace of the *algebraic* dual space V^* consisting of *all* linear functionals on V. We say that V' *separates points* in V if for any two $v \neq w$ in V there exists $\alpha \in V'$ with $\alpha(v) \neq \alpha(w)$.

In this book we generally deal with vector spaces over $\mathbb{C}$. Sometimes it is convenient, however, to use vector spaces over $\mathbb{R}$ instead. Every complex vector spaces naturally is a vector space over the reals. We will now show that in this case every real linear functional is the real part of a complex linear functional.

Lemma C.1.1 *Let W be a complex vector space and W^* be its dual space. The map* $\Psi : \alpha \mapsto \mathrm{Re}(\alpha)$ *is a bijection of W^* to $W_{\mathbb{R}}^* = \mathrm{Hom}_{\mathbb{R}}(W, \mathbb{R})$.*

Proof The map Ψ is real linear. Let α be in its kernel. This implies that α takes only purely imaginary values. Since $\alpha(iv) = i\alpha(v)$ this implies that $\alpha = 0$. Now let $u : W \to \mathbb{R}$ be real linear. Set $\alpha(x) = u(x) - iu(ix)$. Then α is $\mathbb{C}$-linear and $u = \mathrm{Re}(\alpha)$. □

A. Deitmar, S. Echterhoff, *Principles of Harmonic Analysis*, Universitext,
DOI 10.1007/978-3-319-05792-7, © Springer International Publishing Switzerland 2014

Recall that for a linear operator $T : V \to W$ on normed spaces V, W the *operator norm* is defined by

$$\|T\| = \|T\|_{\mathrm{op}} \stackrel{\mathrm{def}}{=} \sup_{v \neq 0} \frac{\|Tv\|}{\|v\|} = \sup_{\|v\|=1} \|Tv\|,$$

and that T is called a *bounded operator* if $\|T\| < \infty$. For every vector v we have $\|Tv\| \leq \|T\|\|v\|$.

Lemma C.1.2 *A linear operator T is bounded if and only if it is continuous as a map from $V \to W$.*

Proof Let T be bounded, and let $v_j \to v$ be a convergent sequence in V. Then $\|Tv_j - Tv\| = \|T(v_j - v)\| \leq \|T\|\|v_j - v\|$ tends to zero, which means that Tv_j tends to Tv, so T is continuous.

For the converse direction assume that T is continuous, but $\|T\| = \infty$. Then there is a sequence (v_j) of vectors in V of norm one such that $\|Tv_j\|$ tends to infinity. It follows that the sequence $\frac{1}{\|Tv_j\|}v_j$ tends to zero. As T is continuous, the sequence

$$T\left(\frac{1}{\|Tv_j\|}v_j\right) = \frac{1}{\|Tv_j\|}Tv_j$$

also tends to zero, but this is a sequence of vectors of norm one. A contradiction. □

Theorem C.1.3 (Hahn-Banach Theorem). *Let M be a subspace of a Banach space V, and let $\alpha : M \to \mathbb{C}$ be linear with $|\alpha(x)| \leq \|x\|$ for every $x \in M$. Then α extends to a linear functional $V \to \mathbb{C}$ such that $|\alpha(x)| \leq \|x\|$ holds for every $x \in V$.*

Proof Let $u = \mathrm{Re}(\alpha)$. Suppose we can show that u extends to a real linear functional with $|u(x)| \leq \|x\|$, then α extends to a complex linear functional whose real part is u. For given $x \in V$ there exists $\theta \in \mathbb{R}$ such that $e^{i\theta}\alpha(x)$ is real. So, $|\alpha(x)| = |e^{i\theta}\alpha(x)| = |\alpha(e^{i\theta}x)| = |u(e^{i\theta}x)| \leq \|e^{i\theta}x\| = \|x\|$.

So it remains to show that u extends to V. What we really show is this: Let M be any real vector subspace of V, and let $u : M \to \mathbb{R}$ be real linear with $|u(x)| \leq \|x\|$ for every $x \in M$, then u can be extended to a linear map from V to $\mathbb{R}$ with $|u(x)| \leq \|x\|$ for every $x \in V$. Assume $M \neq V$, and let $v_1 \in V \smallsetminus M$. Let $M_1 = M \oplus \mathbb{R}v_1$. Then the inequality

$$u(x) + u(y) = u(x+y) \leq \|x+y\| \leq \|x - v_1\| + \|v_1 + y\|$$

implies

$$u(x) - \|x - v_1\| \leq \|y + v_1\| - u(y)$$

for $x, y \in M$. Let $\alpha \in \mathbb{R}$ be any value between the least upper bound of the left hand side as x ranges over M, and the largest lower bound of the right hand side as y ranges over M. Then

$$u(x) - \alpha \leq \|x - v_1\|$$

and

$$u(y) + \alpha \leq \|y + v_1\|$$

holds for all $x, y \in M$. Define $u_1(x + tv_1) = u(x) + t\alpha$ for $x \in M$ and $t \in \mathbb{R}$. Then u_1 extends u to M_1 and is linear. Further, for $t \neq 0$,

$$|u_1(x + tv_1)| = |u(x) + t\alpha| = |u(\frac{1}{t}x) + \alpha| |t|$$

$$\leq \left\| \frac{1}{t}x + v_1 \right\| |t| = \|x + tv_1\|.$$

This implies that u extends to M_1 satisfying the norm bound. The lemma of Zorn implies that there is a maximal subspace M of V, to which u can be extended satisfying the norm bound and what we just have seen implies that then M must be equal to V.   □

Corollary C.1.4 *Every continuous linear functional α on a subspace E of the Banach space V extends to a continuous linear functional α_1 on V such that $\|\alpha_1\| = \|\alpha\|$.*

The continuous linear functionals on V separate points, so if $v, w \in V$ with $\alpha(v) = \alpha(w)$ for every $\alpha \in V'$, then $v = w$.

Proof Let $\alpha \neq 0$ be a continuous linear functional on E. This means that $\|\alpha\| = \sup_{e \in E \setminus \{0\}} |\alpha(e)|/\|e\|$ is finite. Then $\beta = \frac{1}{\|\alpha\|}\alpha$ satisfies $\|\beta\| = 1$, so, by the Hahn-Banach Theorem, β extends to a linear functional β_1 on V with $\|\beta_1\| = 1$. Hence $\alpha = \|\alpha\|\beta$ extends to $\alpha_1 = \|\alpha\|\beta_1$ with $\|\alpha_1\| = \|\alpha\|$.

If $v \neq w$, then there exists a linear functional α on the space E spanned by v and w such that $\alpha(v) \neq \alpha(w)$. As α extends to an element of V', the claim follows.   □

Theorem C.1.5 (Open Mapping Theorem). *Let $T : E \to F$ be a continuous surjective linear map between Banach spaces. Then T is an open mapping, i.e., $T(U)$ is open in F for every open $U \subset E$.*

In particular, if $T : E \to F$ is continuous and bijective, then $T : F \to E$ is a topological isomorphism.

Proof Let $U \subset E$ be an open neighborhood of 0. We have to show that $T(U)$ contains an open neighborhood of 0. There exists $\varepsilon > 0$ such that U contains the ball $U_1 = B_\varepsilon(0)$ around zero of radius ε. Let $U_n = \frac{1}{2^n}U_1$, then $U_{n+1} + U_{n+1} \subset U_n$ for every n. We will show the existence of an open neighborhood W of 0 in F such that

$$W \subset \overline{T(U_1)} \subset T(U).$$

As $F = T(E) = \bigcup_{k=1}^{\infty} k\overline{T(U_2)}$, and F is a Baire space, one of the spaces $k\overline{T(U_2)}$ has non-empty interior, and as the map $x \mapsto kx$ is a homeomorphism on F, the set

$\overline{T(U_2)}$ has non-empty interior, which implies the existence of an open $W_1 \subset \overline{T(U_2)}$. Then $W_2 = -W_1 \subset \overline{T(U_2)}$ as well, and $W = W_1 + W_2$ is an open neighborhood of 0 contained in $\overline{T(U_1)}$. It remains to show $\overline{T(U_1)} \subset T(U)$. For this let $y_1 \in \overline{T(U_1)}$. Assume that $n \geq 1$ and $y_k \in \overline{T(U_k)}$ has been chosen for all $1 \leq k \leq n$. In the same way as for U_1 it follows that $\overline{T(U_{n+1})}$ contains a neighborhood of 0. Hence

$$\left(y_n + \overline{T(U_{n+1})}\right) \cap T(U_n) \neq \emptyset,$$

so there exists $x_n \in U_n$ such that $T(x_n) \in y_n + \overline{T(U_{n+1})}$. Put $y_{n+1} = y_n - T(x_n)$. Then $y_{n+1} \in \overline{T(U_{n+1})}$ and we iterate the inductive construction. Note that by the continuity of T we have $y_n \to 0$. As $\|x_n\| < \frac{\varepsilon}{2^n}$ the sum $x = \sum_{n=1}^{\infty} x_n$ converges in E and lies in $U_1 \subset U$. Now

$$T(x) = \lim_N T\left(\sum_{n=1}^N x_n\right) = \lim_N \sum_{n=1}^N (y_n - y_{n+1}) = \lim_N (y_1 - y_{N+1}) = y_1.$$

We conclude that $T(x) = y_1$ and so $y_1 \in T(U)$ as claimed. □

In the following we equip the direct product $V \times W$ of two Banach spaces V and W with the norm $\|(v, w)\| = \|v\| + \|w\|$. With this norm $V \times W$ is also a Banach space.

Theorem C.1.6 (Closed Graph Theorem). *Suppose that the map $T : V \to W$ is a linear map between Banach spaces, such that the graph $\mathcal{G}(T) = \{(v, Tv) : v \in V\}$ is a closed subset of $V \times W$. Then T is bounded.*

Proof As T is linear, its graph is a linear subspace of $V \times W$. As $V \times W$ is a Banach space, so is its closed subspace $\mathcal{G}(T)$. Let $\pi : \mathcal{G}(T) \to V$ be the projection onto the first factor. Then π is continuous and bijective, so its inverse π^{-1} is continuous by the Open Mapping Theorem C.1.5. Now T is the composition of π^{-1} with the projection onto the second factor, hence T is continuous. □

Proposition C.1.7 *Let F be a closed subspace of the normed space E. Then the quotient space E/F becomes a normed space with respect to the* quotient norm

$$\|v + F\|_q \stackrel{\text{def}}{=} \inf_{u \in F} \|v + u\|.$$

If E is a Banach space, then so is E/F.

Proof We leave it as an exercise to show that $\| \cdot \|_q$ is well-defined and strictly positive, and that it satisfies $|\lambda| \|v + F\|_q = \|\lambda v + F\|_q$ for every $\lambda \in \mathbb{C}$. To see that it satisfies the triangle inequality, let $v_1, v_2 \in E$ and $\varepsilon > 0$. Then we find $w_1, w_2 \in F$ such that $\|v_i + w_i\| < \|v_i + F\|_q + \varepsilon$ for $i = 1, 2$, and then we get

$$\|v_1 + v_2 + F\|_q \leq \|v_1 + v_2 + w_1 + w_2\|$$
$$\leq \|v_1 + F\|_q + \|v_2 + F\|_q + 2\varepsilon.$$

Since $\varepsilon > 0$ was arbitrary, the triangle equation follows.

Assume now that E is a Banach space. To show completeness of E/F, let $(v_n + F)_{n \in \mathbb{N}}$ be a Cauchy sequence in E/F. By passing to a subsequence, we may assume that $\sum_{n=1}^{\infty} \|v_{n+1} - v_n + F\|_q < \infty$, so that $\sum_{n=1}^{\infty} (v_{n+1} - v_n + F)$ converges absolutely in E/F. If it has a limit, say $v + F$, then $v + F$ is clearly a limit for the original sequence $(v_n + F)_{n \in \mathbb{N}}$. Thus it suffices to show that every series $\sum_{n=1}^{\infty} w_n + F$, which converges absolutely in E/F, converges in E/F. But given such series, we may choose for each $n \in \mathbb{N}$ some $u_n \in F$ such that $\|w_n + u_n\| < \|w_n + F\|_q + \frac{1}{2^n}$. Then $\sum_{n=1}^{\infty} \|w_n + u_n\| < \infty$ and since E is complete, there exists $w \in E$ such that $w = \sum_{n=1}^{\infty} (w_n + u_n)$. Since the quotient map $v \mapsto v + F$ is norm-decreasing, it follows that $\sum_{n=1}^{\infty} w_n + F = \sum_{n=1}^{\infty} (w_n + u_n) + F$ converges to $w + F$ in E/F. $\quad\square$

Corollary C.1.8 *Suppose that $T : E \to F$ is a continuous surjective linear map between Banach spaces. Then the linear map $\tilde{T} : E/\ker T \to F$ defined by $\tilde{T}(v + \ker T) = T(v)$ is a topological isomorphism with $\|\tilde{T}\| = \|T\|$.*

Proof We only have to show the equation $\|\tilde{T}\| = \|T\|$. All other assertions will then follow from the Open Mapping Theorem and Proposition C.1.7. Since $T = \tilde{T} \circ q$, where $q : E \to E/\ker T$ denotes the quotient map, and since $\|q\| \leq 1$, it follows that $\|T\| \leq \|\tilde{T}\|$. On the other hand, if $\varepsilon > 0$ is given, we can choose a unit vector $v + \ker T \in E/\ker T$ with $\|\tilde{T}\| < \|\tilde{T}(v + \ker T)\| + \varepsilon$ and we can choose the vector $v \in E$ such that $\|v\| < 1 + \varepsilon$. Then

$$\|\tilde{T}\| < \|\tilde{T}(v + \ker T)\| + \varepsilon = \|Tv\| + \varepsilon \leq \|T\| \|v\| + \varepsilon < \|T\|(1 + \varepsilon) + \varepsilon.$$

As $\varepsilon > 0$ is arbitrary, we get $\|\tilde{T}\| \leq \|T\|$. $\quad\square$

A subspace F of a vector space E is said to have finite codimension if E/F is finite dimensional. We close this section with

Lemma C.1.9 *Suppose that F is a closed subspace of the normed vector space E with finite codimension. If F is complete, then so is E.*

Proof By induction it suffices to assume that E/F is one-dimensional. Choose $v_0 \in E \setminus F$ and fix the isomorphism $E/F \cong \mathbb{C}$ given by $\lambda(v_0 + F) \mapsto \lambda$. Then the quotient map $q : E \to E/F \cong \mathbb{C}$ is a continuous linear functional on E. We claim that $\Phi : F \oplus \mathbb{C} \to E$ given by $\Phi(w, \lambda) = w + \lambda v_0$ is a topological isomorphism. It is clearly continuous. To see continuity of the inverse assume that $(w_n + \lambda_n v_0) \to (w + \lambda v_0)$. Then $\lambda_n \to \lambda$ by the continuity of the quotient map, which then also implies that $w_n \to w$. The result now follows from completeness of $F \oplus \mathbb{C}$. $\quad\square$

C.2   Seminorms

Occasionally, one wants to consider spaces more general than Banach spaces. Here we introduce topological vector spaces whose topology is induced by a family of seminorms. Let V be a complex vector spaces. A *seminorm* on V is a map $N : V \to [0, \infty)$ such that for $v, w \in V$ and $\lambda \in \mathbb{C}$ one has

- $N(\lambda v) = |\lambda| N(v)$, and

- $N(v + w) \le N(v) + N(w)$.

The second condition is the *triangle inequality*.

Let $(N_i)_{i \in I}$ be a family of seminorms on V. Then the family of open balls

$$B_r^i(v) = \{w \in V : N_i(v - w) < r\},$$

where $r > 0$, $i \in I$, $v \in V$, generates a topology on V that makes V a topological vector space. It is an easy exercise for the reader to check that a net (v_j) converges to v with respect to this topology, if and only if $N_i(v_j - j) \to 0$ for all $i \in I$.

Examples C.2.1

- Consider the Schwartz space $\mathcal{S}(\mathbb{R})$ of all infinitely differentiable complex functions on $\mathbb{R}$ such that $x \mapsto |x^m f^n(x)|$ is bounded for all $n, m \in \mathbb{R}$. Then

$$N_{m,n}(f) = \sup_{x \in \mathbb{R}} |x^m f^{(n)}(x)|.$$

  defines a family of seminorms on $\mathcal{S}(\mathbb{R})$, which induces a topology on $\mathcal{S}(\mathbb{R})$.

- Let V be a vector space over the field of complex numbers, and let $E \subset V^*$ be a subset of the dual space of V. Then every $\alpha \in E$ defines a seminorm N_α on V by $N_\alpha(v) = |\alpha(v)|$ for every $v \in V$. This means that every subset $E \subset V^*$ determines a topology on V. If V is a normed space and V' its topological dual space, then the topology on V generated by $V' \subset V^*$ is called the *weak topology* on V. Similarly (and more important for this book) the *weak-* topology* on V' is generated by V viewed as a subspace of $(V')^*$: each $v \in V$ determines a seminorm $N_v : V' \to [0, \infty)$ by $N_v(\alpha) = |\alpha(v)|$. It follows that a net (α_j) in V converges to α in the weak-* topology if and only if $\alpha_j(v) \to \alpha(v)$ for all $v \in V$, i.e., the weak-* topology is the topology of pointwise convergence.

C.3   Hilbert Spaces

Recall that a *Hilbert space* is a complex vector space V with an inner product $\langle \cdot, \cdot \rangle : V \times V \to \mathbb{C}$, such that V is complete in the ensuing norm

$$\|v\| = \sqrt{\langle v, v \rangle}.$$

For basics on Hilbert spaces see Chap. 2 of [Dei05]. The following is often used.

Proposition C.3.1 *Let V be a Hilbert space. Choose an orthonormal basis (e_i). The continuous dual V' is a Hilbert space with the inner product*

$$\langle \alpha, \beta \rangle \overset{\text{def}}{=} \sum_i \alpha(e_i)\overline{\beta(e_i)},$$

which does not depend on the choice of the orthonormal basis.

For a given continuous linear map $\alpha : V \to \mathbb{C}$ there exists a unique vector $w_\alpha \in V$ such that $\alpha(v) = \langle v, w_\alpha \rangle$ holds for every $v \in V$. One has $\|\alpha\| = \|w_\alpha\|$, so the map $\alpha \to w_\alpha$ is an antilinear norm-preserving isomorphism $V' \to V$.

Proof The properties of an inner product are easily verified. Let (f_k) be another orthonormal basis, then there are numbers a_{ik} such that $e_i = \sum_k a_{ik} f_k$ and these satisfy $\sum_k a_{ik}\overline{a_{jk}} = \delta_{ij}$ as well as $\sum_k a_{ki}\overline{a_{kj}} = \delta_{ij}$. Then

$$\sum_i \alpha(e_i)\overline{\beta(e_i)} = \sum_{i,k,l} a_{ik}\overline{a_{il}}\alpha(f_k)\overline{\beta(f_l)} = \sum_k \alpha(f_k)\overline{\beta(f_k)}.$$

The fact that this inner product gives the norm on V' follows from the second part. If $\alpha = 0$, then set $w = 0$. Otherwise, let U be the kernel of α. Then U is closed and so $V = U \oplus U^\perp$. The map α induces an isomorphism $U^\perp \to \mathbb{C}$, so $U^\perp$ is one-dimensional. Let $v_0 \in U^\perp$ of norm one and set $w = \overline{\alpha(v_0)}v_0$. A given $v \in V$ can be written uniquely as $v = u + \lambda v_0$ for $u \in U$ and $\lambda \in \mathbb{C}$. Then

$$\langle v, w \rangle = \langle \lambda v_0, \overline{\alpha(v_0)}v_0 \rangle = \lambda\alpha(v_0) = \alpha(v).$$

The uniqueness of w is clear. Finally, $\|\alpha\| = \sup_{\|v\|=1} |\lambda(v)| = \alpha(v_0) = \|w\|$. $\qquad\square$

It is a bit disturbing that the canonical isomorphism $V' \to V$ is antilinear. It is possible to give linear isomorphisms, but not a canonical one. For this choose an orthonormal basis (e_i) of V and define $\psi : V' \to V$ by $\psi(\alpha) = \sum_I \alpha(e_i)e_i$. It is easy to see that this indeed is an isomorphism.

Definition. Let I be an index set and for each $i \in I$ fix a Hilbert space V_i. The algebraic direct sum $\bigoplus_{i \in I} V_i$ has a natural inner product

$$\langle v, w \rangle \overset{\text{def}}{=} \sum_{i \in I} \langle v_i, w_i \rangle,$$

for $v = (v_i), w = (w_i)$. If $V_i \neq 0$ for infinitely many $i \in I$, then this space will not be complete. We denote by $V = \widehat{\bigoplus}_{i \in I} V_i$ its completion, called the *Hilbert direct sum* of the spaces V_i. In practice, when no confusion can arise, it is convenient to leave out the hat, i.e., to write $\bigoplus_{i \in I} V_i$ for the completed direct sum as well. One can identify this space with the set of all $v \in \prod_{i \in I} V_i$ such that

$$\sum_{i \in I} \|v_i\|^2 < \infty,$$

where the inner product is given by the same formula as above, only now with possibly infinitely many non-zero summands. One can show that the sum defining the inner product always converges.

Likewise, the algebraic tensor product $V \otimes W$ of two Hilbert spaces has a natural inner product, given on simple tensors by

$$\langle v \otimes w, v' \otimes w' \rangle \overset{\text{def}}{=} \langle v, v' \rangle \langle w, w' \rangle.$$

If both spaces are infinite dimensional, then this pre-Hilbert space will not be complete. We denote its completion by $V \hat{\otimes} W$ and call it the *Hilbert tensor product* of the spaces V and W. If (e_i) is an orthonormal basis of V, and (f_j) is an orthonormal basis of W, then the family $(e_i \otimes f_j)_{i,j}$ is an orthonormal basis of $V \hat{\otimes} W$.

If $S : V \to V'$ and $T : W \to W$ are bounded linear operators on Hilbert spaces, then they give rise to a natural bounded linear operator

$$S \otimes T : V \otimes W \to V' \otimes W',$$

which on simple tensors is given by

$$S \otimes T(v \otimes w) = S(v) \otimes T(w).$$

Lemma C.3.2 *Let $T : V \to W$ be a bounded linear operator between Hilbert spaces. Then there exists a bounded linear operator $T^* : W \to V$ such that $\langle Tv, w \rangle = \langle v, T^*w \rangle$ holds for every $v \in V$ and every $w \in W$. We have $(T^*)^* = T$ and $\|T^*\| = \|T\|$.*

Proof For given $w \in W$ consider the linear functional $v \mapsto \langle Tv, w \rangle$. As T is continuous, this functional is, and so there exists a unique vector $T^*w \in V$ such that $\langle Tv, w \rangle = \langle v, T^*w \rangle$ holds for every $v \in V$. The map T^* is easily seen to be linear. For $w \in W$ we can apply this to $v = T^*w$ and, using the Cauchy-Schwarz inequality we get $\|T^*w\|^2 = |\langle TT^*w, w \rangle| \le \|w\| \|T\| \|T^*w\|$. Dividing by $\|T^*w\|$ shows that $\|T^*w\| \le \|T\| \|w\|$, so T^* is indeed bounded with $\|T^*\| \le \|T\|$. Since $\langle T^*w, v \rangle = \overline{\langle v, T^*w \rangle} = \overline{\langle Tv, w \rangle} = \langle w, Tv \rangle$ for all $w \in W, v \in V$ we have $(T^*)^* = T$ and then $\|T\| = \|(T^*)^*\| \le \|T^*\|$. $\qquad\square$

Recall that a bounded operator $T : V \to V$ is *normal* if $T^*T = TT^*$.

Lemma C.3.3 *Suppose that $v \in V$ is an eigenvector for the eigenvalue $\lambda \in \mathbb{C}$ of the normal operator $T \in \mathcal{B}(V)$. Then v is an eigenvector for the eigenvalue $\bar{\lambda}$ of the operator T^*. Moreover, if $\lambda \neq \mu$ are two eigenvalues of T, then $\mathrm{Eig}(T, \lambda) \perp \mathrm{Eig}(T, \mu)$.*

Proof Since $(T - \lambda 1)v = 0$ we have

$$
\begin{aligned}
\|(T^* - \bar{\lambda}1)v\|^2 &= \langle (T^* - \bar{\lambda}1)v, (T^* - \bar{\lambda}1)v \rangle \\
&= \langle v, (T - \lambda 1)(T^* - \bar{\lambda}1)v \rangle \\
&= \langle v, (T^* - \bar{\lambda}1)(T - \lambda 1)v \rangle \\
&= \|(T - \lambda 1)v\|^2 = 0,
\end{aligned}
$$

from which the first assertion follows. For the second assertion let $v \in \mathrm{Eig}(T, \lambda)$ and $w \in \mathrm{Eig}(T, \mu)$. Then

$$
\lambda \langle v, w \rangle = \langle Tv, w \rangle = \langle v, T^*w \rangle = \langle v, \bar{\mu}w \rangle = \mu \langle v, w \rangle.
$$

Since $\lambda \neq \mu$ it follows that $\langle v, w \rangle = 0$. $\qquad\square$

Definition A bounded operator P on a Hilbert space V is called a *projection* if $P = P^2$. The operator is called an *orthogonal projection* if P is a projection, which is self-adjoint.

Lemma C.3.4 *A bounded operator P on a Hilbert space V is an orthogonal projection if and only if the space V is a direct orthogonal sum $V = U \oplus W$, where U and W are closed subspaces and for every $u + w \in V$ with $u \in U$ and $w \in W$ one has $P(u + w) = u$.*

Proof Let P be an orthogonal projection. If x lies in the image of P, say $x = P(y)$, then $P(x) = P^2(y) = P(y) = x$. By continuity, the identity $x = P(x)$ extends to the closure U of $\mathrm{Im}(P)$. This however means that $U = P(U)$ equals the image of P. Let W be its orthogonal complement. For $w \in W$ one has $\langle P(w), P(w) \rangle = 0$, so W lies in the kernel of P. As clearly $V = U \oplus W$, the claim follows. The converse is trivial. $\qquad\square$

Definition In general, if $W \subset V$ is any closed subspace of V, there is a unique orthogonal projection $P_W \in \mathcal{B}(V)$ such that $P_W(V) = W$. Just write $V = U \oplus W$ with $U = W^\perp$ and define $P_W(u + w) = w$. We call this the *orthogonal projection onto W*. Thus there is a one-to-one correspondence between orthogonal projections in $\mathcal{B}(V)$ and closed subspaces of V.

If v is a unit vector in the Hilbert space V, then the orthogonal projection P_v onto $\mathbb{C}v$ is given by $P_v(u) = \langle u, v \rangle v$. More general, if W is a closed subspace of V and

$\{e_i : i \in I\}$ is an orthonormal basis for W, then the orthogonal projection P_W onto W is given by

$$P_W(v) = \sum_{i \in I} \langle v, e_i \rangle e_i.$$

C.4 Unbounded Operators

In this section we give some background on unbounded operators on Hilbert spaces. The content of this section is needed in our study of continuous wavelet transforms. We shall only restrict to those topics we need in this book, so the content of this section cannot be regarded as a full introduction to the theory. In particular, we will not touch on spectral theory of unbounded operators, which can be found in any good text book on Functional Analysis.

Let V and W be Hilbert spaces. If $D \subset V$ is a linear subspace and if $T : D \to W$ is a linear map, we say that T is a linear operator from $V \to W$ with domain D. A linear operator $T : D \subset V \to W$ is *densely defined* if D is dense in V.

The operator $T : D \subset V \to W$ is *closed* if the graph

$$\mathcal{G}(T) = \{(v, Tv) : v \in D\}$$

is closed in $V \oplus W$. It follows from the closed graph theorem, that a densely defined closed operator $T : D \subset V \to W$ is bounded if and only if $D = V$. Since the map $D \to \mathcal{G}(T)$ given by $v \mapsto (v, Tv)$ is clearly a bijection, it follows that T is closed if and only if D becomes a Hilbert space when equipped with the modified inner product

$$\langle\langle v, w \rangle\rangle \overset{\text{def}}{=} \langle v, w \rangle + \langle Tv, Tw \rangle.$$

If $T : D \subset V \to W$ is a densely defined linear operator, then the *adjoint operator* $T^* : D^* \subset W \to V$ is defined as follows: The domain D^* of T^* consists of all vectors $w \in W$ such that $v \mapsto \langle Tv, w \rangle$ is a continuos linear form on D, and hence extends uniquely to a continuous linear functional on all of V. By the Riesz representation theorem, there exists therefore a unique vector, which we call T^*w, in V such that

$$\langle Tv, w \rangle = \langle v, T^*w \rangle$$

for every $v \in D$. It is straightforward to check that $T^* : D^* \subset W \to V$ is a linear operator. Although we don't need it here, we mention the following fact

Proposition C.4.1 *Suppose that $T : D \subset V \to W$ is a densely defined closed operator. Then $T^* : D^* \subset W \to V$ is also a densely defined closed operator and we have $T = T^{**}$.*

For a proof of this and many other useful results on unbounded operators we refer to [RS80]. The following easy observation will be used frequently in this section. We leave the straightforward proof as an exercise

Lemma C.4.2 *Suppose that $T : D \subset V \to W$ is a closed densely defined operator. Then $T(D)^{\perp} = \ker(T^*) \subset D^*$.*

A densely defined operator $T : D \subset V \to V$ is called *self-adjoint* if $D = D^*$ and $T = T^*$. Note that in general this is not the same as being *symmetric*, which means that

$$\langle Tv, w \rangle = \langle v, Tw \rangle$$

for all $v, w \in D$, since one can construct symmetric operators such that the domain D^* of T^* is strictly larger than D (See [RS80, p 258] for an example of a symmetric densely defined closed operator, which is not self-adjoint). So an operator T is self-adjoint if and only if T is symmetric and $D = D^*$. We shall also need the following criterion for self-adjointness:

Proposition C.4.3 *Suppose that $T : D \subset V \to V$ is a closed, densely defined, symmetric operator. Then the following are equivalent:*

(a) *T is self-adjoint.*

(b) $\ker(T \pm i\mathrm{Id}) = \{0\}$.

(c) $(T \pm i\mathrm{Id})(D) = V$.

Proof If T is self-adjoint then it follows easily from the symmetry of T that i and $-i$ cannot be eigenvalues for $T = T^*$, which proves (a) $\Rightarrow$ (b). The assertion (c) $\Rightarrow$ (b) follows from Lemma C.4.2 after checking that $(T + \lambda\mathrm{Id})^* = T^* + \bar{\lambda}\mathrm{Id}$ for every $\lambda \in \mathbb{C}$, which we leave as an exercise for the reader.

Also by Lemma C.4.2 we see that (b) implies that $(T \pm i\mathrm{Id})(D)$ is dense in V. So, in order to get (b) $\Rightarrow$ (c) we only have to check that $(T \pm i\mathrm{Id})(D)$ is closed in V if (b) holds. For this we restrict our attention to the operator $T - i\mathrm{Id}$. Suppose that $(v_n)_{n \in \mathbb{N}}$ is a sequence in D such that $Tv_n - iv_n \to w$ for some w in V. Using the equation

$$\|Tv - iv\|^2 = \|Tv\|^2 + \|v\|^2$$

for every $v \in D$, we observe that $(Tv_n)_{n \in \mathbb{N}}$ and $(v_n)_{n \in \mathbb{N}}$ are Cauchy sequences and there exist $v, u \in V$ such that $v_n \to v$ and $Tv_n \to u$. Since T is closed, it follows that $v \in D$ and $Tv = u$, and then $w = (T - i\mathrm{Id})v \in (T - i\mathrm{Id})(D)$.

To show (c) $\Rightarrow$ (a) let $w \in D^*$. Then it follows from (c) that there exists $v \in D$ with $(T^* - i\mathrm{Id})w = (T - i\mathrm{Id})v$. Since $D \subset D^*$, we see that $w - v \in \ker(T^* - i\mathrm{Id})$, and the latter is zero by (c) $\Rightarrow$ (b). So $D^* = D$ and $T^* = T$. $\square$

A self-adjoint operator $T : D \subset V \to V$ is called *positive* if $\langle Tv, v \rangle \geq 0$ for every $v \in D$, and it is called *positive definite* if, in addition,

$$\langle Tv, v \rangle = 0 \implies v = 0.$$

We shall need the following observations:

Lemma C.4.4 *If $T : D \subset V \to V$ is self-adjoint and injective, then $T(D)$ is dense in V and $T^{-1} : T(D) \subset V \to V$ is also self-adjoint. In particular, if $T : D \subset V \to V$ is positive definite, then T is injective and $T^{-1} : T(D) \subset V \to V$ is also positive definite.*

Proof It follows from Lemma C.4.2 that a selfadjoint operator has dense image if and only if it is injective. Since $\mathcal{G}(T) = \mathcal{G}(T^{-1})$, it follows that T^{-1} is also closed, and it is clearly symmetric. To see that it is self-adjoint, we simply use Proposition C.4.3 to see that

$$\left(T^{-1} - i \mathrm{Id} \right) (T(D)) = -i(i\mathrm{Id} + T)(D) = V$$

since T is self-adjoint, and similarly $\left(T^{-1} + i\mathrm{Id} \right) (D) = V$. Thus T^{-1} is self-adjoint by Proposition C.4.3. $\qquad\qquad\square$

We are now ready for the proof of the main technical result of this section. We first recall the well-known result that any continuous positive (definite) hermitian form $B : V \times V \to \mathbb{C}$ on a Hilbert space V is induced by a positive (definite) operator $T : V \to V$ in the sense that

$$B(v, w) = \langle Tv, w \rangle$$

for every $v, w \in V$. Indeed, the proof of this fact simply uses the Riesz representation theorem: If we fix $w \in V$, then $v \mapsto B(v, w)$ is a positive linear functional on V, and therefore there exists a unique element, which we call Tw, in V such that

$$B(v, w) = \langle v, Tw \rangle.$$

It is quite straightforward to check that T is a bounded linear self-adjoint operator, which is positive (definite) since B is positive (definite). Using the spectral theorem, we may also consider the square-route $C = T^{1/2}$ of T, which is also a positive operator that satisfies the equation

$$B(v, w) = \langle Cv, Cw \rangle$$

for every $v, w \in V$.

We now want to extend these results to certain unbounded hermitian forms. Suppose that D is a dense linear subspace of the Hilbert space V. A hermitian form

$$B : D \times D \to \mathbb{C}$$

is called *closed* if, whenever $(v_n)_{n \in \mathbb{N}}$ is a sequence in D, which converges to some $v \in V$ and such that

$$B(v_n - v_m, v_n - v_m) \to 0$$

for $n, m \to \infty$, then $v \in D$ and $B(v_n - v, v_n - v) \to 0$ for $n \to \infty$.

Theorem C.4.5 *Suppose that D is a dense linear subspace of the Hilbert space V and that*

$$B : D \times D \to \mathbb{C}$$

is a positive semi-definite closed hermitian form on D. Then there exists a closed operator $C : D \subset V \to V$ such that

$$B(v, w) = \langle Cv, Cw \rangle$$

for all $v, w \in D$. If B is positive definite, then C is injective.

Proof We equip D with the inner product

$$\langle\langle v, w \rangle\rangle = \langle v, w \rangle + B(v, w)$$

and claim that D is complete with respect to this inner product. Indeed, if $(v_n)_{n \in \mathbb{N}}$ is a sequence in D, which is Cauchy with respect to $\langle \cdot, \cdot \rangle$, then it follows that $(v_n)_{\in \mathbb{N}}$ is Cauchy in V, hence converges to some $v \in V$, and that $B(v_n - v_m, v_n - v_m) \to 0$ for $n, m \to \infty$. Since B is closed we get $v \in D$ and $v_n \to v$ with respect to $\langle \cdot, \cdot \rangle$.

We first regard the restriction of the inner product on V to D as a positive definite hermitian form on $(D, \langle\langle \cdot, \cdot \rangle\rangle)$. This is clearly bounded, and we obtain a positive definite operator $C_1 : D \to D$ such that

$$\langle v, w \rangle = \langle C_1 v, C_1 w \rangle + B(C_1 v, C_1 w)$$

for all $v, w \in D$. Since C_1 is positive definite, it is injective and has dense image in D with respect to $\langle\langle \cdot, \cdot \rangle\rangle$, and then certainly with respect to the given inner product on V. Let $C_1^{-1} : C_1(D) \to D$ denote the inverse of C_1. It then satisfies

$$\langle C_1^{-1} v, C_1^{-1} w \rangle = \langle v, w \rangle + B(v, w)$$

for all $v, w \in C_1(D)$.

Consider now the hermitian form $B : D \times D \to \mathbb{C}$. Since it is continuous with respect to $\langle\langle \cdot, \cdot \rangle\rangle$ there exists a positive operator $C_2 : D \to D$ with

$$B(v, w) = \langle C_2 v, C_2 w \rangle + B(C_2 v, C_2 w)$$

for all $v, w \in D$.

Put $\tilde{D} = C_2^{-1}(C_1(D))$. We want to define $C : D \to V$ as the composition $C_1^{-1} \circ C_2$. A priori, this operator is only defined on the dense subset $\tilde{D} \subset D$, so we have to show that it extends to a positive (and in particular self adjoint) operator on D. We first observe that we have the equation

$$B(v, w) = \langle Cv, Cw \rangle$$

for all $v, w \in \tilde{D}$. Thus, if we regard C as a map from $(D, \langle\langle \cdot, \cdot \rangle\rangle)$ to $(V, \langle \cdot, \cdot \rangle)$, we see that C becomes bounded, and therefore extends uniquely to all of D, and it follows then from continuity that $B(v, w) = \langle Cv, Cw \rangle$ holds for all $v, w \in D$. If we now regard $C : D \subset V \to V$ as a densely defined operator on V with respect to $\langle \cdot, \cdot \rangle$, then C is closed since D is complete with respect to the inner product

$$\langle v, w \rangle + \langle Cv, Cw \rangle = \langle\langle v, w \rangle\rangle. \qquad \square$$

Remark C.4.6 Let C be as in the above theorem. Using functional calculus for unbounded operators (see [Rud91, Chap. 13]) we may consider the operator $|C| \stackrel{\text{def}}{=} \sqrt{C^*C}$. Then $|C|$ is a positive self-adjoint operator such that

$$\langle |C|v, |C|w \rangle = \langle |C|^2 v, w \rangle = \langle C^*Cv, w \rangle = \langle Cv, Cw \rangle$$

for all $v, w \in D$. Thus the operator C in the above theorem can always be chosen to be positive or even positive definite if $B(\cdot, \cdot)$ is positive definite. Indeed there exists a *unique* positive operator $C : D \subset V \to V$ such that $B(v, w) = \langle Cv, Cw \rangle$, for if S would be another such operator, then

$$\langle (C^2 - S^2) v, w \rangle = \langle C^2 v, w \rangle - \langle S^2 v, w \rangle = 0$$

for all $v, w \in D$, from which it follows that $C^2 = S^2$, and hence $C = S$.

Bibliography

Bea95. Beardon, A.F.: The Geometry of Discrete Groups, Graduate Texts in Mathematics, vol. 91. Springer-Verlag, New York (1995) (Corrected reprint of the 1983 original. MR1393195 (97d:22011))

Ben97. Benedetto, J.J.: Harmonic Analysis and Applications, Studies in Advanced Mathematics. CRC Press, Boca Raton (1997) (MR1400886 (97m:42001))

BCD+72. Bernat, P., Conze, N., Duflo, M., Lévy-Nahas, M., Raïs, M., Renouard, P., Vergne, M.: Représentations des groupes de Lie résolubles. Dunod, Paris (1972) (Monographies de la Société Mathématique de France, No. 4, MR0444836 (56 #3183))

Bla98. Blatter, C.: Wavelets. A K Peters Ltd., Natick (1998) (A primer. MR1654297 (99f:42062b))

BtD95. Bröcker, T., tom Dieck, T.: Representations of Compact Lie Groups, Graduate Texts in Mathematics, vol. 98. Springer-Verlag, New York (1995) (Translated from the German manuscript; Corrected reprint of the 1985 translation. MR1410059 (97i:22005))

Cha68. Chandrasekharan, K.: Introduction to Analytic Number Theory, Grundlehren 148. Springer-Verlag, New York (1968) (MR0249348 (40 #2593))

Cha11. Chandrasekharan, K.: A Course on Integration Theory, Texts and Readings in Mathematics, vol. 8. Hindustan Book Agency, New Delhi (2011) (Reprint of the 1996 edition. MR2858258 (2012h:28001))

Coh93. Donald, L.C.: Measure Theory. Birkhäuser Boston Inc., Boston (1993) (Reprint of the 1980 original. MR1454121 (98b:28001))

Dei05. Deitmar, A.: A First Course in Harmonic Analysis, 2nd ed. Universitext, Springer-Verlag, New York (2005) (MR2121678 (2006a:42001))

DM78. Dixmier, J., Malliavin, P.: Factorisations de fonctions et de vecteurs indéfiniment différentiables. Bull. Sci. Math. **102**(2), 307–330 (1978) (no.4, (French, with English summary). (MR517765 (80f:22005))

Dix96. Dixmier, J.: Les C^*-algèbres et leurs représentations, Les, Grands Classiques Gauthier-Villars. ([Gauthier-Villars, Great Classics] Éditions. Jacques Gabay, Paris, 1996 (French). Reprint of the second (1969) edition. (MR1452364 (98a:46066))

DM76. Duflo, M., Moore, C.C.: On the regular representation of a nonunimodular locally compact group. J. Funct. Anal. **21**(2), 209–243 (1976) (MR0393335 (52 #14145))

Edw82. Edwards, R.E.: Fourier Series, vol. 2, 2nd ed., Graduate Texts in Mathematics, vol. 85. Springer-Verlag, New York (1982) (A modern introduction. MR667519 (83k:42001))

Fol95. Folland, G.B.: A Course in Abstract Harmonic Analysis, Studies in Advanced Mathematics. CRC Press, Boca Raton (1995) (MR1397028 (98c:43001))

Fuehr. Führ, H.: Abstract Harmonic Analysis of Continuous Wavelet Transforms, Lecture Notes in Mathematics, vol. 1863. Springer-Verlag, Berlin (2005) (MR2130226 (2006m:43003))

Gaa64. Gaal, S.A.: Point Set Topology, Pure and Applied Mathematics, vol. XVI. Academic Press, New York (1964) (MR0171253 (30 #1484))

GMP85. Grossmann, A., Morlet, J., Paul, T.: Transforms associated to square integrable group representations. I. general results. J. Math. Phys. **26**(10) 2473–2479 (1985). doi:10.1063/1.526761 (MR803788 (86k:22013))

HC76. Harish-Chandra: Harmonic analysis on real reductive groups. III. The Maass-Selberg relations and the plancherel formula. Ann. Math. **104**(2), 117–201 (1976) (no. 1 MR0439994 (55 #12875)

Hel01. Helgason, S.: Differential Geometry, Lie Groups, and Symmetric Spaces, Graduate Studies in Mathematics, vol. 34. American Mathematical Society, Providence (2001) (Corrected reprint of the 1978 original)

HM06. Hofmann, K.H., Morris, S.A.: The Structure of Compact Groups, Second revised and Augmented edition, de Gruyter Studies in Mathematics, vol. 25. Walter de Gruyter, Berlin (2006) (A primer for the student–a handbook for the expert)

How79. Howe, R.: θ-series and Invariant Theory, Automorphic Forms, Representations and L-functions. Proceedings Sympos. Pure Math., Oregon State Univ., Corvallis, Ore., 1977), Part 1, Proceedings Sympos. Pure Math., XXXIII, Amer. Math. Soc., Providence, R.I., 1979, pp. 275–285. MR546602 (81f:22034)

Kan09. Kaniuth, E.: A Course in Commutative Banach Algebras, Graduate Texts in Mathematics, vol. 246. Springer, New York (2009) (MR2458901 (2010d:46064))

KH95. Katok, A., Hasselblatt, B.: Introduction to the Modern Theory of Dynamical Systems, Encyclopedia of Mathematics and its Applications, vol. 54. Cambridge University Press, Cambridge (1995) (With a supplementary chapter by Katok and Leonardo Mendoza. MR1326374 (96c:58055))

Kna01. Knapp, A.W.: Representation Theory of Semisimple Groups, Princeton Landmarks in Mathematics. Princeton University Press, Princeton (2001) (An overview based on examples; Reprint of the 1986 original. MR1880691 (2002k:22011))

Lan02. Lang, S.: Algebra, Graduate Texts in Mathematics, vol. 211, 3rd ed. Springer-Verlag, New York (2002)

LV80. Lion, G., Vergne, M.: The Weil Representation, Maslov Index and Theta Series, Progress in Mathematics, vol. 6. Birkhäuser Boston, Mass (1980) (MR573448 (81j:58075))

MZ55. Montgomery, D., Zippin, L.: Topological Transformation Groups. Interscience Publishers, New York (1955) (MR0073104 (17,383b))

Ped89. Pedersen, G.K.: Analysis Now, Graduate Texts in Mathematics, vol. 118. Springer-Verlag, New York (1989) (MR971256 (90f:46001))

Pon34. Pontrjagin, L.: The theory of topological commutative groups. Ann. Math. **35**(2), 361–388 (1934). doi:10.2307/1968438.

RS80. Reed, M., Simon, B.: Methods of Modern Mathematical Physics. I, 2nd ed. Academic Press, New York (1980) (Functional analysis. MR751959 (85e:46002))

Rud87. Rudin, W.: Real and Complex Analysis, 3rd ed. McGraw-Hill Book, New York (1987) (MR924157 (88k:00002))

Rud91. Rudin, W.: Functional Analysis, International Series in Pure and Applied Mathematics, 2nd ed. McGraw-Hill, New York (1991) (MR1157815 (92k:46001))

Sel56. Selberg, A.: Harmonic analysis and discontinuous groups in weakly symmetric Riemannian spaces with applications to Dirichlet series. J. Indian Math. Soc. (N.S.) **20**, 47–87 (1956) (MR0088511 (19,531g))

Ser73. Serre, J.-P.: A Course in Arithmetic. Springer-Verlag, New York (1973) (Translated from the French; Graduate Texts in Mathematics, No. 7. MR0344216 (49 #8956))

Tho68. Thoma, E.: Eine charakterisierung diskreter Gruppen vom Typ I. Invent. Math. **6**, 190–196 (1968) ((German). MR0248288 (40 #1540))

War83. Warner, F.W.: Foundations of Differentiable Manifolds and Lie Groups, Graduate Texts in Mathematics, vol. 94. Springer-Verlag, New York (1983) (Corrected reprint of the 1971 edition. MR722297 (84k:58001))

Wil07. Williams, D.P.: Crossed Products of C^*-algebras, Mathematical Surveys and Monographs, vol. 134. American Mathematical Society, Providence (2007) (MR2288954 (2007m:46003))

Wil62. Williamson, J.H.: Remarks on the Plancherel and Pontryagin theorems. Topology. **1**, 73–80 (1962) (MR0141739 (25 #5136))

Wei64. Weil, A.: Sur certains groupes d'opérateurs unitaires. Acta. Math. **111**, 143–211 (1964) ((French). MR0165033 (29 #2324))

Index

A. Deitmar, S. Echterhoff, *Principles of Harmonic Analysis,* Universitext,
DOI 10.1007/978-3-319-05792-7, © Springer International Publishing Switzerland 2014

Printed in the United States
By Bookmasters